Modeling, Identification, and Control for Cyber-Physical Systems Towards Industry 4.0

Emerging Methodologies and
Applications in Modelling,
Identification and Control

Modeling, Identification, and Control for Cyber-Physical Systems Towards Industry 4.0

Edited by

Paolo Mercorelli
Institute for Production Technology and Systems
Leuphana University of Lueneburg
Lueneburg, Germany

Weicun Zhang
School of Automation and Electrical Engineering
University of Science and Technology Beijing
Beijing, China

Hamidreza Nemati
Engineering Design and Mathematics Department
University of the West of England
Bristol, United Kingdom

YuMing Zhang
Department of Electrical and Computer Engineering
University of Kentucky
Lexington, KY, United States

Series Editor
Quanmin Zhu

ACADEMIC PRESS
An imprint of Elsevier

Academic Press is an imprint of Elsevier
125 London Wall, London EC2Y 5AS, United Kingdom
525 B Street, Suite 1650, San Diego, CA 92101, United States
50 Hampshire Street, 5th Floor, Cambridge, MA 02139, United States

ISBN: 978-0-323-95207-1

For information on all Academic Press publications
visit our website at https://www.elsevier.com/books-and-journals

Publisher: Matthew Deans
Acquisitions Editor: Sophie Harrison
Editorial Project Manager: Fernanda Oliveira
Production Project Manager: Manju Paramasivam
Cover Designer: Matthew Limbert

Typeset by VTeX

Dedication

In this tremendous controversial historical period, heavy with injustice and oppression, any epochal scientific and technical progressive process is progressive "if and only if" it provides for a contribution in terms of trust and in terms of believing. This progressive process of the science and technology is decisive, never more so than it is today, with a higher and more important level of trust and belief in the form of a rational integral humanism in which science and technology contributes to peace: Si vis pacem, para pacem!

The editors and contributors of this book

"Look at me!" said the Fox; "because of my passion for studying I have lost a leg."
"Look at me!" cried the Cat; "because of my love for studying I have lost both eyes."

From *Le avventure di Pinocchio* by C. Collodi

Contents

3. Temperature control in Peltier cells comparing sliding
    mode control and PID controllers

Alexandra Mironova, Paolo Mercorelli, and Andreas Zedler

4. A Digital Twin for part quality prediction and control in
    plastic injection molding

Alexander Rehmer, Marco Klute, Hans-Peter Heim, and
Andreas Kroll

Part II
Motion control and autonomous robots as a challenge in Industry 4.0 process
Paolo Mercorelli, Hamidreza Nemati, and Quanmin Zhu

5.   SLAM algorithms for autonomous mobile robots
Sufang Wang and Weicun Zhang

6.   Optimization of motion control smoothness based on Eband algorithm
Sufang Wang, Chuanxu An, and Weicun Zhang

10. Safety automotive sensors and actuators with end-to-end protection (E2E) in the context of AUTOSAR embedded applications

Horia V. Căpriță and Dan Selișteanu

Part III
Motion and control of autonomous unmanned aerial systems as a challenge in Industry 4.0 process

Paolo Mercorelli, Hamidreza Nemati, and Quanmin Zhu

11. Multibody simulations of distributed flight arrays for Industry 4.0 applications

Lewis Yip, Hamidreza Nemati, Paolo Mercorelli, and Quanmin Zhu

12. Recent advancements in multi-objective pigeon inspired optimization (MPIO) for autonomous unmanned aerial systems

Muhammad Aamir khan, Quanmin Zhu, Zain Anwar Ali, and Muhammad Shafiq

13. U-model-based dynamic inversion control for quadrotor UAV systems

Ahtisham Aziz Lone, Hamidreza Nemati, Quanmin Zhu, Paolo Mercorelli, and Pritesh Narayan

14. Nonlinear control allocation applied on a QTR: the influence of the frequency variation

Murillo Ferreira dos Santos, Leonardo de Mello Honório, Mathaus Ferreira da Silva, Vinícius Ferreira Vidal, and Paolo Mercorelli

Part IV
Theoretical and methodological advancements in disturbance rejection and robust control

Paolo Mercorelli, Hamidreza Nemati, and Quanmin Zhu

15. Active disturbance rejection control of systems with large uncertainties

Weicun Zhang and Hui Wang

16. Gain scheduling design based on active disturbance rejection control for thermal power plant under full operating conditions

Zhenlong Wu, Donghai Li, Yali Xue, and YangQuan Chen

17. Active disturbance rejection control of large-scale coal fired plant process for flexible operation

He Ting, Zhenlong Wu, Donghai Li, and Wang Jihong

18. Desired dynamic equational proportional-integral-derivative controller design based on probabilistic robustness

Gengjin Shi, Donghai Li, Yanjun Ding, and YangQuan Chen

Contributors

Chuanxu An, Institute of Software Chinese Academy of Sciences, Beijing, China

Zain Anwar Ali, Department of Electronic Engineering, Sir Syed University of Engineering and Technology, Karachi, Pakistan

Ahtisham Aziz Lone, Department of Engineering Design and Mathematics, University of the West of England, Bristol, United Kingdom

Udo Becker, Ostfalia University of Applied Sciences, Wolfsburg, Germany

Horia V. Căpriță, Continental Automotive Systems Sibiu, Sibiu, Romania University of Craiova, Craiova, Romania

YangQuan Chen, Mechatronics, Embedded Systems and Automation (MESA) Lab, School of Engineering, University of California, Merced, CA, United States

Leonardo de Mello Honório, Faculty of Engineering, UFJF, Juiz de Fora, Brazil

Yanjun Ding, State Key Lab of Power Systems, Department of Energy and Power Engineering, Tsinghua University, Beijing, China

Mathaus Ferreira da Silva, Faculty of Engineering, UFJF, Juiz de Fora, Brazil

Murillo Ferreira dos Santos, Department of Electroelectronics, CEFET-MG, Leopoldina, Brazil

Vinícius Ferreira Vidal, Faculty of Engineering, UFJF, Juiz de Fora, Brazil

C.L. Gilfillan, Faculty of Engineering, Built Environment & Information Technology (EBEIT) at the Nelson Mandela University, Port Elizabeth, South Africa

Hans-Peter Heim, Institute of Material Engineering - Polymer Engineering, University of Kassel, Kassel, Germany

Wang Jihong, University of Warwick, School of Engineering, Coventry, United Kingdom

Muhammad Aamir khan, Department of Electronic Engineering, Sir Syed University of Engineering and Technology, Karachi, Pakistan

Marco Klute, Institute of Material Engineering - Polymer Engineering, University of Kassel, Kassel, Germany

Andreas Kroll, Department of Measurement and Control, University of Kassel, Kassel, Germany

Donghai Li, State Key Lab of Power Systems, Department of Energy and Power Engineering, Tsinghua University, Beijing, China

Alexander B.S. Macfarlane, Faculty of Engineering, Built Environment & Information Technology (EBEIT) at the Nelson Mandela University, Port Elizabeth, South Africa

Paolo Mercorelli, Institute for Production Technology and Systems, Leuphana University of Lueneburg, Lueneburg, Germany

Alexandra Mironova, Institute for Production Technology and Systems, Leuphana University of Lueneburg, Lueneburg, Germany

Pritesh Narayan, Department of Engineering Design and Mathematics, University of the West of England, Bristol, United Kingdom

Hamidreza Nemati, Department of Engineering Design and Mathematics, University of the West of England, Bristol, United Kingdom

Alexander Rehmer, Department of Measurement and Control, University of Kassel, Kassel, Germany

Dan Selişteanu, Department of Automatic Control and Electronics, University of Craiova, Craiova, Romania

Oleg Sergiyenko, Applied Physics Department of Engineering Institute of Baja California Autonomous University, Mexicali, BC, Mexico

Muhammad Shafiq, Department of Electronic Engineering, Sir Syed University of Engineering and Technology, Karachi, Pakistan

Gengjin Shi, State Key Lab of Power Systems, Department of Energy and Power Engineering, Tsinghua University, Beijing, China

He Ting, Jinan University, International Energy College, Zhuhai, China

Theo van Niekerk, Faculty of Engineering, Built Environment & Information Technology (EBEIT) at the Nelson Mandela University, Port Elizabeth, South Africa

Hui Wang, CHINA Metallurgical Group Corporation, Beijing, China

Sufang Wang, Second Academy of China Aerospace Science and Industry Corporation, Institute 706, Beijing, China

Zhenlong Wu, School of Electrical Engineering, Zhengzhou University, Zhengzhou, China

Yali Xue, State Key Lab of Power Systems, Department of Energy and Power Engineering, Tsinghua University, Beijing, China

Lewis Yip, Department of Engineering Design and Mathematics, University of the West of England, Bristol, United Kingdom

Andreas Zedler, Institute for Production Technology and Systems, Leuphana University of Lueneburg, Lueneburg, Germany

Weicun Zhang, University of Science and Technology Beijing, School of Automation and Electrical Engineering, Beijing, China

Quanmin Zhu, Department of Engineering Design and Mathematics, University of the West of England, Bristol, United Kingdom

Biography

Paolo Mercorelli

Paolo Mercorelli (Member, IEEE) received a PhD degree in systems engineering from the University of Bologna, Bologna, Italy, in 1998. In 1997, he was Visiting Researcher for one year with the Department of Mechanical and Environmental Engineering, University of California, Santa Barbara, CA, USA. From 1998 to 2001 he was Postdoctoral Researcher with Asea Brown Boveri Corporate Research, Heidelberg, Germany, and this research activity has brought him 3 patents (dp). From 2002 to 2005 he was Senior Researcher with the Institute of Automation and Informatics, Wernigerode, Germany, where he was the leader of the control group. From 2005 to 2011 he was Associate Professor of Process Informatics with the Ostfalia University of Applied Sciences, Wolfsburg, Germany. Since 2012, he has been Full Professor and Chair of control and drive systems with the Institute for Production Technology and Systems, Leuphana University of Lueneburg, Lueneburg, Germany. Since 2018, he has been International Distinguished Visiting Professor with the Institute of Automatic Control, Lodz University of Technology, Lodz, Poland, where he has been responsible for different courses in the field of control of robotic systems. His current research interests include applications of Kalman filters, robotics, wavelets, geometric control, and sliding mode control. Dr. Mercorelli was the recipient of a three-year scholarship from the Marie Curie Actions Research Fellowship Program, which is one of the most competitive and prestigious European awards sponsored by the European Commission from 1998 to 2001. He received seven best international conferences paper awards: IECON 2013, IECON 2014, CoDIT 2014, ICCC 2017, FedCSIS 2019, ACD 2019, ICCC 2020. In the years 2019, 2020, 2021, 2022 and 2023, he has been on the list of the top 2% scientists by Elsevier Database and (METRICS) University of Stanford (USA) (Ioannidis, J.; Boyack, K. & Baas, J., 2020, Updated science-wide author databases of standardized citation indicators, PLOS, October 16). Since 2022, he has been Editor-in-Chief for the Section "Engineering

Mathematics" in "Mathematics", an Open Access Journal from MDPI, Basel, Switzerland. In summer semester 2023, he has been Visiting Professor with Chandigarh University, India, and responsible for the course of Control Systems.

Prof. Weicun Zhang

Weicun Zhang is Associate Professor at the School of Automation and Electrical Engineering, University of Science and Technology Beijing. He obtained his Bachelor Degree of Engineering in Industrial Automation from Shenyang Institute of Technology, MSc degree in Automatic Control (1989) from Beijing Institute of Technology, and the PhD degree in Control Theory and Applications (1993) from Tsinghua University, P. R. China. From March 1997 to May 1998, he was a visiting research fellow in Industrial and Operations Engineering Department, University of Michigan at Ann Arbor. From September 2006 to August 2007 he was a visiting professor in Department of Electrical and Computer Engineering, Seoul National University, South Korea. His research interests include self-tuning adaptive control, multiple model adaptive control/estimation for both linear and nonlinear dynamic systems. As representative research work, he established a Virtual Equivalent System (VES) theory for unified analysis (stability, convergence, and robustness) of self-tuning control systems, which is independent of specific control strategy and parameter estimation algorithm. With the help of VES, he proved the stability of weighted multiple model adaptive control system.

Hamidreza Nemati

Hamidreza is a highly motivated and skilled researcher, who is currently a lecturer in Aerospace Engineering at the Engineering Design and Mathematics Department of the University of the West of England (UWE Bristol), UK. He has a PhD in Guidance and Control from the Aeronautics and Astronautics Department of the Kyushu University, Japan, postdoctoral experience at the Aerospace Engineering Department of the Amirkabir University of Technology, Iran, and worked as Research Associate at the Engineering Department of the Lancaster University, Lancaster, UK. His work focuses on developing advanced control and navigation techniques for cyber-physical systems with a particular emphasis on robotics and aerospace applications. He has expertise and interest in a wide range of areas, including robust control of uncertain nonlinear systems, disturbance observer-based controllers, modeling and identification of under-actuated systems, autonomous navigation, cooperative control, autonomous control of bio-inspired and insect-sized robots, space debris removal, fault-tolerant control techniques, and ethical implications in robotics and AI. Hamidreza has experience working on a range of projects including the control of a dual-arm hydraulic manipulator, and robust stabilization of unmanned aerial vehicles. He is always interested in applying his skills to real-world problems and has experience collaborating with industries and startups. He is committed to ethical and responsible use of technology and believes that the development of new robotic systems should be guided by principles of safety, fairness, and transparency.

YuMing Zhang

YuMing Zhang has been with the University of Kentucky, Lexington, Kentucky, USA, since 1991, where became Full Professor in 2005. His research focuses on innovative welding processes and intelligent robotic and human-robot collaborative systems and has brought him 12 US patents and over 220 journal publications. His recognition includes (1) Fellow of the American Welding Society (AWS), ASME, the Society of Manufacturing Engineers (SME), IEEE, and Asia-Pacific Artificial Intelligence Association (AAIA); (2) 2002 IFAC World Congress Best Poster Paper; (3) University of Kentucky College of Engineering Dean's Award for Excellence in Research and University of Kentucky Research Professor; and (4) 13 awards from the AWS. In addition, six of his graduate students won IIW (International Welding Institute) Henry Granjon Prize. YuMing Zhang is currently Area Editor for the Journal of Manufacturing Processes published by the SME. He is also a past Chair for the AWS Technical Papers Committee and is currently a member of the AWS Fellow Selection Committee.

Preface

In the variegated scenario of the multi- and inter-disciplinaries in which the concepts of Cyber-Physical Systems (CPS) are located, it is important to take some historical aspects into account. Before and after Cyber-Physical systems, Cyber-Security Systems are of great interest, because of their perception, for controlling and operating of the systems. They are extremely adaptable and changeable and contribute to increasing performance. Processes run autonomously and free up the skilled worker to perform complex tasks. They avoid productivity loss and re-work. On the other hand, they can make better use of cost (time and money). Furthermore, cyber-physical systems can primarily decrease the carbon impact and noise pollution in the local environment. For example, the UK government has been committed to employ carbon neutral operations by 2050. Moreover, United Nations sets out research and development goals into climate action, clean energy, and responsible production and consumption. This implies that "robotification" through renewable and sustainable sources is the required leading technology. In addition, there are dependencies that can paralyze the entire process if individual components or parts of the infrastructure fail. If the systems predominate in autonomy, then it can happen that wrong decisions are made. Another danger arises from the deliberate manipulation of cyber reveal-related items by hackers or administrators. Concerns about network rights require special protection of the systems against attacks or hostile takeovers. This kind of structures is in continuous evolution, and they need to be studied and analyzed continuously, in particular in terms of their modeling and consequently of their control structures. The emphasis on this last crucial point is the task of this book, which represents the crucial role of the algorithms in the context of identification and control of such a system "towards" Industry 4.0. The word "towards" emphasizes that this industrial revolution is not realized yet. The revolution is coming up, and it projects us into a possible future in our life. A scientist who wants to actively operate in this upcoming process should know very well not only the importance of the algorithms and their crucial role for the safety, optimization, and control of such a system, but should be able to generate himself the algorithms taking inspirations from already existing ones. Optimization, control, and identification algorithms as well as fusion safety in the control strategy together with optimal cooperation of different components are central in this book. Machines and humans will cooperate together as well as different components at different levels: actuators, sensors, and their corresponding virtual versions; virtual actuators and sensors play a special role in the context of the functionality, safety in the presence of fault, and optimization aspects. In this context, the book addresses these challenges. The need to emphasize more and more intelligent algorithms is a primary task of this book.

Objectives

In a cyber-physical system, mechanical components are connected to each other via networks and modern information technology. They enable the management and control of complex systems and infrastructures. Cyber-physical systems, often abbreviated as CPS, consist of mechanical components, software, and modern information technology. By networking the individual components can be regulated and controlled via networks such as the Internet and complex infrastructures. The exchange of information from the networked objects and systems can take place in real time, wirelessly, or by cable. The components of CPS include mobile facilities as well as stationary machines, systems, and robots. CPS play a central role in Industry 4.0. The technological basis for the CPS is provided by sciences such as computer science, mathematics, mechanical engineering, electrical engineering, and robotics. The functional principle is based on sensors, actuators, and their virtual versions to ensure safety and continuity of service in the presence of faults or failures. These components are networked with dedicated software. Sensors deliver measurement data from the physical world and report them via networks to software that processes them. This results in the control data that the software forwards to actuators via the network. CPS are characterized by a high level of complexity and are used, for example, for the implementation of intelligent power grids, modern production systems, or in medical technology. All these aspects, which play a crucial role for the co-existence of all these variegated subsystems, are stated by the capability to understand and produce or reproduce intelligent algorithms devoted to the control of such a system. As a conclusion, the objective of this book is to summarize all these pointed out research aspects with their current and future possible long-term significance solutions.

In Chapter 1, "Industry 4.0 more than a challenge in modeling, identification, and control for cyber-physical systems", the authors determine maintenance management and other necessities that businesses need to implement to realize the full potential of Industry 4.0 and to become the perfect smart factory. To accomplish this, two research questions were analyzed. First, we need to figure out how Industry 4.0 and CPS work together to improve preventative maintenance. Second, we want to learn more about the role that this type of integration plays in the future factory of perfect maintenance management. The work employs a case study and a literature review as its research techniques. Empirical data have been gathered for the case study using semi-structured interviews (providing a broad overview of the case company's present maintenance management) and focus groups (generating more particular and detailed information regarding the research issues). Moreover, the example company's relevant documents study improved the overall comprehension of the present concepts applied in the maintenance department. Several challenges and issues related to maintenance are uncovered by the case study. Since not all of these issues can be addressed with just technological means, they are further examined and sorted into distinct categories. Maintenance management is chosen as the primary area of focus in the recommended criteria list for the ideal smart factory. After that, recommendations are made about how to proceed with creating a CPS-based maintenance management system.

In Chapter 2, "Advanced ice-clamping control in the context of Industry 4.0", the authors address an innovative milling system in which the object to be milled is fixed by an ice clamping system. This system allows to obtain "zero deformation" of the object, and thus a precise milling process is obtained. The chapter takes into consideration the control strategy based on Sliding Mode Control (SMC) to ensure robustness even in the absence of the knowledge of the system to be controlled. Measured results demonstrate the effectiveness of the proposed device and its control strategy.

Chapter 3, "Temperature control in Peltier cells comparing sliding mode control and PID controllers", deals with temperature control in Peltier cells comparing SMC and PID controllers. In the context of Industry 4.0 transformation the control strategies play a crucial role, in particular, if the systems to be controlled are innovative and show futuristic techniques of production and transportation and in general new technologies. In this chapter a comparison between a classical PID controller and an SMC is shown. The industrial system considered is an innovative milling system in which the object to be milled is fixed by a clamping-ice system, where the temperature of the ice needs to be controlled to maintain the clamping forces. The measured results are validated and highlight the efficacy of the innovative milling system together with the proposed control strategy.

In Chapter 4, "A Digital Twin for part quality prediction and control in plastic injection molding", the plastic injection molding process has been established as the most widespread manufacturing process in the plastic processing industry. Among the decisive factors contributing to its prevalence are the ability to manufacture parts with intricate geometries and a high degree of automation. Approaches of varying complexity to control part quality to reduce waste and increase the efficiency of the process exist: The industry standard is the control of so-called machine-variables, i.e., process variables that are measured on the machine-side of the process. This does not take into account any variables that reflect the true state of the emerging part. For this reason, the scientific community aims to control process variables that are measured cavity-side, more precisely, the pressure in the mold cavity. However, the implementation of pressure control requires significant control knowledge and is not suitable for large-scale industrial application. The objective of this contribution is therefore to transform an ordinary machine-variable controlled injection molding machine to a Cyber Physical Production System (CPPS) via augmentation by a digital twin (DT). The digital twin predicts part quality from process variables. To this end, a state-of-the-art industrial injection molding machine will be equipped with additional sensors that measure in-cavity process variables. Moreover, an in-line quality measuring cell is added. By doing so, all machine, process, and quality data required for data-driven modeling and prospective control are acquired. Subsequently, an internal dynamics approach for predicting final batch quality from process value trajectories is proposed and compared to the current state-of-the-art modeling approaches in two case studies.

Chapter 5, "SLAM algorithms for autonomous mobile robots", investigates Simultaneous Localization And Mapping (SLAM) for autonomous robots. Due to the limitations of the indoor operating environment, GPS cannot be used to restrict positioning errors, and SLAM opens another door for the development of the indoor robot positioning and guidance. Among different technologies, SLAM based on LiDAR has already become a relatively mature scheme, but the cost-effective problem is still prominent. As a practical solution, the low-cost visual SLAM (VSLAM) has become a research hot spot in recent years. However, no matter which sensor is used alone, there are some obvious defects. The multi-sensor fusion technology based on LiDAR, such as vision sensor and inertial measurement unit, can not only realize the cooperative operation among sensors, but also greatly enhance the robustness of the positioning and guidance. It is believed that the research and applications of multi-sensor fusion technology will bring wider space to driverless vehicles, robotics, augmented reality, and virtual reality. In addition, SLAM can also be combined with deep learning to perform image processing, to generate semantic maps of the environment and improve the human–computer interaction techniques, so that artificial intelligence can be better realized in positioning and guidance.

Chapter 6, "Optimization of motion control smoothness based on Eband algorithm", employs elastic band (Eband), which is a local path planning algorithm similar to timed elastic band (TEB) to improve the robot walking velocity and action consistency. The dynamic obstacle avoidance effect of the local planner is remarkable, and the path to avoid obstacles can be planned in advance, so that the action and response can be made earlier. However, its motion control changes with the fluctuation of "bubble", which sometimes results in the path being not smooth enough. This chapter mainly optimizes the algorithm for the motion control smoothness of Eband and proves the effectiveness of this method through comparative experiments.

Chapters 7 and 8, "Modeling a modular omnidirectional AGV developmental platform with integrated suspension and power-plant" and "Control system strategy of a modular omnidirectional AGV", deal with an omni-directional automatic guided vehicle (AGV), which is a wheeled self-navigation system. Unlike autonomous vehicles, this AGV follows a predetermined "virtual" path. Since factories are space optimized and designed around human ergonomics, it is often the case that a traditional Ackermann-style wheeled robot will have difficulty navigating in such an environment. To combat this, it is necessary to make the vehicle "omni-directional", which means creating a system that is able to drive in any cardinal direction. These two chapters discuss two possible strategies for achieving this goal using both mecanum wheels and a novel swerve drive system. These two chapters also justify the use of AGV in general over human labor both economically and socially, especially within the South African environment.

Chapter 9, "Mecanum wheel slip detection model implemented on velocity-controlled drives", discusses the creation of a slip mitigation controller implemented on a omnidirectional autonomous guided vehicle (AGV). The AGV utilizes four mecanum wheels to achieve its omni-directional capabilities. The algorithm was developed to reduce the negative effects that occurred when a single mecanum wheel in a four-wheel mecanum wheel AGV experienced a loss of traction or "slipped".

Chapter 10, "Safety automotive sensors and actuators with end-to-end protection (E2E) in the context of AUTOSAR embedded applications", studies automotive embedded applications. These applications control vehicle dynamics, environment interpretation (including pedestrian or traffic signs detection), intra- and inter-car communication (which requires strong cyber security algorithms), power control, stability, and so on. Hardware and software supports are needed to fulfill all the requirements requested by the functionality itself and implementing the safety mechanisms specified by the ISO 26262 standard. Automotive software is designed based on Automotive Open System Architecture (AUTOSAR) standard. AUTOSAR also describes a safety mechanism to be used in data transfer between the electronic control units (ECUs) inside the car. This software safety mechanism is called end-to-end communication protection (E2E) and consists in protecting the data through a calculated cyclic redundancy code (CRC) before sending on the communication bus (e.g., LIN, CAN, Flexray, etc.) and checking the data integrity on the receiving side. This chapter presents a method to migrate the software E2E mechanism inside the hardware to improve the model of the basic automotive sensors and actuators. By adding this feature it is possible that, besides increasing the safety level, these modules can be directly connected to the network ECUs via standard communication buses. Modeling, designing, and mapping of the hardware E2E modules inside Field Programmable Gate Arrays (FPGAs) are described. The models are validated also by comparing the output of the proposed E2E hardware against the output provided by AUTOSAR software E2E library.

In Chapter 11, "Multibody simulations of distributed flight arrays for Industry 4.0 applications", the authors introduce distributed flight arrays (DFAs) for the new generation of industries. DFAs are an experimental type of aerial, multirotor, vehicle capable of land-based navigation and cooperative aerial flight involving physically docking with N-number of other agents forming a larger structure with some designs allowing for unassisted solo flight. DFAs would be able to be configured into the most resource efficient structures for achieving a specific logistics operation and be capable of manoeuvring around the warehouse environment in a relatively unrestricted manner. For the application of material, handling a reliable, predictable, and safe mode of transporting a payload is required. Gathering large amounts of data across a large variety of payload systems and DFA formations is an extremely large undertaking when done through real-world experimentation. This large scope is much more suitable for a computer-based physics simulation as it allows for rapid iteration and data gathering without the high resource investment of real-world experimentation. This research finds that the multibody simulation software Simscape is capable of complex control system simulation research for handling the flight and navigation of a DFA and that DFA slung payload systems are highly likely to be compatible with future material handling operations due to developments in automation in Industry 4.0.

Chapter 12, "Recent advancements in multi-objective pigeon inspired optimization (MPIO) for autonomous unmanned aerial systems", deals with Unmanned Aerial Vehicles (UAVs) for high functional complex procedures that can improve the intellectual ability and are considered to be a fast-growing prototype system in a broad range of applications. In a context of communication, aerial systems are an effective strategy to increase the autonomous unmanned systems in terms of modeling, sensing, control, and efficient application. This study presents the technical analysis of multi-objective pigeon-inspired optimization (MPIO), which covers the most recent advancements and developments in key technologies in this demanding area. This work also presents the recent developments in the integrated designs, computing algorithms, and mathematical methodologies for enhancing the guidance, navigation, and control of unmanned autonomous ground, aerial, underwater, and surface vehicles. The proposed study justifies the findings and restrictions to create a more efficient framework for algorithm design applied in the autonomous systems.

In Chapter 13, "U-model-based dynamic inversion control for quadrotor UAV systems", a quadcopter as an example of a cyber-physical system is stabilized using a new control law. Due to the recent advances in the field of electronics, resulting in increased capabilities of quadcopters and improved performances, they have gained popularity and garnered widespread attention for their practical applications, and consequently this has resulted in efforts by researchers to try and further improve their performances and capabilities. In this chapter, the authors consider a Parrot Mambo minidrone as a subject of study, deriving its dynamic model and designing a controller using a Model-Independent Design approach (U-Model). The simulation results are presented with graphical illustration demonstrating the effectiveness of the proposed controller.

Chapter 14, "Nonlinear control allocation applied on a QTR: the influence of the frequency variation", presents a study on the influence of the frequency variation of a nonlinear control allocation technique execution, known as Fast Control Allocation (FCA) for the Quadrotor Tilt-Rotor (QTR) aircraft. Then, through Software-In-The-Loop (SITL) simulation, the proposed work considers the use of Gazebo, QGroundControl, and MATLAB applications, where different frequencies of the FCA can be implemented separated in MATLAB, always analyzing the QTR stability conditions from the virtual environment performed in Gazebo. The results showed that the FCA needs at least 200 Hz of frequency for the QTR safe flight conditions, i.e., two times smaller than the main control loop frequency, 400 Hz. Lower frequencies than this one would cause instability or crashes during the QTR operation.

Chapter 15, "Active disturbance rejection control of systems with large uncertainties", presents a multiple model active disturbance rejection control strategy for a class of systems with large uncertainties. The authors first have a look of some popular control strategies, including PID (proportional-integral-derivative) control, active disturbance rejection control (ADRC), model predictive control (MPC), and parameter identification-based adaptive control, to find their connections, advantages, and disadvantages. The key points include: PID is actually a model reference adaptive control with human in the control loop as an adjuster of controller parameters according to a desired step response curve; ADRC is a reinforced PID control and also a special kind of adaptive control; and finally, there is a common virtual equivalent system (VES) framework for MPC, ADRC, and parameter identification-based adaptive control. In terms of methodology (how authors treat model and model error), different control strategies usually lead to the same end (tracking and disturbance rejection). Different control strategies have their own suitable application situations with consideration of control requirements and cost-effectiveness. Finally, a multiple model ADRC control system is presented with simulation verification, with the purpose to address the control problem of plants with large uncertainties.

In Chapter 16, "Gain scheduling design based on active disturbance rejection control for thermal power plant under full operating conditions", the authors introduce a gain scheduling design based on active disturbance rejection control (ADRC) for thermal power plants under full operating conditions. To integrate more renewable energy into the power grid, thermal power plants have to accelerate the speed of power output and extend their operating ranges, and this can result in great challenges for their safe operations and even safety accidents. The urgency of the proposed control strategy is illustrated by analyzing the control difficulties of coordinated control systems. Then the scheduling parameter selection and the linear switching method for the proposed control strategy are analyzed. Moreover, the qualitative stability analysis based on the Kharitonov theorem and the calculation of quantitative stability regions is carried out to ensure the stability of the closed-loop system. Simulations of the power tracking under different load tracking rates and disturbance rejection under the coal quality variation are carried out. Simulation results show that the tracking and disturbance rejection performance in both power output and throttle pressure loops has been improved simultaneously compared with the regular ADRC and the traditional proportional-integral control strategies. Based on the verified superiority, the proposed gain scheduling design based on ADRC shows a promising potential in industrial applications.

Chapter 17, "Active disturbance rejection control of large-scale coal fired plant process for flexible operation", proposes a quantitative tuning method for parameters available in the active disturbance rejection control (ADRC) law. Laboratory experiments on the water tank and the power plant simulator highlight the feasibility and superiority of the proposed tuning method for high-order industrial processes.

Chapter 18, "Desired dynamic equational proportional-integral-derivative controller design based on probabilistic robustness", deals with existing uncertainties in industrial processes that may lead to many challenges for controller design. To enhance the ability of the closed-loop system to handle the uncertainties, a desired dynamic equational (DDE) proportional-integral-derivative (PID) controller is designed based on probabilistic robustness (PR). The necessity of the proposed design method is demonstrated by introducing the problem formulation. Based on its fundamentals, DDE PID designed based on PR (DDE-PR PID) is proposed for uncertain systems, and the corresponding design procedure is summarized as a flow chart. Then the proposed DDE-PR PID is designed for several typical processes, and simulation results indicate that the proposed DDE-PR PID can not only achieve satisfactory control performance for nominal systems, but also satisfy control requirements for all uncertain systems with the maximum probability. Finally, the proposed DDE-PRPID is applied to the level system of a water tank. Its superiority in robustness is validated by both simulations and experiments, which shows the promising prospect of DDE-PR PID in future power industry.

Many thanks to Massimiliano Dominici for his comprehensive help in suggestions, sharing ideas and concepts, and LaTeX support, but not only limited to those.

Paolo Mercorelli[a]
Hamidreza Nemati[b]
Quanmin Zhu[b]

[a]*Institute for Production Technology and Systems, Leuphana University of Lueneburg, Lueneburg, Germany,* [b]*College of Arts Technology and Environment, Department of Engineering Design and Mathematics, University of the West of England, Bristol, United Kingdom*

Chapter 1

Industry 4.0 more than a challenge in modeling, identification, and control for cyber-physical systems

Paolo Mercorelli[a], Hamidreza Nemati[b], and Quanmin Zhu[b]

[a]Institute for Production Technology and Systems, Leuphana University of Lueneburg, Lueneburg, Germany, [b]Department of Engineering Design and Mathematics, University of the West of England, Bristol, United Kingdom

1.1 Introduction

1.1.1 Background and challenging issues

Globalization has forced firms to cater to customers with widely different preferences and to contend with rivals from around the world. Historically, sellers held more influence in the market, but as the market has developed and matured, buyers have become more influential. As a result, there is a movement away from the factory-style mentality of mass production and toward one that is more subtle and deliberate. This has led most companies to adopt the practice of personalization as a means of increasing profits. To truly realize customization, a corporation needs to have two things: (1) an in-depth understanding of market customers and (2) the capacity to suit those customers' different wants while maintaining economies of scale. Market segmentation based on customer feedback, competition analysis, demand forecasting, etc. is essential for the former [26]. The second criterion focuses on production-related concerns, such as achieving a balance between competing performance goals and ensuring output, and corresponds to consumer demand. Both sets of rules stress the importance of a company's ability to acquire and manage massive amounts of data, both internal and external.

Manufacturers can respond to this demand with the support of a set of practices known as "Industry 4.0," which was coined in a German government project in 2012. Due to the widespread impact, that computerization of manufacturing systems during the third industrial revolution had, the notion of "Industry 4.0," also known as the fourth industrial revolution, has been gaining steam since

1

its introduction in 2012. Industry 4.0, a manufacturing system in which sensors are implanted in all physical items to link the real world with a computer simulation of it, is built on the foundation of Cyber-Physical Systems (CPS) [9]. Big Data is the term used to describe the enormous quantities of data that are being produced on a regular basis as a result of the proliferation of sensors and other forms of networked technology.

It is extremely difficult to understand what is meant by the term "Industry 4.0" in isolation. The fourth industrial revolution ushers in a new way of thinking about the creation and use of CPS, though, so keep that in mind. Additionally, there are some similarities between CPS and Industry 4.0, but the latter word barely touches the surface of what the former entails. Since CPS research is still in its infancy, many questions remain unanswered about its optimal design and implementation. Given that CPS is the bedrock of the fourth industrial revolution, it is important to look at how the two concepts relate to one another to fully grasp Industry 4.0.

Not to mention, as was previously noted, the shift in production priorities has stimulated the development and implementation of innovative manufacturing strategies like Just-In-Time. However, the efficiency of these methods is often impeded by inflexible or broken equipment. Therefore, in a world where everything is always changing, it is crucial for firms to maintain relatively stable output levels. When it comes to the health of a factory, one of the most important metrics to look at is overall equipment efficiency (OEE), which is inversely proportional to the standard of care given to the machines housed there [1]. As a result, the efficiency and efficacy of a manufacturer's maintenance function have a significant impact on the manufacturer's bottom line. Due to its many downsides, including unannounced downtime, excessive damage, and a high incidence of troubleshooting issues, the tried-and-true reactive maintenance method can no longer be relied upon by manufacturers. Therefore firms started implementing innovative maintenance practices, which enhanced the efficiency of their machines but required more time, money, and other resources to implement successfully [26]. The CPS is a cutting-edge tool for real-time asset tracking and forecasting. However, CPS is still in its infancy, and a great deal of research and effort is still required before CPS-based strategies can be developed and implemented. Since the success of maintenance processes is becoming increasingly vital, studying how CPS could be used to maintain operations in line with Industry 4.0 is exciting.

An automobile glass manufacturer collaborated on the study. In this situation the unit under review was factory maintenance at the company's main office. The company has faced increasingly stiff rivalry in recent years. It was considered by the board of directors that leading the charge into the fourth industrial revolution was the best way to regain lost ground in the market. Given the existence of problems, the company decided to look into CPS as a potential source of innovation. Some preventative maintenance tasks, for example, drained resources, were probably unneeded, whereas others had serious negative effects on product quality. Studying the CPS criterion and its potential role in upkeep efforts would have benefited both theories and practice.

1.1.2 Basic concepts of Industry 4.0

Specifically in the manufacturing and industrial sectors, the term "Industry 4.0" has been introduced to denote upcoming changes in the corporate environment. These changes are expected to be brought about by technological advancements. It should come as no surprise that both the business world and the academic world recognize the significance of the fourth industrial revolution. In the context of this discussion, the term "smart manufacturing" refers to the fundamental component of the idea. The term "smart manufacturing" refers to a burgeoning sub-industry within the manufacturing industry that makes use of the integration of production assets with sensors, computer platforms, communication technologies, control, simulation, data-intensive modeling, and predictive engineering.

A new era in manufacturing has been dubbed "Industry 4.0," and it is characterized by the convergence of a wide range of cutting-edge technologies in order to give digital solutions. The new paradigm of Industry 4.0 is based on the two-way communication between humans and machines in a highly networked environment. This communication is made possible by revolutionary automation technologies such as cyber-physical systems (CPS), the Internet of Things (IoT), and cloud computing, among others. The term "computational and physical processes" (CPP) refers to a set of processes that combines computational approaches with physical ones. In other words, CPS stands for computational and physical processes. Controlling electronics typically requires the utilization of embedded computers and networks.

The utilization of feedback loops in Asian techniques is essential to ensuring that things remain stable. Because of the communication-executive network, sensors and actuators are becoming increasingly commonplace. This makes it possible for them to be integrated naturally into our surroundings and for systems to easily share and receive data from one another. As a direct result of this change, businesses have begun distributing the various stages of their products across a number of different sites. By utilizing a remote monitoring and management system for instrumental performance, this organizational structure makes it possible to perform an accurate simulation of real-world conditions. The decentralized and virtualized paradigms are characteristics of the Fourth Industrial Revolution since this is the case.

The ever-expanding domain of Industry 4.0 is a cut-throat marketplace that necessitates the development of cutting-edge production strategies predicated on consistent adaptability and reconfigurability. The world of business has moved on to a new paradigm, and modeling strategies that have been effective in other areas are assisting with the change. In point of fact, when applied to a new setting in which modeling plays a significant part, new production strategies and

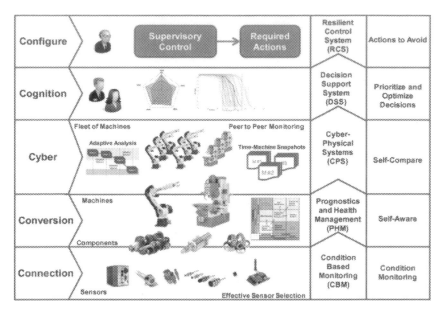

FIGURE 1.1 5C architecture for implementation of CPS [13].

processes become apparent. Fig. 1.1 shows 5C architecture of CPS as follows: connection, conversion, cyber, cognition, and configuration. This new kind of organization has abandoned traditional command-and-control hierarchies in favor of a management style that is more decentralized. This is the defining feature of the new type of organization. The intelligence of connected machinery, which is given by smart devices, is essential to the functioning of Industry 4.0.

The internet's ability to connect seemingly unrelated pieces of everyday life is one of the most important ideas to come out of the fourth industrial revolution. In this envisioned future, sensors that are installed in machines will be able to enable two-way communication across ad hoc networks made up of direct links between individual objects. This would lead to the creation of what is known as "the Internet of Things" (IoT). The various machines are interconnected with one another and function as a unified whole. The IoT also presents substantial commercial and economic potential. The Internet of Things (IoT), which is an essential component of Industry 4.0, is anticipated to provide game-changing insights into the administration and operation of various extant industrial systems within the digital enterprises that will be a part of the complex industrial ecosystems of the future.

Due to the fact that it places emphasis on the confluence of the digital and physical worlds, Industry 4.0 makes extensive use of the Internet of Things. Embedded sensing and acting capabilities, distributed processing, and cloud-based management are the three key components that make up the 5C architecture, which is used throughout the CPS development process. At its heart, the 5C architecture consists of these three primary components.

The idea of "cognitive automation" is a central tenet of the fourth industrial revolution. This is a concept that aims to bridge the gap between the current model of industry, in which robots have largely replaced human labor, and the ideal model, in which data is transmitted quickly and processes are perfectly in sync with one another.

The term "human–machine interaction" refers to the exchange of information in both directions that takes place between humans and machines in an active environment and using a variety of user interfaces. Ever since the discovery of tools, humans and machines have collaborated on many projects. The character of this cooperation has shifted throughout the course of so many years. The genuine technological adaptation of humanity took place prior to World War II when individuals learnt how to correctly employ mechanical equipment. However, as the war progressed, technological advancements made it increasingly difficult to educate personnel. As a result, it was required to conduct an in-depth analysis and synthesis of the interaction between humans and machines. The history of interactions between humans and machines can be loosely divided into four distinct time periods. Beginning in the 1940s and continuing until the middle of the 1950s, researchers investigated the limits of what was possible for humans. All of the highly sophisticated technologies available today are created with the end user in mind. Between the years 1955 and 1970, various enhancements were carried out. The scientists of the time attempted to develop mechanical models of humans in the hope that they could improve the tools that were designed for humans. In the 1970s, advancements were first made in the field of electronics. Plus, beginning in 1970 . . .

As early as 1985, this technology was already being used to replace humans in a variety of tasks across a variety of industries. The previous controller's duties were elevated to those of a supervisor once the position was reclassified. There has been a significant amount of development since the year 1985. During the process of developing robots, it is currently common practice to take into account aspects such as workload, cognitive processes, and emotional interaction models.

As a direct result of this, there has been a significant change over the course of history in the manner in which people interact with machines, and Industry 4.0 is a significant advancement in this particular domain.

1.2 Theoretical background

As stated in the introduction, the study's main focus is on the function of CPS in Industry 4.0 and the connection between CPS and upkeep. The significance of "Industry 4.0" in the context of contemporary production will first be introduced theoretically, allowing the reader to gain a firm footing. The origins of CPS are then elaborated upon, and its modern applications are outlined. Later, we will talk about how maintenance has changed over time and how maintenance tactics have changed over time.

Three prior industrial revolutions might be linked to significant technological advancements. The advent of the water and steam engine towards the end of the 18th century marked the beginning of this new era of industrialization. Because of the widespread availability of electricity and the introduction of the assembly belt, the second industrial revolution began in the latter half of the 19th century. Advances in electronics and information technology in the middle of the 20th century prompted the third industrial revolution. With CPS, the fourth industrial revolution may begin, as data from several sources can be monitored and kept in sync between the physical factory floor and the digital computer environment. After more than two decades of employing CPS in production, the German government coined the term "Industry 4.0." German companies currently face global competition in many areas, but especially in product quality and pricing [8]. Industry 4.0 has evolved in response to this demand, with the expectation that it will strengthen the German economy, foster greater international cooperation, and give rise to innovative online marketplaces. The authors Åström and Eykhoff [3] found more evidence that an Industry 4.0 approach could help Germany keep its competitive edge in the global high-wage economy. The phrase "Industry 4.0" has come to signify these changes, but it is being used inconsistently and has no agreed-upon definition.

1.2.1   Internet of Things and services

Due to the increasing adoption of the concept, the phrase "Internet of Things (IoT)" has been ubiquitous over the past several years. An IoT is "a global infrastructure for the Information Society, enabling advanced services by interconnecting (physical and virtual) things based on, existing and evolving, interoperable information and communication technologies," as defined by the International Telecommunication Union (ITU) in 2012. There are many possible applications for the Internet of Things (IoT), but one of the most prominent is in the smart industry, especially in the form of Industry 4.0, which refers to the development of intelligent production systems and connected production facilities. The Internet of Things is utilized to enhance the primary physical functions of an item by including secondary IT-based digital services accessible on a worldwide scale [10]. For instance, a light bulb serves primarily as a source of illumination. This information technology service could be used in a variety of contexts, such as a security system that detects and responds to individuals in the vicinity by adjusting the lights. That is to say, with the help of digital features made possible by information technology, products can be made smart and remote-controlled to better serve the demands of their owners. The features and capabilities of any product can be enhanced when they are integrated into a larger product system. As an added bonus, Kim [11] claims that IoT technology will allow the integration of previously independent product systems, resulting in systems capable of altering competitive dynamics and expanding the bounds of existing industries.

Despite its many advantages, the IoT cannot be built without first integrating numerous hardware and software components. These needs are in keeping with the current pattern of growth in Industry 4.0, which aims to achieve integration between networks by means of improved information flow. In a report published in 2010, Zuehlke proposed the automation pyramid as part of the so-called "factory-of-things" strategy. A total of four tiers make up this model: device, control, manufacturing execution system (MES), and enterprise resource planning (ERP). Sensors, actuators, programmable logic controllers (PLC), MES, and ERP system are all covered by the system [20]. Signals are captured at the device level, then relayed via messages to the ERP systems, and lastly, shown to the user in some way. Most modern factories use some variation on this automation pyramid, with data from various enterprise resource planning (ERP) systems serving as the backbone for manufacturing management. However, the necessity for differentiated information at levels beyond the ERP level has been disregarded by the one-way information flow that relies on messages.

1.2.2   Smart manufacturing

The current production paradigm is unsustainable, despite the advances in contemporary industrial methods over the past few decades. In addition, adaptability has become a vital performance indicator for most businesses since the need for highly customized goods in small batches remains constant. Like conventional manufacturing and business intelligence, "smart manufacturing" seeks to "do something" with data by transforming it from "information" into "knowledge." The emphasis in smart manufacturing, on the other hand, is on the real-time gathering, integration, and sharing of data between the many different physical and digital processes involved [5].

Nowadays, data analysis plays a crucial role in factories for the purpose of monitoring operations and detecting defects. Implementations of OEE measuring tools at the system level are used to minimize waste and maximize efficiency in production. In contrast, the "Industry 4.0 factory" allows machines and their constituent parts to acquire intelligence and predict their own actions. An Industry 4.0 facility's ability to self-configure, self-maintain, and self-organize at the system level guarantee optimal performance of the factory's complete production system.

Thanks to the abundance of data available online, smart devices can also engage in cross-fleet comparisons of similar equipment. The ability to anticipate issues with the use of comparisons like these is crucial for making decisions regarding just-in-time maintenance that results in minimal downtime. Industry 4.0 aspires to transform organizations into interdependent networks with the ability to perform self-optimization across a wide range of operational metrics, including but not limited to situational awareness, predictive ability, comparative ability, configuration, and maintenance [6].

1.2.3   Vertical integration and networked manufacturing system

Vertical integration happens in the factory setting. Some of the many informational and physical subsystems that can be found at a manufacturing facility include actuators and sensors, control and production management, manufacturing, and corporate planning. When an organization's IT is vertically integrated, all layers of the system communicate and collaborate to deliver superior support to customers at every level. CPS is used to facilitate rapid responses to changes in demand, stock levels, or equipment failure. As a result, changes need to be made in the factory to accommodate CPS. Configuration rules are gradually replacing concrete factory architectures and process definition to automatically construct case-specific topologies [12]. For this reason, "smart factory" emerged as a result of the practice of vertical integration. Smart factories provide significant real-time advantages in quality, time, resources, and costs over conventional production processes. The flexible network of the CPS-based production system is responsible for these advantages, since it streamlines and automates numerous processes in the manufacturing sectors. CPS-based vertical integration optimizes process structure and increases production process adaptability, allowing for the dynamic reallocation of production schedule in response to price discrepancies, order changes, quality fluctuations, etc. [6]. Therefore the modern production techniques we employ are geared at accommodating the specific needs of each individual customer.

1.2.4   Marginalization of the network center: towards horizontal integration

These modern methods of creating money are based on real-time optimized networks that have high degrees of openness and flexibility. Organizations can only maintain their global competitive advantages by concentrating on their core competencies and delegating non-core duties to their network of partners. Networks allow for the multiplication of existing resources with no additional financial investment [24].

According to Navickas, Kuznetsova, and Gruzauskas [17], adapting network's efforts to fluctuating market conditions is a key to maintaining a competitive edge over the long term. To make significant progress in today's ever-increasingly complex world, a wide spectrum of organizations must collaborate. Horizontal integration can encourage transparency because users have access to both aggregated data and detailed information about their own machines. Because of this, deciding which maintenance tasks are most important will be simpler.

However, due to the reality that building new business models involves the combined efforts of various firms with distinct but supplementary sets of skills, new business modes and cooperation models will emerge. To maximize the time-saving potential of collaboration, the supply chain must be adaptable. Monitoring not just commodity flows, but also delivery reliability and customer satisfaction metrics, is essential to supply chain agility. Therefore networked production systems promote transparency and adaptability from procurement to manufacture to retail, and so do horizontal networks that allow networking through CPS.

1.2.5 Cyber-physical system

CPS is the backbone of the Internet of Things, which, along with the Internet of Services, makes Industry 4.0 possible. In the near future, organizations will build CPS-shaped worldwide networks to link all of their facilities, facilities, and equipment. Since the fourth industrial revolution is the subject of this thesis, we provide a brief overview of the term and how it is being applied in the manufacturing sector.

Cyber-physical systems are a set of technologies that are changing the way businesses handle their physical assets and digital data [24]. This new era of fully networked, intelligent, and interconnected physical goods is being heralded by the development of Cyber-Physical Systems, a set of enabling technologies that serve as a bridge between the digital and physical world. This system bridges the gap between the virtual world of computing and communication and the physical one.

Cloud platforms, embedded systems, and sensor networks, as stated by Lin et al. [14], form the basis of CPS, and individuals cannot avoid these components regardless of how they engage with CPS. Embedded systems, real-time systems, distributed sensor networks, and control systems are all components of what Havard et al. [7] call cyber-physical systems (CPS), which is in line with this definition.

1.2.6 Commercialization of CPS

Manufacturing facilities, data centers, and other intelligent machines that can work together without human oversight are all examples of such CPS. The CPS field, however, is just now emerging from its infancy. Consequently, there is no universally applicable CPS structure or architecture. Traditionally, CPS has been split into two parts: hardware and Paula Ferreira et al. [19], who state that CPS has a physical component that perceives the physical environment, collects data, and puts the decisions made by the computational component into action. A more nuanced perspective emerged as a result of CPS. For now, let us start with a high-level framework, which will hopefully provide some light on this system.

1.2.7 Outcomes of application of CPS

According to MacDougall [16], the implementation of CPS into production systems makes it possible to have a smart factory, which is a fundamental idea behind the Industry 4.0 initiative. Vertical integration, with Industry 4.0 serving as a backdrop, requires the implementation of a smart factory that can swiftly adapt changes in design. This is necessary for vertical integration. The finished product is the result of the factory's processing of inputs such as raw materials and intermediate items, among other things. For production and administrative purposes, factories frequently make use of a diverse range of specialized physical and/or electronic subsystems. Traditional businesses frequently make use of hierarchical organizational systems composed of numerous levels and divisions. Therefore concerns at the unit level are typically prioritized over issues that arise at the frontiers and interfaces of systems. Material flow is accomplished through the use of regimented production lines; the flow of information between various subsystems is hampered by a number of bottlenecks, and it is frequently challenging to ensure continuity and uniformity [25].

Unbelievably, there is a lot of hope that intelligent industrial technology will be able to help with this problem. Its exceptional levels of efficiency, agility, and adaptability are made possible by the web of information from interconnected systems, which can be broken down into three primary categories: components, machines, and production systems. Adaptability is the ability to quickly respond to changing conditions and circumstances. The degree to which these items have the ability to increase our factory's visibility and our level of understanding varies. The "smart factory" is an attractive and promising production strategy that has the ability to address global concerns. This model focuses on the manufacturing of individualized and low-volume items. It is an entirely novel method of production, which reduces expenses and increases productivity, and the fact that it is networked means that it opens up a great deal of new doors for conducting business [4].

1.3   Method and implementation

Generally speaking, when people talk about "research techniques," they mean a comprehensive approach to answering research issues. According to Oluyisola, Bhalla, Sgarbossa, and Strandhagen [18], validity and reliability are typically used as the primary means of ensuring integrity in qualitative research, mostly because "a good process of evaluating qualitative research methodologies has not been discovered". Moreover, they suggested triangulation as a way to guarantee credibility. Consequently, the thesis employs a wide variety of research techniques in an effort to practice method triangulation, the goal of which is to guarantee that study outcomes are comparable.

An interview is a conversation in which one person asks another questions for a predetermined purpose. Semenkov, Promyslov, Poletykin, Mengazetdinov [21] suggest doing exploratory interviews as a first step in many different types of research projects. This method is frequently seen as highly efficient for collecting information because the interviewer can address specific concerns as they arise. As a result of the one-on-one nature of the interview, respondents are more likely to respond to questions asked in person, making interviews one of the most effective techniques of data collection.

According to Albers et al. [2], structured, unstructured, and semi-structured interviews are the three main varieties of the interview. Initial information on the case firm was gleaned from an informal interview with the production manager, and thanks to the manager's suggestion, key individuals who might be useful in future studies have been uncovered. Then semi-structured interviews were used to extract in-depth comprehension of the research issue at hand, with participants also given the chance to submit supplementary data. It is crucial to conduct both open-ended and in-depth interviews with participants to properly comprehend their points of view.

1.3.1   Research process

Due to lack of familiarity with automotive glass production, researchers in the case study lacked a proper framework for understanding the challenging workplace and process of the case company. Thus the case study research approach comprises of an orientation study and a primary study, the former of which aims to familiarize oneself with the factory, and the latter of which strives to identify key individuals/phenomena for future investigation.

The research strategy kicked off with a complete Gemba walkthrough of the manufacturing line lead by the production manager. After being briefed by management, investigators viewed the whole production process. Both managers were initially reluctant to have their interviews videotaped. Therefore we took turns pitching questions and taking notes to keep the interviews moving along smoothly. With the manager's help during the interviews, we were able to identify possible participants and learn when they were available, laying a solid framework for the subsequent, more extensive study.

1.3.2   Findings and analysis

There is a dearth of research into the potential integration of CPS with Industry 4.0 for use in maintenance. To illustrate how CPS could be used to maintenance in an Industry 4.0 setting, we give here theoretical findings from the literature.

The model's five tiers provide a stepwise framework for introducing and employing CPS-based maintenance. Acquiring precise, up-to-date data from the real world is the first step in the "smart collection" stage. It is possible that the data may be measured in a number of ways, then sent to the centralized server over the internet or some other suitable network.

Level of conversation: In recent years, numerous analytical tools have been created to aid and support this procedure. Prognostics and health management (PHM) algorithms, for instance, have seen rapid development in recent years because of their potential to give machines a degree of intelligence and foresight [13].

When we talk about the "cyber level," we are referring to the digital realm where all of our data reside. On the other hand, the future behavior of machines can be predicted by measuring the degree to which their performance is similar to that of prior assets (historical data).

Knowledge is generated at the cognition level, and reasoning information is provided to correlate the effect of various system components. As a result of having access to data that allows for comparison as well as individual machine state, decisions can be made on the order of maintenance tasks [23].

1.4 Conclusions

Rather than emphasizing how the digital and physical worlds can be brought together, Constant & Proactive Surveillance places an emphasis on the ways in which they already are [22]. Theoretical and empirical studies alike uncovered a number of holes that needed filling before the full potential of CPS-based maintenance could be realized within the context of Industry 4.0.

The initial step in deploying a Cyber-Physical System (CPS) is gathering accurate and up-to-date data from the physical world. An in-depth understanding of the function, design, and consequences of every machine and part used in the production process is essential for success in this endeavor. After that point, data can be either gathered directly from the sensors or extracted through enterprise resource planning (ERP) systems or other types of industrial software.

Next, producers will benefit from having the raw data transformed into insights about the current status of the system. A dependable and potent piece of software is required for the data collection process to properly categorize, normalize, and standardize the data for subsequent usage. Furthermore, a large amount of storage space is necessary for the data to be preserved in a useful form. The data-to-information transition can subsequently be finished with the help of analytical tools. Fault detection and diagnosis (FDD), prognostics and health management (PHM), etc. are only some of the problem-finding methods that can be used across a wide range of strategies and situations. FDD uses input data and a fault logic to locate issues, whereas PHM considers the big picture [15].

A centralized server for storing all information is also necessary. The security and trustworthiness of the server data storage are essential. Now advanced analytic techniques are used to provide a more complete view of the fleet overall and machine-specific health [13]. The future actions of machines can be predicted by looking at how they have previously performed in relation to other assets (historical data). To achieve the desired outcomes, advanced analytic methods are required for determining the causes of failure that are typical for machines of a given type. Further, at this point, you will be creating a digital twin model of the physical machine and all its elements. Maintaining such a twin model in a state of continuous simulation requires a computer with the processing power to deal with the massive amounts of data being created.

Potential benefits may not outweigh the massive cost of developing a CPS-based maintenance management system. Although a maintenance management system based on CPS could be beneficial, it is likely to be useful only for large, well-established organizations with long-term plans. It is possible that fixing problems with human resource management and organizational structure can yield a quicker and higher return on investment than fixing problems caused by technological progress. Filling up these spaces also sets the door for ground-breaking technological advances. To rephrase, using CPS-based maintenance management is more effective and efficient following failures that can be linked to humans.

References

[1] M.H. Ahmed, OEE can be your key, Industrial Engineer 45 (8) (2013) 43–48.

[2] A. Albers, B. Gladysz, T. Pinner, V. Butenko, T. Stürmlinger, Procedure for defining the system of objectives in the initial phase of an Industry 4.0 project focusing on intelligent quality control systems, Procedia CIRP 52 (2016) 262–267.

[3] K.J. Åström, P. Eykhoff, System identification — a survey, Automatica 7 (2) (1971) 123–162.

[4] L. Bakule, Decentralised control: status and outlook, Annual Reviews in Control 38 (1) (2014) 71–80.

[5] D. Cogliati, M. Falchetto, D. Pau, M. Roveri, G. Viscardi, Intelligent cyber-physical systems for Industry 4.0, in: 2018 First International Conference on Artificial Intelligence for Industries (AI4I), IEEE, 2018, September, pp. 19–22.

[6] A.W. Colombo, G.J. Veltink, J. Roa, M.L. Caliusco, Learning industrial cyber-physical systems and Industry 4.0-compliant solutions, in: 2020 IEEE Conference on Industrial Cyber-physical Systems (ICPS), vol. 1, IEEE, 2020, June, pp. 384–390.

[7] V. Havard, M.H. Sahnoun, B. Bettayeb, F. Duval, D. Baudry, Data architecture and model design for Industry 4.0 components integration in cyber-physical production systems, Proceedings of the Institution of Mechanical Engineers, Part B: Journal of Engineering Manufacture 235 (14) (2021) 2338–2349.

[8] T. Huldt, I. Stenius, State-of-practice survey of model-based systems engineering, Systems Engineering 22 (2) (2019) 134–145.

[9] N. Jazdi, Cyber physical systems in the context of Industry 4.0, in: 2014 IEEE International Conference on Automation, Quality and Testing, Robotics, IEEE, 2014, May, pp. 1–4.

[10] M. Khakifirooz, D. Cayard, C.F. Chien, M. Fathi, A system dynamic model for implementation of industry 4.0, in: 2018 International Conference on System Science and Engineering (ICSSE), IEEE, 2018, June, pp. 1–6.

[11] J.H. Kim, A review of cyber-physical system research relevant to the emerging IT trends: industry 4.0, IoT, big data, and cloud computing, Journal of Industrial Integration and Management 2 (03) (2017) 1750011.

[12] M. Krugh, L. Mears, A complementary cyber-human systems framework for industry 4.0 cyber-physical systems, Manufacturing Letters 15 (2018) 89–92.

[13] J. Lee, B. Bagheri, H.A. Kao, A cyber-physical systems architecture for industry 4.0-based manufacturing systems, Manufacturing Letters 3 (2015) 18–23.

[14] W.D. Lin, Y.H. Low, Y.T. Chong, C.L. Teo, Integrated cyber physical simulation modelling environment for manufacturing 4.0, in: 2018 IEEE International Conference on Industrial Engineering and Engineering Management (IEEM), IEEE, 2018, December, pp. 1861–1865.

[15] Y. Lu, Cyber physical system (CPS)-based industry 4.0: a survey, Journal of Industrial Integration and Management 2 (03) (2017) 1750014.

[16] W. MacDougall, Industrie 4.0: Smart Manufacturing for the Future, Germany Trade & Invest, 2014.

[17] V. Navickas, S.A. Kuznetsova, V. Gruzauskas, Cyber-physical systems expression in industry 4.0 context, Financial and Credit Activity Problems of Theory and Practice 2 (23) (2017) 188–197.

[18] O.E. Oluyisola, S. Bhalla, F. Sgarbossa, J.O. Strandhagen, Designing and developing smart production planning and control systems in the industry 4.0 era: a methodology and case study, Journal of Intelligent Manufacturing 33 (1) (2022) 311–332.

[19] W.D. Paula Ferreira, F. Armellini, L.A.D. Santa-Eulalia, C. Rebolledo, Modelling and simulation in industry 4.0, in: Artificial Intelligence in Industry 4.0, Springer, Cham, 2021, pp. 57–72.

[20] D.G. Pivoto, L.F. de Almeida, R. da Rosa Righi, J.J. Rodrigues, A.B. Lugli, A.M. Alberti, Cyber-physical systems architectures for industrial internet of things applications in Industry 4.0: a literature review, Journal of Manufacturing Systems 58 (2021) 176–192.

[21] K. Semenkov, V. Promyslov, A. Poletykin, N. Mengazetdinov, Validation of complex control systems with heterogeneous digital models in industry 4.0 framework, Machines 9 (3) (2021) 62.

[22] F. Tao, Q. Qi, L. Wang, A.Y.C. Nee, Digital twins and cyber-physical systems toward smart manufacturing and industry 4.0: correlation and comparison, Engineering 5 (4) (2019) 653–661.

[23] L.D. Xu, L. Duan, Big data for cyber physical systems in industry 4.0: a survey, Enterprise Information Systems 13 (2) (2019) 148–169.

[24] J. Yan, M. Zhang, Z. Fu, An intralogistics-oriented Cyber-Physical System for workshop in the context of Industry 4.0, Procedia Manufacturing 35 (2019) 1178–1183.

[25] Q. Zhu, Complete model-free sliding mode control (CMFSMC), Scientific Reports 11 (1) (2021) 1–15.

[26] Q. Zhu, Y. Wang, D. Zhao, S. Li, S.A. Billings, Review of rational (total) nonlinear dynamic system modelling, identification, and control, International Journal of Systems Science 46 (12) (2015) 2122–2133.

Part I

Manufacturing as a challenge in Industry 4.0 process

Paolo Mercorelli[a], Hamidreza Nemati[b], and Quanmin Zhu[b]
[a]*Institute for Production Technology and Systems, Leuphana University of Lueneburg, Lueneburg, Germany,* [b]*Department of Engineering Design and Mathematics, University of the West of England, Bristol, United Kingdom*

Today, manufacturing is changing faster than ever before and the drivers for this include globalization, individualization, time to market and sustainability

—Brian Holliday
Managing Director, Digital Factory

The term "industry 4.0" refers to the fourth industrial revolution. This revolution is characterized by a number of new technologies and trends that are transforming manufacturing and other industries. Some key technologies and trends associated with Industry 4.0 include big data and analytics, the Internet of Things, 3D printing, robotics, and artificial intelligence. These technologies are enabling manufacturers to achieve unprecedented levels of efficiency and productivity. Industry 4.0 is already having a profound impact on manufacturing and is poised to transform other industries in the future. Manufacturing is the process of converting raw materials into finished products. It can be done by hand, using machine tools, or using various industrial processes. Manufacturing is a critical part of the economy, as it is responsible for the production of essential goods and services. The manufacturing sector is also a significant source of employment, providing jobs for millions of people around the world. Therefore industry 4.0 is a concept that impacts manufacturing, shows its importance, and faces challenges in manufacturing.

Industry 4.0 is going to transform manufacturing. By integrating digital and physical technologies Industry 4.0 enables factories to become more connected, agile, and responsive to changing market demands. One of the most significant impacts of Industry 4.0 is the way it is changing the role of human workers. With the rise of intelligent machines and artificial intelligence, many tasks that human workers once performed are now automated. This is freeing up workers to focus on more strategic and value-added tasks, such as problem solving and innovation. Although there is still some debate about the exact definition of Industry 4.0, there is no doubt that it has a major impact on manufacturing. As technology continues to evolve, we can expect even more transformative changes in the years to come.

Industry 4.0 is essential because it represents a major shift in the way that manufacturing is done. In the past, manufacturing was largely manual, with workers performing tasks using tools and machines. With Industry 4.0, manufacturing is becoming increasingly automated, with machines and robots performing many tasks previously done by human workers. This shift is essential because it can help improve the efficiency of manufacturing processes, and it can also help decrease the cost of manufacturing products. It allows the industry to innovate and work together with greater ease and efficiency. The creation of Industry 4.0 has helped companies innovate new processes, machines, and systems that create a more seamless workflow, which results in continuous processes that are easier to execute, thus allowing them to become more competitive.

Manufacturing has always been challenging, but with the advent of Industry 4.0, it has become even more complex. With the increasing use of data and technology in manufacturing, there are more opportunities for things to go wrong. For example, real-time data acquisition, storage, transmission, distribution, and data analysis are predominant challenges in Industry 4.0 [1]. The problem is more challenging due to high uncertainty and fluctuation in incoming and future product returns [2], which may require further intelligent monitoring and control [3]. In addition, the increasing complexity of products means more potential areas for errors. It involves a lot of innovation, investment, and technology that shifts from the traditional physical manufacturing in industrial space to virtual space. This shift requires new skill sets like business engineering, cloud computing, big data analytics, and artificial intelligence (AI). Moreover, companies need to achieve complete customer satisfaction with the right supply chain management. Similarly, they also need to reduce waste over their product lifecycle (PLC). Despite the challenges, manufacturing is still a vital part of the economy and plays a crucial role in the production of many products. To stay competitive, manufacturers must continue to invest in data and technology to improve their processes. By doing so they can stay ahead of the curve and remain a key player in the global economy.

References

[1] D. Mourtzis, Introduction to cloud technology and Industry 4.0, in: D. Mourtzis (Ed.), Design and Operation of Production Networks for Mass Personalization in the Era of Cloud Technology, Elsevier, 2022, pp. 1–12.

[2] A.-L. Andersen, T.D. Brunoe, M.T. Bockholt, A. Napoleone, J.H. Kristensen, M. Colli, B.V. Wæhrens, K. Nielsen, Changeable closed-loop manufacturing systems: challenges in product take-back and evaluation of reconfigurable solutions, International Journal of Production Research (2022) 1–20.

[3] B. Dafflon, N. Moalla, Y. Ouzrout, The challenges, approaches, and used techniques of CPS for manufacturing in Industry 4.0: a literature review, The International Journal of Advanced Manufacturing Technology 113 (2021) 2395–2412.

Chapter 2

Advanced ice-clamping control in the context of Industry 4.0

Alexandra Mironova, Paolo Mercorelli, and Andreas Zedler

Institute for Production Technology and Systems, Leuphana University of Lueneburg, Lueneburg, Germany

2.1 Introduction

Ice clamping as a technique to fixate fragile workpieces during machining was already introduced in various papers; see [1], [2], [3], or [4]. The precondition for the choice of a phase changing clamping technique such as clamping with frozen water is its property to clamp workpieces without introducing local contact deformations and, especially, without irreversible structural plastic deformations. In contrast to conventional, mechanical clamping methods, the ice-clamping method addresses this issue with its force- and form-fitted character. The conformity and uniformity of the clamping forces introduce the force-fitted character of ice clamping without local force peaks. Note that the term *clamping force* is used as an analogy to the conventional mechanical clamping, to facilitate a better correspondence and comparability to this common expression. In fact, there are no real clamping forces in the classical sense, acting on the workpiece. The terms adhesion and cohesion are more accurate.

Furthermore, the additional cooling through the clamping device can be advantageous with regard to the machined workpiece quality and tool life. A quantity of research effort was applied to cryogenic assisted machining operations, investigating the positive effects of additional cooling to part quality, tool life, and machinability, as reported in [5–8]. Common cryogenic techniques nowadays use gases like argon or nitrogen, which introduce temperatures between $-100°C$ and $-200°C$. However, as recorded in [9], the temperature of interest is rather the material embrittlement temperature, corresponding to the transition temperature were the material changes from ductile to brittle. A brittle, or hard, state is desired for machining (material cutting) to avoid long and continuous chip formation [10] and thus facilitating and optimizing the chip breakability. This is especially beneficial for plastics, as elastomers with low Young's modulus, which are hard to machine [11]. Since the embrittlement temperature is material dependent, it does not has to be necessarily as low as stated before, as for instance, a temperature of $-55°C$ for steel (AISI/SAE 1008)

Modeling, Identification, and Control for Cyber-Physical Systems Towards Industry 4.0
https://doi.org/10.1016/B978-0-32-395207-1.00012-3

19

was recorded to be appropriate for a high strain rate process, such as high-speed milling [9]. For plastics, even warmer temperatures are sufficient for embrittlement, such as $-15°C$ for polyoxymethylene (wide area of application in engineering, e.g., for gears, valves, spring elements) or $-20°C$ for polypropylene (e.g., used for interiors in domestic or automobiles, the textile industry, etc.) [12]. Even if the current work was primarily focused on an exemplary operational setpoint of $-10°C$, this does not inhibit lower ice-clamping temperatures; see Fig. 2.1 for an overview of the proposed system. In fact, lowering the operational temperature would bring further advantages, such as increased ice adhesion strength, hence higher clamping forces, and thus more security to the process. Furthermore, a steep temperature gradient would facilitate the heat removal from the cutting zone [13]. This chapter is organized as follows. Section 2.2 deals with the model of the system. Section 2.3 introduces the advanced clamping structure. Section 2.4 shows the results and their discussion. Section 2.5 introduces the concept of intelligent clamping. The final Section 2.6 indicates the most important issues for the integration of the proposed technology into Industry 4.0 process. Conclusions close the chapter.

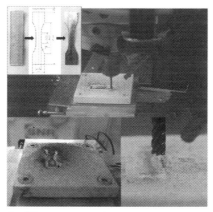

FIGURE 2.1 High-speed milling operation of an aluminum workpiece clamped onto an ice-clamping device.

2.2 Model

Form the physical point of view, Fig. 2.2 shows a detailed view of a positive and negative doped semiconductor combined to a thermocouple in real shown in (a) and as a schematic view with current, heat, and electron flow depicted in (b). Both semiconductors are connected by a metal, usually soldered on copper, to the so-called thermocouple or dice, which is depicted in Fig. 2.2(a). Since one thermocouple produces only some millivolts, multiple couples have to be connected in series to enhance the electric potential in terms of a generator application, but as well for the reversal effect of a heating/cooling application.

Hence a set of thermocouples can be seen as a battery, which also cause a larger voltage drop if a number of them are connected in a row.

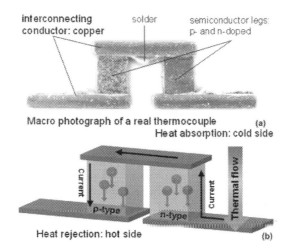

FIGURE 2.2   Scheme of a thermocouple, [2].

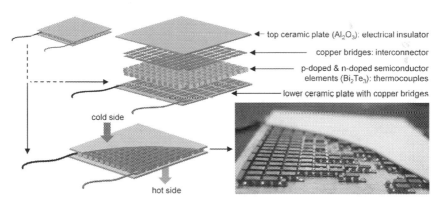

FIGURE 2.3   Peltier cells, [2].

To enhance the thermal performance, the couples are connected thermally in parallel to decrease the thermal resistance. For a firm mounting and electrical insulation, the thermocouples and their interconnects are sandwiched between two thin ceramic (Al_2O_3) substrates; see Fig. 2.3. The emitted heat Q_h towards the hot side and the absorbed heat Q_c from the cold side, driven by the PC, is illustrated in Fig. 2.4 and is formulated as

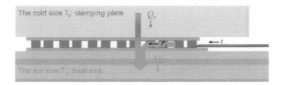

FIGURE 2.4 Scheme of ice-clamping system.

$$Q_h(t) = \underbrace{i(t)S(T_h(t) - T_c(t) + Ri^2(t)}_{P_q(t)} + \underbrace{Si(t)T_c(t) - R\frac{1}{2}i^2(t) - K(T_h(t) - T_c(t))}_{Q_c(t)}.$$

$$(2.1)$$

The cold and hot side temperatures change according to the following nonlinear differential equations:

$$\frac{dT_c(t)}{dt} = \frac{1}{m_a c_a}\left(\frac{1}{2}Ri(t)^2 - ST_c(t)i(t) + K\left(T_h(t) - T_c(t)\right)\right), \qquad (2.2)$$

$$\frac{dT_h(t)}{dt} = \frac{1}{m_k c_k}\left(\frac{1}{2}Ri(t)^2 + ST_h(t)i(t) - K_k\left(T_h(t) - T_a\right) - K\left(T_h(t) - T_c(t)\right)\right),$$

$$(2.3)$$

where S is the Seebeck coefficient, $T_h(t)$ and $T_c(t)$ are the hot and cold side temperatures, R is the electric resistance, and $i(t)$ is the input electric current. The ambient temperature (which can be assumed to be constant) is represented by T_a, m_a is the mass of the cooling plate placed on top of the cold side, m_k is the mass of the heat sink situated below the hot side, c_a and c_k are the corresponding specific heat capacities of the plate and heat sink, K_k is the cooling power coefficient of the heat sink, and K is a parameter quantifying the internal heat transfer between the hot and cold sides. The input power

$$P_q(t) = u_q(t)i(t) \qquad (2.4)$$

has in turn, besides its electric part, also a thermal one, expressed as the voltage generated due to a temperature gradient. Then from Eq. (2.1) we derive that operating the Peltier cell (PC) under heating conditions is more efficient than under cooling, since the applied power has a positive effect on the heat pump in the T_h direction, whereas the cooling performance is affected negatively with the increase of input power, and the absorbed heat remains lower than the released one ($Q_c = Q_h - P_q$). Thus the utilization of a heat sink for an accurate heat removal on the hot side is significant when the application is focused on cooling.

2.3 Advanced ice-camping structure

The control strategy was proposed in [4], in which the asymptotical and robust stability was proved. In this work, we design the first control loop using the *equivalent input current* $i_{eq}(t)$ and the *corrective input current* $i_c(t)$ to obtain a controlling closed-loop current $i_{cl}(t)$ as follows:

$$i_{cl}(t) = \frac{m_a c_a}{ST_c(t)} \underbrace{\left[a(T_a - T_{cd})e^{-at} - k_{ec}(T_{cd}(t) - T_c(t))}_{i_{eq}(t)} \underbrace{-\lambda_c(s_c(t)) - \beta_c \operatorname{sgn}(s_c(t))\right]}_{i_c(t)}.$$

(2.5)

In a similar way, we calculate a controlling closed-loop input $\varkappa_{cl}(t)$ consisting of the *equivalent conduction input* $\varkappa_{eq}(t)$ and the *corrective conduction input* $\varkappa_c(t)$ for the hot side as follows:

$$\varkappa_{cl}(t) = \frac{m_k c_k}{(T_h(t) - T_a)} \underbrace{\left[a(T_a - T_{hd})e^{-at} - \frac{K(T_h(t) - T_c(t))}{m_k c_k} - k_{eh}(T_{hd}(t) - T_h(t))\right]}_{\varkappa_{eq}(t)}$$

$$\underbrace{-\frac{m_k c_k}{(T_h(t) - T_a)}\left[\lambda_h(s_h(t)) + \beta_h \operatorname{sgn}(s_h(t))\right]}_{\varkappa_c(t)}.$$

(2.6)

The variable $\varkappa_{cl}(t)$ is treated analogously to the value of K_k. It indirectly represents the forced convection due to the use of a fan to increase the effect of the heat removal rate and the relationship between forced convention and the heat sink conductance coefficient K_k. A sketch of demonstration of the asymptotical stability of the control laws is shown in the Appendix.

To fulfill the requirements for a clamping system in an industrial, commercial, and intelligent manner (smart manufacturing), we propose an embedded system for an optimized ice-clamping system based on the gained results. The number of thermoelectric coolers (TECs) has to be increased to maintain a more flexible controllability through the enhancement of granularity. A new distributed and robust control system, based on the sliding mode control approach, should ensure separate control of each TEC to better adjust to unequal environmental conditions and external disturbances.

To sum up, the ice clamping device should address the needs of the new hardware configuration as follows:

- increase number of TECs to twelve elements (enhancement of the granularity), but maintaining the total output power, such that a qualitative comparison to the original system (of 6 TECs) can be still obtained: 1070 W total maximal electric power and 680 W total maximal cooling power
- Controller: high performance, controllability of minimum every single TEC
- usability: optimized and simplified operation by means of buttons and switches
- Output of status messages via a LCD

- Record data: cold temperature of each TEC, mean hot temperature, ambient temperature
- possibility of switching between two modi: controllability of each single TEC via H-bridges (PWM) and via power transistors (DC)

In Fig. 2.5, the schematic 3D view of the proposed system is shown with twelve TECs.

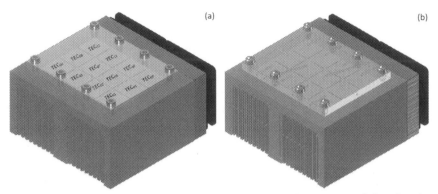

FIGURE 2.5 Isometric 3D CAD view of the proposed ice-clamping system consisting of twelve TECs with numbered position in (a) and the location of thermo sensor inputs inside the clamping plate in (b), [2].

2.4 Measured results and performance evaluation

Before showing the results and discussing them, we give some technical aspects related to the measurements in the lab. Most measurements were performed within a 16 m² laboratory room. The first error to be assessed is the atmospheric environment, classified into the ambient temperature and the relative air humidity. Especially for measurements over a long period, which are subject to thermodynamics, it has to be guaranteed that the surrounding climate is constant over the entire time. For this purpose, the climate in the laboratory was controlled with the air conditioning system Argo of type AUR213CL. The *stability of the ambient temperature* $= f_1$ was permanently kept constant at $20°C \pm 1.0°C$. The *stability of the air humidity* $= f_2$ was controlled to be $50\% \pm 10\%$ rel. humidity. For the crosscheck, the values were additionally monitored with a hand-held measuring device from Testo. According to the chosen PC as well as to the heat sink's performance, high temperature differences can be achieved, which make freezing the plate under $-10°C$ possible, representing an operating point where high adhesives strengths of ice can guarantee a secure grip during machining operations [14] by encapsulating the workpiece force- and form-fitted. For a more detailed description of this method, we refer to [4,15,16]. According to the theory, the sliding mode control is a highly robust approach, but its main limitation is the so-called *chattering phenomenon*, which is especially observable in real measurements [17]. The chattering is represented by high-frequency oscillations around the sliding surface. This chattering not only reduces the control performance but also affects the hardware electronic devices. High oscillations of the system input from the controller forces the input electric current to switch with high amplitudes, as shown in Fig. 2.7(b), which can lead to premature wear of electronic parts.

The presence of the sign function in Eq. (2.6), which induces the system chattering, can be smoothened by defining the general control law

$$u(t) = \hat{u}(t) + \beta_s sat\left(\frac{s(t)}{\delta}\right), \tag{2.7}$$

where $\hat{u}(t)$ represents the first terms of the right side of Eq. (2.6), being the equivalent input, and δ is a constant that defines the thickness of the boundary layer and is devoted to the reduction of the chattering phenomenon, with

$$sat\left(\frac{s(t)}{\delta}\right) = \begin{cases} 1 & \text{if } \frac{s(t)}{\delta} \geq 1, \\ \frac{s(t)}{\delta} & \text{if } -1 < \frac{s(t)}{\delta} < 1, \\ -1 & \text{if } \frac{s(t)}{\delta} \leq -1. \end{cases} \tag{2.8}$$

Increasing parameter δ, the chattering phenomenon can be reduced. A similar approach was found to be advantageous for the use in a thermoelectric system in [18]. The choice of the integral-type sliding surface was made to also limit the chattering characteristics. Without using integral action, the chattering behavior can be increased additionally, in particular, in the final part of the tracking. This aspect is known in the classical literature (see [19]), and there Fillipov [20] and Slotine [21] are cited as fundamental background emphasizing the importance of the integral presence to reduce chattering.

Moreover, neglecting the sign function would suppress the chattering completely, since in [14], it was shown that the convergence is still guaranteed just using the parameter λ. However, this contradicts the theory of the classical approach, and the robustness can no longer be guaranteed in the sense of SMC. In fact, the system presented in [14] is rather inspired by the SMC but does not explicitly corresponds to the purpose of a sliding mode approach, so that actually the approach described in [14] belongs rather to a Lyapunov-based controller, as presented in [16,22]. Therefore the presence of the sign function is justified by the fact that it is possible that for different working points (or parameters), the switching function is explicitly needed.

The block diagram of the controller with chattering reduction for the new system consisting of twelve TECs is shown in Fig. 2.6.

In Fig. 2.7(a) the cold side temperature results for all twelve TECs in a detailed view are shown for the SM control *without* chattering suppression. In Fig. 2.7(b) the corresponding input currents show high oscillation characteristic. Due to the chattering action, the error of the cold side temperatures

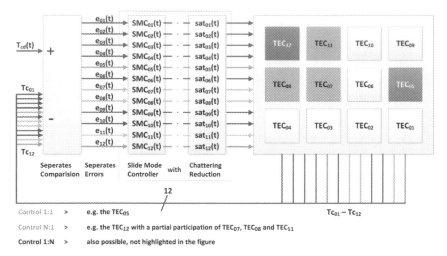

FIGURE 2.6 Block diagram of the new controller logic, [2].

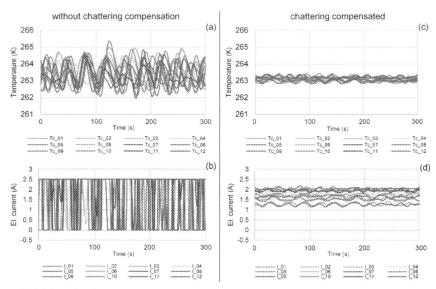

FIGURE 2.7 SMC chattering reduction: control without chattering reduction is shown for the cold side temperatures in (a), and the corresponding input currents are shown in (b); temperature results with reduced chattering are shown in (c), and the chatter reduced input currents are shown in (d), [2].

is high, having a maximal absolute overshoot of $+2.2$ K and a maximal undershoot of -1.2 K. In Fig. 2.7(c)–(d) the temperature results and the corresponding currents are shown for chattering compensated SM control according to Eqs. (2.7)–(2.8). The high oscillated switching of the input currents is reduced significantly, so that the chattering and thus the temperature profile are smoothened noticeably. The maximal absolute overshoot is $+0.28$ K, and the maximal absolute undershoot is -0.34 K, which is near the resolution limit of the sensors and can thus be disregarded. Two types of disturbances were applied on the middle of the clamping plate, as shown in Fig. 2.16(a). The disturbances were selected such that 600 J (40 W over 15 s) were applied first, shown in detail in Fig. 2.8(b), and the second disturbance was a permanent one, hence representing the same heating power applied to the plate over the entire final measuring time; see Fig. 2.8(c). Both represent extra high thermal load to evaluate the controller's performance at maximum stress. The two largest temperature peaks belong to the TEC_{06} and TEC_{07} (temperatures Tc_{06} and Tc_{07}, respectively). This is due to the fact that these TECs are both placed in the center of the twelve TECs, located right under the external disturbance, the heated specimen; see Figs. 2.5 and 2.16(a).

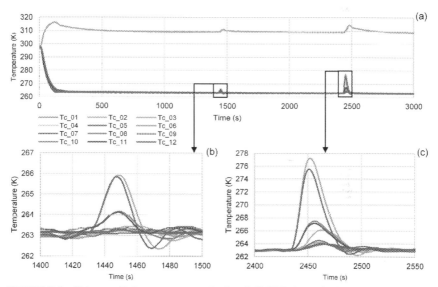

FIGURE 2.8   SM-controlled temperature results for all 12 TECs in (a), in the presence of a short disturbance in (b), and a permanent disturbance in (c), [2].

2.4.1   Performance comparison

To compare the performance of the new optimized ice-clamping approach with a traditional one, measurements with the infrared camera InfraCAM (Flir) were performed for the steady state and when a thermal disturbance is applied. In Fig. 2.9, measured infrared results of the temperature profile at the defined operating temperature point of $-10°C$ are shown in Fig. 2.9(a) for the old system and in Fig. 2.9(c) for the new system. The homogeneity of the temperature distribution is significantly optimized in Fig. 2.9(c). Furthermore, the disturbance attenuation, as shown in Figs. 2.9(b) and 2.9(d), is noticeably better for the new ice-clamping system. All pictures were recorded after a certain time of period at stable and thus steady states.

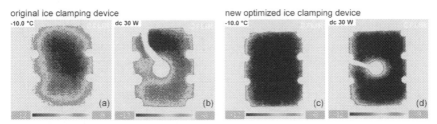

FIGURE 2.9 Performance comparison: temperature distribution of the clamping plate of the original system in (a), in the presence of a thermal disturbance in (b), and of the optimized clamping system in (c) and (d), respectively, [2].

2.5 Towards intelligent clamping

As for an intelligent clamping system, the ice-clamping device can be used to detect workpieces on the plate, such that the controller is capable of adapting its controller action as a function of changed states or the environment. By introducing a new state, which is referred to as the *stand-by mode*, the ice-clamping plate is cooled to a quasi-operating state between 0°C and +1°C. When a workpiece at ambient temperature is put onto the plate, the change in temperature is recognized as a *deliberated disturbance*, by taking advantage of the thermal conductance, which functions as an *activation point* to launch the controlled freezing action to the desired operating temperature point (e.g., −10°C). Also, with the TEC's ability to be used as a generator, it is also possible to exploit the phenomenon of voltage generation (due to a temperature difference) as another detection method. After detection of a workpiece, the controller is now *focused to control* only a certain area of the plate, where the workpiece is *effectively* located by deciding autonomously which TECs have to be activated to reach the desired setpoint. This phase is called adaption. As a result, either a single *spot* is controlled, and hence only one TEC is powered, which is true for small workpieces, or an *area* is controlled, and hence several TECs are actuated. This happens when the contact surface area of the placed workpiece is large or when the workpiece is placed over more than one TEC, soch that several TECs are triggered at once (or in case the cooling power needs to be enhanced, the controller can decide to add neighboring TECs). The remaining area (thus the remaining TECs) is either kept in the stand-by mode or can be used to support cooling of the effective area, in case the cooling power is not sufficient enough, as in the case of high ambient temperature. But in the most cases, the remaining TECs will be switched off by the controller, since the thermal conductance will inevitably lead to an indirect cooling of the remaining area. This, however, is very desirable, since the consumption of electric power is reduced in this way, which meets the requirement of an ecological and economical clamping. Then it is also possible to refer to this approach as *green clamping*, in line with green manufacturing.

Note that for all control actions, a sliding mode controller is still used. In Fig. 2.10, infrared measurement results are shown. For this measurement, the workpiece was placed on the lower right-hand corner of the plate in the stand-by mode, and the controller detects the piece and adaptively starts the cooling, so that only the corresponding area, where the workpiece was located, was controlled to the desired setpoint temperature ($-10°C$). As is visible in Fig. 2.10, the adaptive feature of the controller decides to cool an area actuated by four TECs, since the workpiece was placed in such a way that several TECs were triggered.

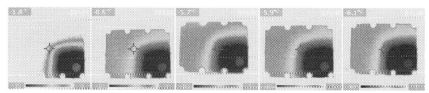

FIGURE 2.10   Infrared measurement results of the detection phase and the adaptive clamping of a workpiece, [2].

The described concept can also be traced in the block diagram of Fig. 2.6. Different control configurations are represented in colored arrows and TECs. The configuration *Control 1:N* means a basic control strategy. One measured error is generally valid for the entire system, so that one system input is valid for all TECs of the system, no matter how many system outputs are present. The control configuration *Control 1:1* is a distributed control, in which a single error signal is mapped to a single system input, but in its entirety, every TEC is addressed individually. The configuration *Control N:1* is the core of the new designed controller approach. With this configuration, an evaluation of the situation takes place, in which the controller is capable of deciding to connect or disconnect further neighboring TECs to enhance or decrease the cooling power, depending on the desired setpoint and environmental conditions.

The proposed approach allows the clamping of workpieces successively to simultaneously process multiple parts during the operating state. This means that after detecting and adaptively clamping the first workpiece, the algorithm within the controller constantly updates and checks the *detection* state, so that the placement of a second workpiece onto the plate is recognized as a new piece and the adaptive feature decides which TECs have to be powered on to start the second control action. In Fig. 2.11 a photo of the setup is shown for two workpieces placed successively on the plate. After the clamping of the first workpiece reaches steady state ($-10°C$) (see the first infrared picture of Fig. 2.12), a second workpiece was placed onto the opposite corner of the plate. The next eleven pictures of Fig. 2.12 illustrate the second controller action stepwise. It is interesting

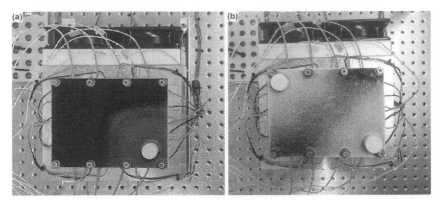

FIGURE 2.11 Intelligent clamping concept autonomously detects workpieces on the clamping plate and adaptively controls the clamping: in (a), first one workpiece is placed, detected, and controlled to a desired setpoint, and in (b) a second workpiece is placed during active clamping process onto the plate, [2].

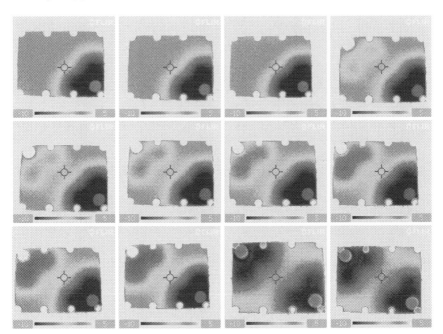

FIGURE 2.12 Infrared measurements of temperature distribution of intelligent clamping during operation with regard to the setup presented in Fig. 2.11, [2].

to note that in this case, only three TECs on each corner were active to control the clamping, whereas in the previous measuring example, four TECs were used to actuate the freezing and thus the clamping (compare with Figs. 2.10 and 2.12). This is due to the adaptive feature of the controller, which decides autonomously how many actuators are needed to reach the desired setpoint.

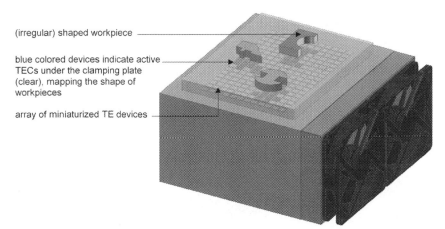

(irregular) shaped workpiece

blue colored devices indicate active
TECs under the clamping plate
(clear), mapping the shape of
workpieces

array of miniaturized TE devices

FIGURE 2.13 By enhancing the granularity of the TEC array the subpixel approach allows a more exact position detection of workpieces and a geometrical shape recognition, [2].

2.6 Towards Industry 4.0

The proposed approach has two features, to detect a workpiece and to adapt the right control decision, which includes recognizing a correct positioning and maintaining accurate tracking of the desired trajectory. As described above, the detective and adaptive characteristics of this approach are based on identifying either the measured temperature change of the clamping plate or the measured voltage change of a TEC in the presence of a temperature change. Assuming that the dimensions of the ice clamping device are maintained, miniaturized TE devices can be used to enhance the granularity in such way that a shape recognition of the workpiece's contact surface area is possible. This means that in future works, a *subpixel* approach can be applied to detect not only the position of the workpiece but simultaneously the geometrical form, thus being more effective in terms of energy-saving clamping. The smaller the TE devices are chosen,[1] the more precise the shape can be recognized. A shape recognition might be useful to determine workpiece characteristics and lead to a clamping device-to-machine communication. By detecting the exact position and recognizing the shape through the clamping device this information can be used as feedback passed to the CNC processing machine. This data can be then stored and used to automatically match the CAD data, convert it to a CNC code, and execute the motion. Especially for irregularly shaped workpieces, this might be advantageous to facilitate the machine setup (reduce manual setting actions), as autonomous reference and boundary points as well as zero position detections. In Fig. 2.13 a schematic view of the proposed subpixel concept for the shape recognition of an ice-clamping device is shown.

---

[1] It is also possible to assume the smallest possible unit being one dice, hence one controllable thermocouple.

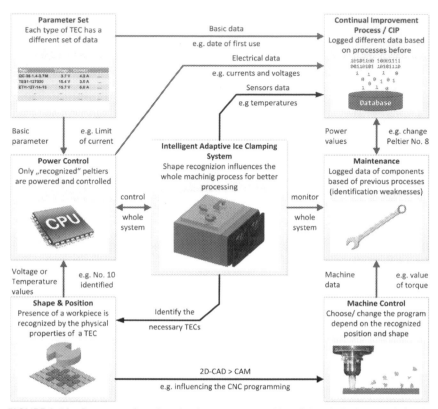

FIGURE 2.14   Smart manufacturing: development opportunities of the subpixel approach for the position and shape recognition of the ice-clamping device in order to meet the requirements of Industry 4.0, [2].

Future trends, especially with regard to Industry 4.0, seek the integration of multiple technologies in one processing step [23]. The combination of different methods and production steps is a key issue to shorten the process chain with the aim of a *complete machining* of complex parts, as deduced in a recent research study of 2017 in [23], analyzing the future trends of the machine tool industry. To meet these demands, the proposed method fulfills the integration of multiple technologies and combines several processing steps, as represented by Fig. 2.14.

The hardware is configured to power 12 TECs [TEC01]–[TEC12] of the type QC–241–1.0–3.0M. A safety margin of +25% has been taken into account for the electric power. The TECs are placed onto a heat sink [HS01] with two fans [FAN01]. The Controller [C01] is an Arduino DUE with a 32-bit ARM, 84 MHz Atmel CPU core architecture with 512 kB ARM for fast computational tasks (fast algorithm calculations). The supply voltage for the processor is reduced to 3.3 V to compensate energy losses due to an increased computing power (in contrast to 5 V for similar Controller with lower performance). For the communication, the DUE board has 54 I/O ports, with 14 PWM ports and 14 analog-IN-ports. The connections and communication are controlled via bus systems as SPI and i^2C. The HMI is represented by a group of eight buttons [B01], subdivided in a three- and five-button subgroup, and a LCD [DO1] in accordance with the HD44780 standard with 4×20 characters. The LCD displays, e.g., status messages, measured data, or the setup menu. With the three-button subgroup, the basic functions are controlled, such as the *start of cooling process* or *start of unclamping (heating) process*. With the five-button subgroup, the navigation of the menu structure is controlled, in which the change of displayed values, or the change of input of parameter values is possible. With the proposed installation structure, it is possible to switch [S01] between two operation modes: H-bridges power the TECs with PWM and power transistors supply the TECs with DC. The switching is performed by the relay group [R01]–[R12]. In the case of the PWM operation mode, twelve H-bridges [H01]–[H12] of type L6203 (STMicroelectronics) are used, powered up to $U = 48$ V and $I = 5$ A, so that a sufficient power reserve capacity is ensured for all TECs. The H-bridges are supplied via the power supply unit [PSU1]. For the cooling operation, the bridges are controlled via the PWM output ports [PWM01]–[PWM12] of the [C01]. For the heating (unclamping) operation, the controller does not provide enough output ports, so that the controller is extended by the board [IC01] (PCA9685 from NXP Semiconductors), which provides further twelve PWM output ports [PWM13]–[PWM24] connected by i^2C. The H-bridges are mounted in a the heat sink [HS02] with one fan [FAN02]. In the case of the DC operation mode, the TECs are powered by twelve transistors [T01]–[T12] of type BD249C with a total power of $U = 125$ V and $I = 25$ A, hence also having a sufficient reserve capacity. The transistors are controlled via a DAC [IC02] of type MAX521, which, in turn, is controlled via a i^2C bus. As a protective measure for the output ports of the DAC, impedance converters [OP01]–[OP12] of type LM741 (Texas Instruments) are used. The transistors are mounted together with the bridges to the heat sink [HS02]. Temperature data for each TEC element are obtained from twelve cold-junction compensated type-k thermocouples with the converter boards MAX31855 [TC01]–[TC12] (measured data is provided via SPI bus) and for the hot side, defined as [TH01]–[TH04]. The hot side temperature is recorded for safety monitoring purposes only, since it was shown and experimentally proven that for the chosen controller, the hot side temperature value is not necessarily needed. Nevertheless, the hardware application is designed in a way to provide the possibility of an extension for further sensors (e.g., observer application). All internal low voltages from 3.3 V, 5.5 V, 12 V, and ±15 V are provided by several power supply units [PS02]–[PS06]. The described structure is implemented into the embedded controller system unit of the ice-clamping device shown in Fig. 2.16, whereas in Fig. 2.16(b) a detailed top view of the electronic is shown as depicted in the schematic circuit diagram of Fig. 2.15.

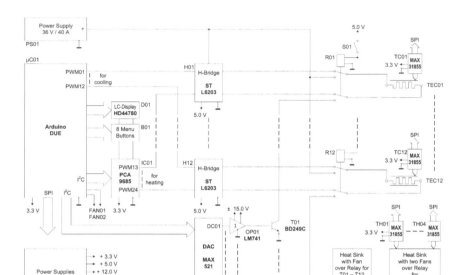

FIGURE 2.15   Circuit diagram of electronic components of the prosed controller unit for the optimized ice clamping, [2].

2.7   Conclusions

This chapter is dedicated to an innovative milling system in which the object to be milled is fixed by an ice-clamping system. One of the most important issues in manufacturing systems is to obtain "zero deformation". In particular, the chapter takes into consideration the control strategy is based on sliding mode control to obtain robust results even in the absence of the knowledge of the system to be controlled. Measured results are shown to demonstrate the effectiveness of the proposed device and its control strategy. Moreover, the chapter indicates a direction to integrate this new technology into the Industry 4.0 process.

Appendix

Theorem 1.   *Given the system represented by Eqs. (2.1), (2.2), and (2.3), there exists an asymptotically stable controller characterized by the controlling closed-loop input current $i_{cl}(t)$ and by the real constants $\lambda_c > 0$, $\lambda_h > 0$, and $\beta_c > 0$, $\beta_h > 0$ together with two real constants $k_{ec} > 0$ and $k_{eh} > 0$ such that*

$$\lim_{t \to \infty} T_c(t) = T_{cd}(t) = T_{cd} \quad \text{and} \quad \lim_{t \to \infty} T_h(t) = T_{hd}(t) = T_{hd}. \tag{2.9}$$

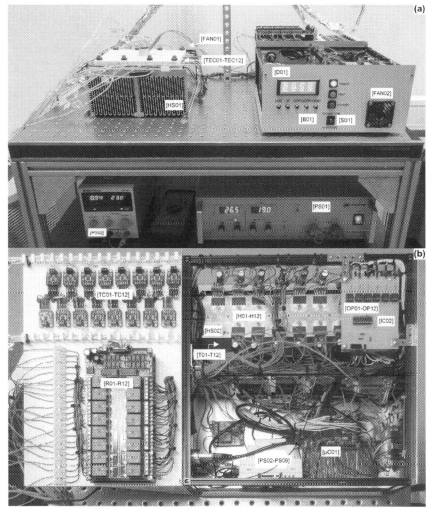

FIGURE 2.16   Embedded system: ice-clamping device next to its embedded control unit in (a) and a detailed top view of the electronic components in (b), [2].

We define the following two sliding mode surfaces:

$$s_c(t) = T_{cd}(t) - T_c(t) + k_{ec} \int_0^t \Big(T_{cd}(\tau) - T_c(\tau)\Big) d\tau - s_c(0), \qquad (2.10)$$

$$s_h(t) = T_{hd}(t) - T_h(t) + k_{eh} \int_0^t \Big(T_{hd}(\tau) - T_h(\tau)\Big) d\tau - s_h(0), \qquad (2.11)$$

where the presence of terms $s_c(0)$ and $s_h(0)$ reduces to zero the reaching phase ($s_c(0) = 0$ and $s_h(0) = 0$) regardless of initial condition $T_c(0)$ and $T_h(0)$; T_{cd}

and T_{hd} can either represent the *constant* desired cold and hot side temperatures, or these setpoints can be described in a *dynamic* way as $T_{cd}(t)$ and $T_{hd}(t)$ as follows:

$$T_{cd}(t) = (T_a - T_{cd})e^{-at} + T_{cd} \qquad (2.12)$$

and

$$T_{hd}(t) = (T_a - T_{hd})e^{-at} + T_{hd}, \qquad (2.13)$$

where the constant a states the velocity of the dynamics of the desired temperature profile. Let

$$e_{ec}(t) = T_{cd}(t) - T_c(t), \qquad (2.14)$$
$$e_{eh}(t) = T_{hd}(t) - T_h(t) \qquad (2.15)$$

be the temperature errors. It is then possible to define two sliding temperature surfaces, one for the cold and one for the hot side as follows:

$$\dot{s}_c(t) = \dot{e}_{ec}(t) + k_{ec}e_{ec}(t), \qquad (2.16)$$
$$\dot{s}_h(t) = \dot{e}_{eh}(t) + k_{eh}e_{eh}(t). \qquad (2.17)$$

Eqs. (2.16) and (2.17) with $\dot{s}_c(t) = 0$ and $\dot{s}_h(t) = 0$ state the tracking dynamics of the controlled system in sliding mode for the cold and hot temperatures, respectively. The constants k_{ec} and k_{eh} state their rates of convergence. At $s_c(t) = 0$ the controlled trajectory approaches the sliding surfaces, and with $s_h(t) = 0$, the trajectory remains in the sliding surface. Imposing $s_h(t) = 0$, we obtain

$$\dot{e}_{ec}(t) + k_{ec}e_{ec}(t) = 0, \qquad (2.18)$$
$$\dot{e}_{eh}(t) + k_{eh}e_{eh}(t) = 0, \qquad (2.19)$$

which admit the following solutions:

$$e_{ec}(t) = \eta_c \exp^{-k_{ec}t}, \qquad (2.20)$$

and

$$e_{eh}(t) = \eta_h \exp^{-k_{eh}t}, \qquad (2.21)$$

where η_c and η_h are constants determined by the initial conditions. To realize a region of attraction around this sliding surface, we can consider a Lyapunov approach. Starting by defining two $\Phi_c(t)$ and $\Phi_h(t)$ functions such that

$$\Phi_c(t)\dot{\Phi}_c(t) = \frac{R}{2m_a c_a}i^2(t) + \frac{K}{m_a c_a}\big(T_h(t) - T_c(t)\big), \qquad (2.22)$$

and

$$\Phi_h(t)\dot{\Phi}_h(t) = \frac{S}{m_k c_k}i(t)T_h(t) + \frac{R}{2m_k c_k}i^2(t), \qquad (2.23)$$

it follows that

$$\Phi_c(t)d\Phi_c(t) = \left(\frac{R}{2m_ac_a}i^2(t) + \frac{K}{m_ac_a}(T_h(t) - T_c(t))\right)dt, \quad (2.24)$$

$$\int \Phi_c(t)d\Phi_c(t) = \int \left(\frac{R}{2m_ac_a}i^2(t) + \frac{K}{m_ac_a}(T_h(t) - T_c(t))\right)dt, \quad (2.25)$$

$$\frac{1}{2}(\Phi_c(t))^2 + C_c = \int_0^t \left(\frac{R}{2m_ac_a}i^2(\tau) + \frac{K}{m_ac_a}(T_h(\tau) - T_c(\tau))\right)d\tau \quad (2.26)$$

with $C_c = 0$, and through similar consideration it follows that

$$\frac{1}{2}(\Phi_h(t))^2 + C_h = \int_0^t \left(\frac{S}{m_kc_k}i(\tau)T_h(\tau) + \frac{R}{2m_kc_k}i^2(\tau)\right)d\tau \quad (2.27)$$

with $C_h = 0$. In [4], it was shown that $T_c(t) < T_h(t)$ if $i(t) > 0$, and considering that $T_c(t) > 0$ and $T_h(t) > 0$, we have

$$\frac{R}{2m_ac_a}i^2(t) > 0, \quad \frac{K}{m_ac_a}(T_h(t) - T_c(t)) > 0, \quad \frac{S}{m_kc_k}i(t)T_h(t) > 0, \quad (2.28)$$

and thus the right parts of Eqs. (2.26) and (2.27) are positive, and thus positions (2.26) and (2.27) are consistent. Denote

$$V(T_h, T_c, t) = \frac{s_c^2(t) + \Phi_c^2(t) + s_h^2(t) + \Phi_h^2(t)}{2}. \quad (2.29)$$

Then

$$\dot{V}(T_h, T_c, t) = s_c(t)\dot{s}_c(t) + \Phi_c(t)\dot{\Phi}_c(t) + s_h(t)\dot{s}_h(t) + \Phi_h(t)\dot{\Phi}_h(t). \quad (2.30)$$

Using the model of Eqs. (2.1)–(2.3), it follows that

$$\dot{V}(T_h, T_c, t) =$$

$$s_c(t)\left[-a(T_a - T_{cd})e^{-at} + \frac{S}{m_ac_a}i(t)T_c(t) - \frac{R}{2m_ac_a}i^2(t) - \frac{K}{m_ac_a}(T_h(t) - T_c(t))\right.$$

$$\left. - \frac{d_c(t)}{m_ac_a} + k_{ec}(T_{cd}(t) - T_c(t)) + \Phi_c(t)\dot{\Phi}_c(t)\right] + s_h(t)\left[-a(T_a - T_{hd})e^{-at}\right.$$

$$- \frac{S}{m_kc_k}i(t)T_h(t) - \frac{R}{2m_kc_k}i^2(t) + \frac{K}{m_kc_k}(T_h(t) - T_c(t)) + \frac{K_k}{m_kc_k}(T_h(t) - T_a)$$

$$\left. - \frac{d_h(t)}{m_kc_k} + k_{eh}(T_{hd}(t) - T_h(t)) + \Phi_h(t)\dot{\Phi}_h(t)\right]. \quad (2.31)$$

Considering the definition of $\Phi_c(t)$ and $\Phi_h(t)$ and in particular Eqs. (2.22) and (2.23), it follows that

$$\dot{V}(T_h, T_c, t) = s_c(t)\left[-a(T_a - T_{cd})e^{-at} + \frac{S}{m_ac_a}i(t)T_c(t) - \frac{d_c(t)}{m_ac_a}\right.$$

$$+ ke_c(T_{cd}(t) - T_c(t))\Big] + s_h(t)\Big[-a\big(T_a - T_{hd}\big)e^{-at} + \frac{K}{m_k c_k}(T_h(t) - T_c(t))$$

$$+ \frac{K_k}{m_k c_k}(T_h(t) - T_a) - \frac{d_h(t)}{m_k c_k} + ke_h(T_{hd}(t) - T_h(t))\Big]. \quad (2.32)$$

The first control loop can be designed using the *equivalent input current* $i_{eq}(t)$ and the *corrective input current* $i_c(t)$ to obtain a controlling closed loop current $i_{cl}(t)$ as follows:

$$i_{cl}(t) = \frac{m_a c_a}{S T_c(t)} \underbrace{\Big[a\big(T_a - T_{cd}\big)e^{-at} - ke_c\big(T_{cd}(t) - T_c(t)\big)}_{i_{eq}(t)} \underbrace{-\lambda_c\big(s_c(t)\big) - \beta_c \, \mathrm{sgn}\big(s_c(t)\big)\Big]}_{i_c(t)}.$$

$$(2.33)$$

In a similar way, we calculate a controlling closed-loop input $\varkappa_{cl}(t)$ consisting of the *equivalent conduction input* $\varkappa_{eq}(t)$ and the *corrective conduction input* $\varkappa_c(t)$ for the hot side:

$$\varkappa_{cl}(t) = \frac{m_k c_k}{\big(T_h(t) - T_a\big)} \underbrace{\Big[a\big(T_a - T_{hd}\big)e^{-at} - \frac{K\big(T_h(t) - T_c(t)\big)}{m_k c_k} - ke_h\big(T_{hd}(t) - T_h(t)\big)\Big]}_{\varkappa_{eq}(t)}$$

$$\underbrace{- \frac{m_k c_k}{\big(T_h(t) - T_a\big)}\Big[\lambda_h\big(s_h(t)\big) + \beta_h \, \mathrm{sgn}\big(s_h(t)\big)\Big]}_{\varkappa_c(t)}.$$

$$(2.34)$$

In Eqs. (2.33) and (2.34) the $sgn(s_c, s_h)$ represents the sgn function,

$$\mathrm{sgn}(s_c, s_h) = \begin{cases} +1 & \text{if } s_c, s_h > 0, \\ -1 & \text{if } s_c, s_h < 0. \end{cases} \quad (2.35)$$

As is well known, the equivalent control makes the derivative of the sliding surface equal zero to stay on the sliding surface, and the corrective control compensates the deviations from the sliding surface to reach the sliding surface. The feedback control laws in Eqs. (2.33) and (2.34) are such that $V(T_h, T_c, t)$ remains a Lyapunov-like function of the closed loop system.[2] It is worth noting that if input Eqs. (2.33) and (2.34) are inserted into Eq. (2.32), then

$$\dot{V}(T_h, T_c, t) \leq s_c(t)\dot{s}_c(t) + s_h(t)\dot{s}_h(t) = s_c(t)\Big(-\frac{d_c(t)}{m_a c_a} - \lambda_c s_c(t) - \beta_c \, \mathrm{sgn}(s_c(t))\Big)$$

$$+ s_h(t)\Big(-\frac{d_h(t)}{m_k c_k} - \lambda_h s_h(t) - \beta_h \, \mathrm{sgn}(s_h(t))\Big), \quad (2.36)$$

---

[2] We refer for details to [21], p. 282.

and since $\lambda_c > 0$ and $\lambda_h > 0$, it follows that

$$\dot{V}(T_h, T_c, t) \leq s_c(t)\dot{s}_c(t) + s_h(t)\dot{s}_h(t) =$$
$$s_c(t)\left(-\frac{d_c(t)}{m_a c_a} - \beta_c \operatorname{sgn}(s_c(t))\right) + s_h(t)\left(-\frac{d_h(t)}{m_k c_k} - \beta_h \operatorname{sgn}(s_h(t))\right). \quad (2.37)$$

The sufficient condition

$$\dot{V}(T_h, T_c, t) \leq 0 \qquad (2.38)$$

is guaranteed if

$$\beta_c \geq \max\left|\frac{d_c(t)}{m_a c_a}\right| \qquad (2.39)$$

and

$$\beta_h \geq \max\left|\frac{d_h(t)}{m_k c_k}\right|. \qquad (2.40)$$

To show the asymptotic stability, a Lyapunov-like lemma is invoked. In fact, the condition $\dot{V}(T_h, T_c, t) \leq 0$, together with the fact that assuming that the disturbances $d_c(t)$ and $d_h(t)$ are uniformly continuous functions, implies that $\dot{V}(T_h, T_c, t)$ is a uniformly continuous function because it consists of a sum of uniformly continuous functions. These two facts, together with $V(T_h, T_c, t)$ being lower bounded, guarantee that the hypotheses of Lyapunov-like lemma are satisfied and thus $\lim_{t \to +\infty} \dot{V}(T_h, T_c, t) = 0$, and we obtain the following result:

$$\lim_{t \to +\infty} T_c(t) = T_{cd}(t) = T_{cd} \quad \text{and} \quad \lim_{t \to +\infty} T_h(t) = T_{hd}(t) = T_{hd}. \quad (2.41)$$

□

References

[1] A. Mironova, Effects of the influence factors in adhesive workpiece clamping with ice: experimental study and performance evaluation for industrial manufacturing applications, The International Journal of Advanced Manufacturing Technology 99 (1–4) (August 2018) 137–160.

[2] A. Mironova, Industrial Application Study and Development of a Control Strategy for an Ice Clamping Device, PhD thesis, Leuphana University of Lueneburg, Institute for Production Tehnology and Systems, ISBN 13: 978-3-96548-007-0 (Print), Sierke Verlag, 2018.

[3] A. Mironova, B. Haus, A. Zedler, P. Mercorelli, Extended Kalman filter for temperature estimation and control of peltier cells in a novel industrial milling process, IEEE Transactions on Industry Applications 56 (2) (March 2020) 1670–1678.

[4] A. Mironova, P. Mercorelli, A. Zedler, A multi input sliding mode control for peltier cells using a cold–hot sliding surface, Journal of the Franklin Institute 355 (18) (December 2018) 9351–9373.

[5] S.Y. Hong, Economical and ecological cryogenic machining, Journal of Manufacturing Science and Engineering 123 (2) (1999) 331–338.

[6] Y. Yildiz, M. Nalbant, A review of cryogenic cooling in machining processes, International Journal of Machine Tools and Manufacture 48 (9) (2008) 947–964.

[7] R. Nayak, R. Shetty, Cutting force and surface roughness in cryogenic machining of elastomer, International Journal of Mechanical Engineering and Technology 5 (2014) 151.

[8] G. Manimaran, M.P. Kumar, R. Venkatasamy, Influence of cryogenic cooling on surface grinding of stainless steel 316, Cryogenics 59 (2014) 76–83.

[9] S.Y. Hong, Y. Ding, Micro-temperature manipulation in cryogenic machining of low carbon steel, Journal of Materials Processing Technology 116 (1) (2001) 22–30. Containing papers selected from the 2nd international conference on Advanced Manufacturing Technology.

[10] Y. Ding, S.Y. Hong, Improvement of chip breaking in machining low carbon steel by cryogenically precooling the workpiece, Journal of Manufacturing Science and Engineering 120 (02 1998).

[11] M. Putz, M. Dix, M. Neubert, G. Schmidt, R. Wertheim, Investigation of turning elastomers assisted with cryogenic cooling, in: 13th Global Conference on Sustainable Manufacturing-Decoupling Growth from Resource Use, Procedia CIRP 40 (2016) 631–636.

[12] Inc. Zeus Industrial Products. Low temperature properties of polymers, technical white paper, http://www.appstate.edu/~clementsjs/polymerproperties/plastics_low_temp.pdf, 2005. (Accessed 25 April 2018).

[13] R. Ghosh, Z. Zurecki, J. Frey, Cryogenic machining with brittle tools and effects on tool life, in: American Society of Mechanical Engineers, ASME 2003 International Mechanical Engineering Congress and Exposition, vol. 14, 01 2003.

[14] A. Mironova, P. Mercorelli, A. Zedler, Robust control using sliding mode approach for ice-clamping device activated by thermoelectric coolers, Ifac Papersonline 49 (25) (2016) 470–475.

[15] A. Mironova, P. Mercorelli, A. Zedler, E. Karaman, A model based feedforward regulator improving pi control of an ice-clamping device activated by thermoelectric cooler, Number 8014064, 2017, pp. 484–489.

[16] A. Mironova, P. Mercorelli, A. Zedler, Control of a two-thermoelectric-cooler system for ice-clamping application using Lyapunov based approach, Number 7976183, 2017, pp. 24–29.

[17] K. Széll, P. Korondi, Mathematical basis of sliding mode control of an uninterruptible power supply, Acta Polytechnica Hungarica 11 (2014) 87–106.

[18] T. Hatano, M. Deng, S. Wakitani, Operator and sliding mode based nonlinear control for cooling and heat-retention system actuated by Peltier devices, in: 2015 International Conference on Advanced Mechatronic Systems (ICAMechS), Aug 2015, pp. 343–348.

[19] P. Kachroo, M. Tomizuka, Integral action for chattering reduction and error convergence in sliding mode control, in: 1992 American Control Conference, June 1992, pp. 867–870.

[20] A.F. Fillipov, Differential Equations with Discontinuous Righthand Sides, 1 edition, Springer Netherlands, Springer Science+Business Media B.V., Netherlands, 1988. Originally published in Russian.

[21] J.-J.E. Slotine, W. Li, Applied Nonlinear Control, Pearson, Upper Saddle River, NJ, 1991.

[22] A. Mironova, P. Mercorelli, A. Zedler, Thermal disturbances attenuation using a Lyapunov controller for ice-clamping device actuated by thermoelectric coolers, Thermal Science and Engineering Progress (2018).

[23] J. Dispan, Development trends of the machine tool industry 2017 (original title: Entwicklungstrends im Werkzeugmaschinenbau 2017), January 2017, Working paper research funding, IGM.

Chapter 3

Temperature control in Peltier cells comparing sliding mode control and PID controllers

Alexandra Mironova, Paolo Mercorelli, and Andreas Zedler

Institute for Production Technology and Systems, Leuphana University of Lueneburg, Lueneburg, Germany

3.1 Introduction

To face the assumed model simplification, the signal d_c represents any kind of lumped model uncertainty that is not considered in the cold dynamics equation and, in particular, lumped thermal disturbances on the cold side due to any kind of external heating factor. Any kind of lumped thermal disturbance and model uncertainty, regardless of the hot dynamics equation, is represented by d_h. To a large extent, the external heating factor is associated with the machining process, which generates process (frictional) heat through the working tool acting on the workpiece. The majority of the mechanical energy in common machining operations is converted into heat [3,4]. Heat is generated at the tool–chip interface [4]. An exemplary examination of process heat is performed further in Section 3.3, whereas initial experiments already showed the tendency of heat generation during various machining operation steps as shown in [1]. Another external disturbance may also be due to the change of *ambient temperature*, which is normally assumed to be *constant*. However, in the real production environment, this condition cannot be guaranteed, except if an accurately controlled air-conditioning system is available. The same applies to the moisture condensation on the clamping plate due to the ambient humidity.

3.1.1 Sliding mode controller

According to the results of the previous chapter, any uncertainties, whether they are related to model simplifications or unknown parameters, have to be addressed by an appropriate control law formulation. Furthermore, the derived controller has to be stable against any external disturbances. For these objectives, it was chosen to design a Sliding Mode-based Controller (SMC).

The classic control approach, which is widely available on the market due to its simplicity and has been widely investigated [5–7], is the model-free control

Modeling, Identification, and Control for Cyber-Physical Systems Towards Industry 4.0
https://doi.org/10.1016/B978-0-32-395207-1.00013-5
41

strategy using the proportional-integral-derivative (PID) controller. However, the drawbacks of this strategy include oscillations, slow settling time characteristics or overshoots, and amplifications of noise [8,9]. With changes in operating conditions, the model-free PID control strategy shows limitations, and although gain-tuning strategies might be applied to enhance accuracy, they represent an additional effort. Therefore model-based control strategies are recorded to be more beneficial, especially for nonlinear systems [10,11], since they already encompass the relevant physics of the thermal system. However, incorporate model dynamics is accompanied by the challenge of representing the nonlinearities of the thermoelectric system. To deal with this, linear approximations, i.e., neglecting the Joule heat or the assumption of fixed values such as for the hot side temperature, are typical approaches, which are used to simplify the mathematical model of thermoelectric coolers (TECs), as shown in [12–16]. Another novel and control-oriented modeling approach for TEHs is introduced by the study [17], in which the method of integro-differential relations (MIDR) using parabolic partial differential equations is applied and analyzed by measurements for a multivariable heating temperature control. Although the latter research presents promising and accurate results, it is governed by high mathematical complexity, which involves a disproportionate effort of control law derivation. Since the present work is focused on reducing mathematical complexity while simultaneously avoiding model reduction and linearizing efforts, the SMC is a highly promising choice, since it encompasses the properties of robustness, high accuracy, and ease of tuning and implementation. Sliding mode is seen as *"one of the most significant discoveries in modern control theory"* [18], since one of its fundamental advantages is a high insensitivity to uncertainties and parameter variations and a complete rejection of disturbances [2,19–24]. Therefore, in terms of a nonlinear thermal system control, Sliding Mode is a highly promising tool when it comes to the present thermal system with unknown thermal parameters, since no parameter measurements or complex derivations of temperature dependencies have to be performed.

3.1.2   Multi-input multi-output control motivation

In control practice, normally one side of a TEC is controlled, so that the focus lies either on a cooling or a heating action. Controlling both sides simultaneously means controlling mainly the temperature difference. The upcoming section is focused on modeling both sides with their own dynamics simultaneously, using multiple input characteristics based on *Lyapunov* expressions for this approach. Despite the advantages of the robustness using Sliding Mode, controlling both sides of a TEC concurrently, the Sliding Mode (or another control) approach, to the current knowledge, has not been yet applied. Research activities like in [15,22,25,26] identify and verify the advantages of SMC for TECs, but still they are focused on controlling one side. The reasons for this lie in the difficulty of modeling the nonlinear dynamics and in finding an appropriate control law, allowing the control of both interacting parts of a TEC dependently. Facing this gap, the following derivation of the control law presents a new perspective of dealing with the input dynamics, allowing an interaction between the control algorithm and in this way representing the corresponding electro-thermal interaction of the physics in TECs more realistically.

Another reason why a simultaneous control of both temperatures of a TEC has not been yet derived in research is due to the fact that, primarily, one objective is pursued, whether heating *or* cooling. The application of TECs for ice clamping, however, shows new perspectives, which do not occur to this extent in any other application, since a major contribution of including an additional hot-side-control is focused on the zero point displacement of the clamping plate in the z-axis, which was discussed in detail in [1]. Especially in CNC-controlled precise machining, positioning errors in fixture systems have strictly to be avoided when fastening a workpiece in the processing machine. Due to the freezing process, the thermal stress, caused by high temperature differences between the hot and cold sides, lead to thermal expansion or shrinking of the clamping device materials. Whether the chilled plate is shrinking or extending (and thus whether the zero point offset is negative or positive) depends on the applied current and thus on the emerging hot side temperature and the heat sink performance. In particular, it depends on the combination of mass and the type of materials used for the clamping system, the heat sink, the TEC(s), and all utilized fastening elements of the whole device, having different coefficients of thermal expansion. Knowing this and taking the environment temperature into account, a control strategy focused on controlling the temperature delta, rather than a single temperature setpoint, represents an important upside in particular with respect to the field of the presented industrial application. As shown in the previous chapter, the cold and hot side dynamics are highly coupled to the thermal effects, so that a decoupling of both temperature states for control can only be achieved by two inputs. A detailed model, which also includes the dynamical change of the zero point displacement as a function of input values such as current and fan action, is difficult to achieve in terms of lumped components, since the complexity of the heterogeneous system belongs to the type of distributed components. Although the thermal expansion coefficients of most of the materials used to build the presented clamping device are known, the difficulty lies in the complexity of all interacting mechanical, thermal and electrical parts. Moreover, the difficulty of building a model also lies in describing the expansion of the TEC element itself, due to electrical and thermal stress, representing a subsystem of a conglomerate of different materials. With the given ideal equation, an exact physical representation and thus *desired* hot side temperature, which is associated with the spatial expansion, cannot be derived explicitly. However, an *implicit* relationship can be derived, which is associated with the thermal conductance of the heat sink K_k from Eq. (3.2), such that a multi-input control law can be formulated to control two target states simultaneously. In the area of multi-input systems combined with an SMC, the following two contributions analyze two very important problems. In [7] a robust adaptive sliding mode controller is proposed for a class of uncertain nonlinear multi-input multi-output (MIMO) systems. The upper bounds of the uncertainties are not needed in the procedure of the controller design, and the controller is continuous, which guarantees that the tracking error can converge to a small residual set. In [27] a logarithmic sliding surface obtained from a systematic procedure has been provided, in which this sliding surface without an initial constraint on its structure is proposed. Operational characteristics, such as sensitivity and control effort, are regulated. The proposed procedure ensures the global asymptotic stability of the closed-loop system. Nevertheless, sliding mode control approaches suffer from chattering problems, and in the case of high-precision velocity control, including soft landing, it is not always possible to obtain good results. In [28] a systematic design method of full order sliding mode control for nonlinear systems is presented, which allows both the chattering and singularity problems to be solved. More interesting in the context of sliding mode control is the MIMO case, which recently received particular attention. In fact, in [29] an adaptive scheme of designing sliding mode control for affine class of MIMO nonlinear systems with uncertainty in the system dynamics and control distribution gain is proposed. Typically, the closed-loop stability conditions are derived based on Lyapunov theory. The chapter is organized in the following way. Section 3.2 takes into consideration the structure of the proposed SMC. Section 3.3 is dedicated to the experimental validation of the results obtained using the SMC and PI controller. In Section 3.5, conclusions on the controller comparison close the chapter.

3.2 Sliding mode control law derivation

Considering the model of the previous chapter, it follows that the model is extended and reformulated as

$$\frac{dT_c(t)}{dt} = \frac{1}{m_a c_a}\left[-Si(t)T_c(t) + \frac{1}{2}Ri^2(t) + K\left(T_h(t) - T_c(t)\right) + d_c(t)\right],$$
(3.1)

$$\frac{dT_h(t)}{dt} = \frac{1}{m_k c_k}\left[Si(t)T_h(t) + \frac{1}{2}Ri^2(t) - K(T_h(t) - T_c(t))\right.$$
$$\left. - K_k(T_h(t) - T_a) + d_h(t)\right].$$
(3.2)

To derive a control law based on the sliding mode approach, it is common to analyze its finite time stability using the method of Lyapunov functions [30]. In 1892, Alexander Lyapunov in his doctoral thesis [31,32] described *"the existence of positive definite function[s]"* [33], which demonstrates sufficient conditions for nonlinear stability of a system. By this method the asymptotic stability of a system is determined by the criteria that the energy of the system is decreasing and thus converges towards its minimum (along its trajectory). At this minimum (point of equilibrium where the Lyapunov function reaches its local minimum), no more increase is possible and hence no instability, and thus a stability in the sense of Lyapunov is present.

The SMC is based on the idea of controlling and forcing the states onto a defined path in the state space, which is called the *sliding surface*. Hence, first and before the derivation of the controller input law, the sliding surface has to be defined.

For the sliding surface, which is shown in the following theorem, an integral-type surface was chosen. This is due to the fact that this action can limit the chattering behavior. Following this, without using an integral term, the chattering could be increased, in particular, in the final part of the tracking. This issue is a well-known and discussed aspect, which can be found in classic SMC-related research, emphasizing the importance of the integral term to reduce chattering; see [34].

Theorem 2. *Given the system represented by Eqs.* (3.1) *and* (3.2), *there exists an asymptotically stable controller characterized by the controlling closed-loop input current $i_{cl}(t)$ and by the real constants $\lambda_c > 0$, $\lambda_h > 0$, $\beta_c > 0$, and $\beta_h > 0$ together with two real constants $k_{ec} > 0$ and $k_{eh} > 0$ such that*

$$\lim_{t \to \infty} T_c(t) = T_{cd}(t) = T_{cd} \quad \text{and} \quad \lim_{t \to \infty} T_h(t) = T_{hd}(t) = T_{hd}. \tag{3.3}$$

We define the following two sliding mode surfaces:

$$s_c(t) = T_{cd}(t) - T_c(t) + k_{ec} \int_0^t \Big(T_{cd}(\tau) - T_c(\tau) \Big) d\tau - s_c(0), \tag{3.4}$$

$$s_h(t) = T_{hd}(t) - T_h(t) + k_{eh} \int_0^t \Big(T_{hd}(\tau) - T_h(\tau) \Big) d\tau - s_h(0), \tag{3.5}$$

where the presence of terms $s_c(0)$ and $s_h(0)$ reduces to zero the reaching phase ($s_c(0) = 0$ and $s_h(0) = 0$) regardless of the initial conditions $T_c(0)$ and $T_h(0)$; T_{cd} and T_{hd} can either represent the *constant* desired cold and hot side temperatures, or these setpoints can be described in a *dynamic* way as

$$T_{cd}(t) = (T_a - T_{cd})e^{-at} + T_{cd} \tag{3.6}$$

and

$$T_{hd}(t) = (T_a - T_{hd})e^{-at} + T_{hd}, \tag{3.7}$$

in which the constant a states the velocity of the dynamics of the desired temperature profile. There are a few reasons, which may be valuable and advantageous, to treat the desired setpoints dynamically. A dynamic expression can minimize the initial start energy of the controller output. This can generate smooth trajectories, prevent overshoots in the tracking, and represent temperature distribution realistically. Due to the thermal inertia of the system (caused by the system intrinsic properties, e.g., the mass of plate and heat sink, which resembles capacitor behavior), the velocity of the reaching phase is, comparably to classic mechanical clamping systems, in any case significantly longer, so that there is no special need to try to enhance the velocity, especially with regard to its physical limitations. Moreover, as pointed out in [1], with slower cooling rates, the cohesive bonding of ice may be enhanced. Following this, the overall clamping forces can be increased.

Let

$$e_{ec}(t) = T_{cd}(t) - T_c(t), \tag{3.8}$$
$$e_{eh}(t) = T_{hd}(t) - T_h(t) \tag{3.9}$$

be the temperature errors. It is then possible to define two sliding temperature surfaces, one for the cold and one for the hot side, as follows:

$$\dot{s}_c(t) = \dot{e}_{ec}(t) + k_{ec}e_{ec}(t), \tag{3.10}$$
$$\dot{s}_h(t) = \dot{e}_{eh}(t) + k_{eh}e_{eh}(t). \tag{3.11}$$

Eqs. (3.10) and (3.11) with $\dot{s}_c(t) = 0$ and $\dot{s}_h(t) = 0$ state the tracking dynamics of the controlled system in sliding mode for the cold and hot temperatures, respectively. The constants k_{ec} and k_{eh} state their rates of convergence. By solving $\dot{s}_c(t) = 0$ and $\dot{s}_h(t) = 0$ formally for the control input we obtain an expression for the inputs called *equivalent control*, which can be interpreted as the continuous control law that would maintain $\dot{s}_c(t) = 0$ and $\dot{s}_h(t) = 0$ if the dynamics were known exactly.

At $s_c(t) = 0$ and $s_h(t) = 0$, this means that the sliding mode system approaches the sliding surfaces, and we obtain

$$\dot{e}_{ec}(t) + k_{ec}e_{ec}(t) = 0, \tag{3.12}$$
$$\dot{e}_{eh}(t) + k_{eh}e_{eh}(t) = 0, \tag{3.13}$$

which admit the following solutions:

$$e_{ec}(t) = \eta_c \exp^{-k_{ec}t} \tag{3.14}$$

and

$$e_{eh}(t) = \eta_h \exp^{-k_{eh}t}, \tag{3.15}$$

where η_c and η_h are constants determined by the initial conditions. To realize a region of attraction around this sliding surface, we can consider a Lyapunov approach. Starting by defining two functions $\Phi_c(t)$ and $\Phi_h(t)$ such that

$$\Phi_c(t)\dot{\Phi}_c(t) = \frac{R}{2m_ac_a}i^2(t) + \frac{K}{m_ac_a}(T_h(t) - T_c(t)) \tag{3.16}$$

and

$$\Phi_h(t)\dot{\Phi}_h(t) = \frac{S}{m_k c_k}i(t)T_h(t) + \frac{R}{2m_k c_k}i^2(t), \qquad (3.17)$$

it follows that

$$\Phi_c(t)d\Phi_c(t) = \left(\frac{R}{2m_a c_a}i^2(t) + \frac{K}{m_a c_a}\big(T_h(t) - T_c(t)\big)\right)dt, \qquad (3.18)$$

$$\int \Phi_c(t)d\Phi_c(t) = \int \left(\frac{R}{2m_a c_a}i^2(t) + \frac{K}{m_a c_a}\big(T_h(t) - T_c(t)\big)\right)dt, \qquad (3.19)$$

$$\frac{1}{2}\big(\Phi_c(t)\big)^2 + C_c = \int_0^t \left(\frac{R}{2m_a c_a}i^2(\tau) + \frac{K}{m_a c_a}\big(T_h(\tau) - T_c(\tau)\big)\right)d\tau \qquad (3.20)$$

with $C_c = 0$, and through similar consideration it follows that

$$\frac{1}{2}\big(\Phi_h(t)\big)^2 + C_h = \int_0^t \left(\frac{S}{m_k c_k}i(\tau)T_h(\tau) + \frac{R}{2m_k c_k}i^2(\tau)\right)d\tau \qquad (3.21)$$

with $C_h = 0$. According to Remark 3, $T_c(t) < T_h(t)$ if $i(t) > 0$, and considering that $T_c(t) > 0$ and $T_h(t) > 0$, we have

$$\frac{R}{2m_a c_a}i^2(t) > 0, \quad \frac{K}{m_a c_a}(T_h(t) - T_c(t)) > 0, \quad \frac{S}{m_k c_k}i(t)T_h(t) > 0, \qquad (3.22)$$

thus the right parts of Eqs. (3.20) and (3.21) are positive, and thus positions (3.20) and (3.21) are consistent. According to a Lyapunov-like lemma, which is used for the asymptotic stability analysis for non-autonomous systems,[1][2] the following *lower bounded scalar function* can be chosen to design the sliding mode controller:

$$V(T_h, T_c, t) = \frac{s_c^2(t) + \Phi_c^2(t) + s_h^2(t) + \Phi_h^2(t)}{2}. \qquad (3.23)$$

Then

$$\dot{V}(T_h, T_c, t) = s_c(t)\dot{s}_c(t) + \Phi_c(t)\dot{\Phi}_c(t) + s_h(t)\dot{s}_h(t) + \Phi_h(t)\dot{\Phi}_h(t). \qquad (3.24)$$

Using the model of Eqs. (3.1) and (3.2), it follows that

$$\dot{V}(T_h, T_c, t) =$$
$$s_c(t)\left[-a\big(T_a - T_{cd}\big)e^{-at} + \frac{S}{m_a c_a}i(t)T_c(t) - \frac{R}{2m_a c_a}i^2(t) - \frac{K}{m_a c_a}(T_h(t) - T_c(t))\right]$$

---

[1] It is worth noting that a system $\dot{x}(t) = f(x(t), u(t))$ under time-varying feedback $u(x(t), t)$ represents a non-autonomous system. This is needed, for instance, also in the typical case in which $x(t)$ will be required to track a time-varying trajectory.

[2] See pg. 125 and pg. 126 of [35].

$$- \frac{d_c(t)}{m_a c_a} + k_{ec}(T_{cd}(t) - T_c(t)) + \Phi_c(t)\dot{\Phi}_c(t)\Big] + s_h(t)\Big[-a(T_a - T_{hd})e^{-at}$$

$$- \frac{S}{m_k c_k}i(t)T_h(t) - \frac{R}{2m_k c_k}i^2(t) + \frac{K}{m_k c_k}(T_h(t) - T_c(t)) + \frac{K_k}{m_k c_k}(T_h(t) - T_a)$$

$$- \frac{d_h(t)}{m_k c_k} + k_{eh}(T_{hd}(t) - T_h(t)) + \Phi_h(t)\dot{\Phi}_h(t)\Big]. \quad (3.25)$$

Considering the definition of $\Phi_c(t)$ and $\Phi_h(t)$ and in particular Eqs. (3.16) and (3.17), it follows that

$$\dot{V}(T_h, T_c, t) = s_c(t)\Big[-a(T_a - T_{cd})e^{-at} + \frac{S}{m_a c_a}i(t)T_c(t) - \frac{d_c(t)}{m_a c_a}$$

$$+ ke_c(T_{cd}(t) - T_c(t))\Big] + s_h(t)\Big[-a(T_a - T_{hd})e^{-at} + \frac{K}{m_k c_k}(T_h(t) - T_c(t))$$

$$+ \frac{K_k}{m_k c_k}(T_h(t) - T_a) - \frac{d_h(t)}{m_k c_k} + ke_h(T_{hd}(t) - T_h(t))\Big]. \quad (3.26)$$

The first control loop can be designed using the *equivalent input current* $i_{eq}(t)$ and the *corrective input current* $i_c(t)$ to obtain a controlling closed loop current $i_{cl}(t)$ as follows:

$$i_{cl}(t) = \frac{m_a c_a}{S T_c(t)} \underbrace{\Big[a(T_a - T_{cd})e^{-at} - k_{ec}(T_{cd}(t) - T_c(t))}_{i_{eq}(t)} \underbrace{- \lambda_c(s_c(t)) - \beta_c \operatorname{sgn}(s_c(t))\Big]}_{i_c(t)}.$$

$$(3.27)$$

In a similar way, we calculate a controlling closed loop input $\varkappa_{cl}(t)$ consisting of the *equivalent conduction input* $\varkappa_{eq}(t)$ and the *corrective conduction input* $\varkappa_c(t)$ for the hot side:

$$\varkappa_{cl}(t) = \frac{m_k c_k}{(T_h(t) - T_a)} \underbrace{\Big[a(T_a - T_{hd})e^{-at} - \frac{K(T_h(t) - T_c(t))}{m_k c_k} - k_{eh}(T_{hd}(t) - T_h(t))\Big]}_{\varkappa_{eq}(t)}$$

$$\underbrace{- \frac{m_k c_k}{(T_h(t) - T_a)}\Big[\lambda_h(s_h(t)) + \beta_h \operatorname{sgn}(s_h(t))\Big]}_{\varkappa_c(t)}. \quad (3.28)$$

The variable $\varkappa_{cl}(t)$ is treated analogously to the value of K_k. It indirectly represents the forced convection due to the use of a fan to increase the effect of the heat removal rate. The relationship between forced convention and the heat sink conductance coefficient K_k was explained in [1]. Since the aluminum heat sink dimensions and material properties remain unchanged, the only way to actively change the performance of the heat sink and thus the temperature of the hot side is obtained by controlling the ventilation. With $\varkappa_{cl}(t)$, the air volume flow rate is controlled in terms of the fan speed, which is controlled by the input power to the DC external rotor motor of the fan. The contribution of the fan power to the total conductance of the heat sink can be obtained from manufacturer data sheets; more generally, we can refer to [36–38], where the mathematical descriptions are shown more comprehensively. Hence $\varkappa_{cl}(t)$ can be seen as an effective input value responsible for the heat sink total performance.

In Eqs. (3.27) and (3.28) the $sgn(s_c, s_h)$ represents the sign function

$$sgn(s_c, s_h) = \begin{cases} +1 & \text{if } s_c, s_h > 0, \\ -1 & \text{if } s_c, s_h < 0. \end{cases} \tag{3.29}$$

As is well known, the equivalent control makes the derivative of the sliding surface equal zero to stay on the sliding surface, and the corrective control compensates the deviations from the sliding surface to reach the sliding surface. The feedback control laws in Eqs. (3.27) and (3.28) are such that $V(T_h, T_c, t)$ remains a Lyapunov-like function of the closed loop system.[3] It is worth noting that if input Eqs. (3.27) and (3.28) are inserted into Eq. (3.26), then

$$\dot{V}(T_h, T_c, t) \leq s_c(t)\dot{s}_c(t) + s_h(t)\dot{s}_h(t) = s_c(t)\left(-\frac{d_c(t)}{m_a c_a} - \lambda_c s_c(t) - \beta_c \, sgn(s_c(t))\right)$$
$$+ s_h(t)\left(-\frac{d_h(t)}{m_k c_k} - \lambda_h s_h(t) - \beta_h \, sgn(s_h(t))\right), \tag{3.30}$$

and since $\lambda_c > 0$ and $\lambda_h > 0$, it follows that

$$\dot{V}(T_h, T_c, t) \leq s_c(t)\dot{s}_c(t) + s_h(t)\dot{s}_h(t) = s_c(t)\left(-\frac{d_c(t)}{m_a c_a} - \beta_c \, sgn(s_c(t))\right)$$
$$+ s_h(t)\left(-\frac{d_h(t)}{m_k c_k} - \beta_h \, sgn(s_h(t))\right). \tag{3.31}$$

The sufficient condition

$$\dot{V}(T_h, T_c, t) \leq 0 \tag{3.32}$$

is guaranteed if

$$\beta_c \geq \max\left|\frac{d_c(t)}{m_a c_a}\right| \tag{3.33}$$

and

$$\beta_h \geq \max\left|\frac{d_h(t)}{m_k c_k}\right|. \tag{3.34}$$

---

[3] Please refer in detail to [35] on pg. 282.

To show the asymptotic stability, a Lyapunov-like lemma is invoked. In fact, the condition $\dot{V}(T_h, T_c, t) \leq 0$, together with the fact that assuming that disturbance $d_c(t)$ and $d_h(t)$ are uniformly continuous functions, implies that $\dot{V}(T_h, T_c, t)$ is a uniformly continuous function because it consists of a sum of uniformly continuous functions. These two facts, together with $V(T_h, T_c, t)$ being lower bounded, guarantee that the hypotheses of Lyapunov-like lemma are satisfied and thus $\lim_{t \to +\infty} \dot{V}(T_h, T_c, t) = 0$, and we obtain the following result:

$$\lim_{t \to +\infty} T_c(t) = T_{cd}(t) = T_{cd} \quad \text{and} \quad \lim_{t \to +\infty} T_h(t) = T_{hd}(t) = T_{hd}. \quad \square \quad (3.35)$$

Remark 1. The fact that $\dot{V}(T_h, T_c, t) \leq 0$ guarantees the boundedness of the controlling closed-loop input current $i_{cl}(t)$ and the controlling closed-loop conduction input $\varkappa_{cl}(t)$. Moreover, the scalar function (3.23) is radially unbounded, which guarantees the global asymptotic stability of the controlled system. $\quad \square$

Remark 2. In case of parametric uncertainties, conditions (3.33) and (3.34) become

$$\beta_c \geq \max \left| \frac{d_c(t)}{m_a c_a} \right| + \max |\Delta_c(t)| \quad (3.36)$$

and

$$\beta_h \geq \max \left| \frac{d_h(t)}{m_k c_k} \right| + \max |\Delta_h(t)| \quad (3.37)$$

with the assumptions

$$\mathbf{A_1}: \quad \max |\Delta_c(t)| < \delta_c \quad (3.38)$$

and

$$\mathbf{A_2}: \quad \max |\Delta_h(t)| < \delta_h, \quad (3.39)$$

where δ_c and δ_h are known bounded real quantities. With conditions $\mathbf{A_1}$ and $\mathbf{A_2}$, the system described in Eqs. (3.1) and (3.2) is assumed to be stable. The functions $\Delta_c(t)$ and $\Delta_h(t)$ represent the estimated maximal margin of uncertainties due to the variations of the parameters, which generate a non-perfect cancelation through the equivalent inputs $i_{eq}(t)$ and $\varkappa_{eq}(t)$. $\quad \square$

Remark 3. We can see that the cooling phase is guaranteed for $i(t) > 0$ for all t such that $T_h(t) > T_c(t)$ for all t and all considered TECs. This is subject to the Peltier effect, which governs the directional dependency of the electric current applied. Changing polarity leads to a redirection of the heat flow, so that for an unclamping and thus ice thawing action, heating is guaranteed for $i(t) < 0$ with the consequence that $T_h(t) \leq T_c(t)$. In terms of ice clamping, the last condition does not necessarily have to be $T_h(t) < T_c(t)$, since thawing conditions only depend on $T_c(t) > 273 \, \text{K}$, so that for a sufficient time period

T_{th}, the thawing is even guaranteed for $i(T_{th}) = 0$. Whereas the time T_{th} is then a function of thermal conductance, natural convection and radiation, and the heat quantities of all involving matter. To facilitate fast thawing, however, it is a highly recommended practice to force the input current to $i(t) < 0$. ☐

Structural stability

The Lyapunov-like lemma was introduced to mathematically prove the stability of the derived Sliding Mode Control law. A further assumption can be formulated to prove the existence of an additional *structural stability*. The structural stability of the desired cold side temperature is defined to depend on the coefficient K_k, representing the heat removal through forced convection and conduction on the hot side. The higher the thermal conductance, the lower the thermal resistance of the heat sink. Thus the higher the possible temperature difference can be achieved, the lower the T_{cd} can be chosen. So the stability of $T_h(t)$ can be seen as a function of the term K_k, which must guarantee enough cooling of the hot side. Assuming that due to the action of the fan the variation of $T_h(t)$ is small enough, the hot side temperature change of Eq. (3.2) can be rewritten in a steady-state form considering the following assumption.

A1: There exists a value of variable $K_k(t)$ such that

$$0 = Si(t)T_h(t) + \frac{1}{2}Ri^2(t) - K\left(T_h(t) - T_c(t)\right) - K_k\left(T_h(t) - T_a\right). \quad (3.40)$$

Eq. (3.40) can be guaranteed by the existence of an input current, which can achieve a *maximal heating power* Q_{hmax} by defining

$$\frac{dQ_h(t)}{di(t)} = 0 \quad (3.41)$$

with $Q_h(t)$ corresponding to Eq. (3.2). By differentiating Eq. (3.2), in view of Eq. (3.41), we have

$$ST_h(t) + Ri_{Qhmax}(t) = 0, \quad (3.42)$$

so that

$$i_{Qhmax}(t) = \frac{T_h(t)S}{R}. \quad (3.43)$$

Condition (3.40) can be guaranteed if we prove that $T_h(t)$ consists of small variations with respect to the right terms of (3.2). Considering Eqs. (3.40) and (3.43), this can be guaranteed by

$$K_k(t) = \left(T_h(t) - T_a\right)^{-1}\left(\left(R^{-1}T_h^2(t)S^2\right) + \left((2R)^{-1}T_h^2(t)S^2\right) - K\left(T_h(t) - T_c(t)\right)\right). \quad (3.44)$$

To be more practical, it is possible to consider the maximal and feasible value of $K_k(t) = \hat{K}_k$. Eq. (3.44) can be taken to calculate the needed parameter beforehand or to evaluate the thermal performance of an existing heat sink; thus it is a good tool for designing and sizing proper thermal systems with high temperature differences. Furthermore, Eq. (3.44) proves the existence of a value stabilizing the thermal system.

Remark 4. – Case Study With reference to the practical application, the boundary values, which determine the minimal and maximal heat removals, are represented by the reciprocal of the term K_k, since this value can be gathered from the manufacturer's data sheet. The value of interest is the thermal resistance R_{th}. For example, by following Eq. (3.44) and assuming constant steady state values (e.g., maximal desired temperature delta), we can derive that:

$$\hat{R}_{th} = \frac{T_h - T_a}{T_h^2 S^2 R^{-1} + T_h^2 S^2 (2R^{-1}) - K(T_h - T_{cd})} \tag{3.45}$$

with $\hat{R}_{th}$ representing the thermal resistance that guarantees the stability in the presence of maximal heating power when the corresponding $i_{Qh_{max}}$ is applied to the element. An exemplary calculation can be then derived, considering a desired setpoint at $T_{cd} = 263$ K with a constant ambient temperature at $T_a = 295$ K and formulating the boundary condition that T_h should not be more than 10 degrees over the ambient temperature. Then, to attain the structural stability, a heat sink should be chosen with thermal performance of $\hat{R}_{th} < 0.19 \frac{K}{W}$ or $\hat{K}_k > 5.3 \frac{W}{K}$. In fact, according to the data sheet, the performance metric of the chosen heat sink for measurements (see Table 3.1 in Section 3.2.1) is $R_{th} = 0.125 \frac{K}{W}$, describing the maximal achievable cooling performance (in the presence of maximum ventilation performance). Furthermore, if the chosen boundary condition is revised, then the structural stability can no longer be guaranteed. Assuming a new case that the temperature difference between the ambient and hot sides is lower than 5 degrees, the chosen heat sink performance would not be sufficient, as, for stability, it must be then guaranteed that $\hat{R}_{th} < 0.095 \frac{K}{W}$. Note that this calculation is for exemplary use only.

3.2.1 Simulation results of MIMO SM controller

For the simulation of the proposed controllers, we use constant parameters for the Seebeck coefficient, thermal conductance, and the electric resistance shown in Table 3.1. For the cold side temperature, a value of $-10°C$ (corresponding to 263.15 K) was chosen as the desired setpoint, as it represents a sufficiently good operation temperature for generating high adhesive ice strength [1]. The hot side temperature setpoint was chosen to be $+32°C$ (corresponding to 305.15 K), such that in total a temperature delta of $\triangle T = 42$ K should be controlled.

Fig. 3.1 shows the conceptual scheme of the designed control law. In Fig. 3.2 the tracking results for the cold and hot sides are shown. Whereas the rising time

TABLE 3.1 Constant coefficient values used for simulation of MIMO SMC controller. Values referring to the TEC type QC–450–0.8–3.0, heat sink LA6250 and fan ebmpapst 24VDC.

Coefficient	Value	Unit	Coefficient	Value	Unit	Coefficient	Value	Unit
m_a	0.1323	kg	K	0.976	W/K	λ_c	0.03	s^{-1}
m_k	1.360	kg	S	0.17	V/K	β_c	0.001	Ks^{-1}
c_a	898	$Jkg^{-1}K^{-1}$	R	20	Ω	λ_h	0.0003	s^{-1}
c_k	898	$Jkg^{-1}K^{-1}$	a	0.05	s^{-1}	β_h	0.0001	Ks^{-1}
A	0.026	m^2	k_{ec}	0.01	s^{-1}	k_{eh}	1	s^{-1}

Technical data of fan	
thermal resistance at maximal input voltage (DC)	$0.125\,KW^{-1}$
maximal input voltage (DC)	$24\,V$
maximal power	$3\,W$
maximal air volume	$56\,m^3\,h^{-1}$
maximal speed	$6.850\,min^{-1}$

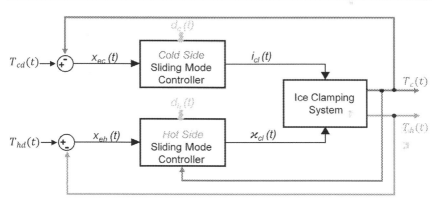

FIGURE 3.1 Block diagram of conceptual control scheme.

dynamic of the desired hot side shows fast response, taking into account an error tolerance of ±0.5 K, the cooling velocity on the cold side of the system is much slower. This is because the temperature difference of the desired cold side with regard to the ambient temperature is much higher. However, it is important to note that the settling time dynamics of the cold side is 2.5 times faster than that of the hot side, being subject to the higher λ_c and β_c values, which were chosen to emphasize the cold side dynamics with regard to the purposed freezing application. It should be pointed out that the rising time seems to be long, concerning a manufacturing system, but we have to consider that this represents the launch time, when the system has to cool down starting at room temperature. After reaching the operational setpoint, the unclamping procedure is achieved through a reversal of the polarity of the applied current, which leads to a heating of the

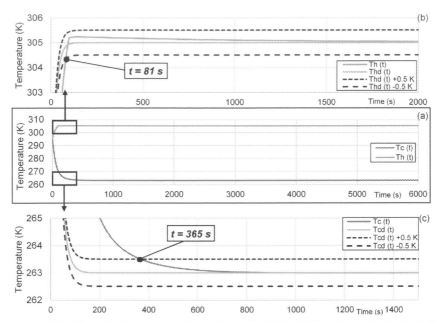

FIGURE 3.2   Hot and cold temperature simulation results in (a) with (b) and (c) being the detailed views of cold and hot side temperatures, showing their rising time dynamics depending on the tolerance set as ±0.5 K.

cold side and thus to a fast thawing of the ice. For this, the system is heated to 0°C, which in turn speeds up the velocity of cooling in the next clamping phase.

In Fig. 3.3 the corresponding input current and the input value for the conduction are shown. As shown in Fig. 3.3(b), the boundaries are set to 2 W/K and 8 W/K, describing the minimum and maximum of the heat sink performance. The minimum heat conductance is accounted for the heat sink itself, without fan action, in the presence of natural convection, which is always present, and thus this value has to be taken as the lowest constant limit. The maximum value is related to the maximal power output of the fan at maximum voltage input. In conclusion, this range defines the operational area of the designed controller and is derived from the electric power consumption of the applied fan.

From the simulation results we see that the controller is capable to control both sides stable and for a long time run. It leads to the conclusion that an accurate temperature delta control can be obtained using the proposed control strategy. A precise *temperature difference controller* shows not only advantages towards the avoidance of zero point displacement, but it may also be interesting regarding other application areas, such as measurement or calibration technologies.

However, as discussed in Section 3.1.2 and with consideration of the results presented in [1], it is not possible to deduce the exact value of the hot side tem-

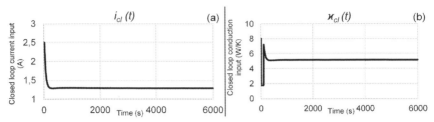

FIGURE 3.3 Simulation results of controlling closed loop input current in a) and controlling closed loop conduction input in b).

perature, which would guarantee a non-displacement, without further modeling. For this purpose, a more complex physical model representation needs to be derived, including material properties of each component of the ice-clamping system. Therefore it was decided to keep the fan action at its maximum for the upcoming simulations and measurements. Hence a single-input single-output (SISO) system is considered. Moreover, with regard to an economical, ecological, and thus sustainable manufacturing, at maximal cooling performance, this choice is advantageous, since the electric power consumption of the TEC can be minimized when powering the fan on its maximum. The performance metrics of the TEC show a high voltage at a high current (see Table 3.1), whereas the power consumption of the fan is, in contrast, much lower. Through simulation it is possible to derive that almost 11 W can be saved at maximum fan action (at the same setpoint $T_{cd}(t)$).

3.3 Experimental validation

Measurements were performed to validate the simulations and experimentally validate the stability of the controller in the presence of model and parameter uncertainties when external thermal disturbances are applied to the clamping plate. Controller gains were taken from simulations, summarized in Table 3.1. The conceptual structure of the experimental setup is shown in Fig. 3.4.

The temperature is measured with type-k thermocouples. The data is logged within the μController Arduino Due, which controls the magnitude of the output current and drives the TEC by forcing the *closed-loop input current* to track the desired temperature values via DC. Furthermore, Fig. 3.4 shows the scheme of how an external lumped thermal disturbance is implemented on the cold side of the clamping device, simulating a machining action. For this, a specimen made of AlMgSi0.5, similar to the reference specimen used for measurements in [1], is heated with a heating element till thermal inertia. After reaching thermal inertia, it is put on the clamping plate for a certain amount of time, simulating high thermal stress of an operation with a worn tool and forcing the system to its

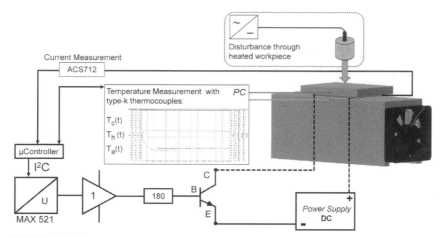

FIGURE 3.4   Structure of the experimental setup, showing the basic overview of the electronic hardware.

limits, namely nearly to the melting point of the ice. The properties of the TEC, heat sink, and fan remain the same, as used for the simulations; see Table 3.1.

Definition of thermal disturbance

To test the controller performance when thermal disturbance is applied, pre-measurements were performed. External lumped thermal disturbances depend on different factors, such that their determination cannot be generalized. Beginning with the material of the workpiece, its specific heat coefficient, conductivity, and mass have to be taken into account. The operation parameters, like the machining mode, whether a new or worn tool is used, cutting forces, feed speed, and the cutting depth are also decisive factors, which need to be taken into consideration. Therefore, to better interpret the presented results for the introduced system, experiments with the reference specimen were performed to attain some exemplary disturbance values, based on common machining operations, such as drilling and milling. Since a milling action was already discussed in the previous chapter, an exemplary drilling action is introduced in addition. Drilling was performed without cooling at room temperature, by thermally isolating the specimen, to attain the input thermal power and energy. The evolving temperature was measured with a k-type thermocouple sensor, which was placed inside the specimen. The time to derive the thermal energy input was monitored. A worn tool was used to simulate extra high thermal stress and to illustrate a common machining procedure, since in the industry the tool would not be replaced with each new machining procedure. The average feed rate was 17.2 mm/min, and the drilling hole depth was 10.5 mm. The results plotted in Fig. 3.5 show that the maximal thermal power for the largest (worn) tool was recorded to be 56 W, corresponding 2300 J.

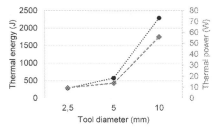

FIGURE 3.5 Introduced power and energy by a drilling action with three different drill diameters at an average feed rate of 17.2 mm/min and with a drill hole depth of 10.5 mm.

Measured controller performance

The tracking results, in the presence of an external lumped disturbance, are shown in Fig. 3.6, comparing measured (subscript *meas*) and simulated results (subscript *sim*). The simulations of the cold side dynamics are much in line with the experiments, having a steady state error $< 0.01\%$. However, the simulated current deviates from the real measurements (see Fig. 3.6(b)), which can be accounted for model uncertainties due to the use of a compact model with constant parameters. The corresponding mean absolute error is 280 mA; see Fig. 3.6(c). We can see from the results that the disturbance attenuation of the controller shows fast and robust behavior, having a small undershoot in the response, which is highly desirable for a homogeneous temperature distribution on the clamping plate. A uniform heat allocation equals constant clamping forces. We can conclude that even among uncertainties and model simplifications, the proposed controller is characterized by a strong and robust performance.

According to Fig. 3.5, the applied thermal disturbance represents the second drilling action (with a tool diameter of 5 mm), and hence 600 J were applied to the cold clamping plate. In fact, the chosen thermal disturbance led the temperature of the cold side to exceed the zero point, and hence a thawing process was already present. Consequently, applying 600 J to the system shows clearly its upper limit (at least for the considered system). Despite the quick response of the controller, the thermal inertia of the system lowers its efficiency. To counter this, it is therefore advisable to lower the setpoint to increase the safety margin. Nevertheless, it should be noted that this kind of thermal disturbance represents an especially high thermal stress, since low feed rates with worn tools were used to determine exemplary values. From the real performance test in [1] it was shown that even without a controller action the clamping plate temperature on the cold side did not exceed the dew point.

In Fig. 3.7 the performance of the controller was tested, when, next to a short time disturbance (400 J), a permanent thermal noise was applied to the cold side of the system. The results show a good disturbance compensation and accurate tracking and prove a longtime stability of the controller when a continuous thermal load is applied to the system (here again, 600 J were applied on the cold side to simulate extra high thermal load), representing a long machining opera-

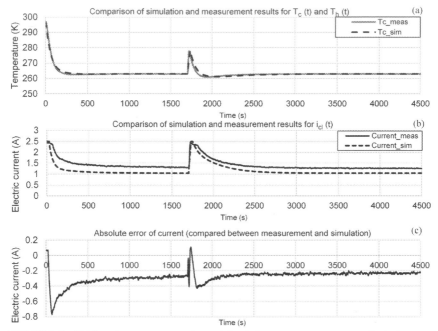

FIGURE 3.6   SMC performance: comparison of simulation and measurement results in the presence of an external lumped disturbance; the cold side temperatures are shown in (a), the input currents in (b), and the absolute error of input current compared to the experiments in (c).

tion. In Fig. 3.7(b) the corresponding controlling input current is shown, which shows a quick response to the applied thermal disturbance, emphasizing a good sensitivity to temperature changes as soon as they occur. This sensitivity is due to the appropriate values of the parameters λ and β obtained from simulation.

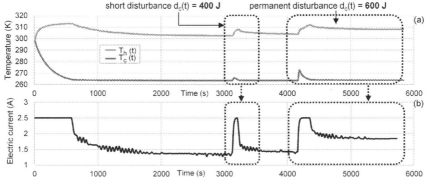

FIGURE 3.7   (a) Sliding Mode controller performance in the presence of a short and permanent (longtime) disturbances applied to the cold side of the clamping system; (b) the corresponding controlling closed-loop input current of the controller.

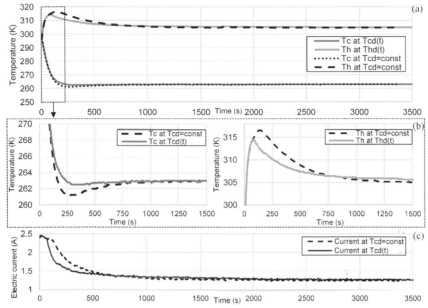

FIGURE 3.8 Comparative measurement of using fixed values or a dynamic expression for the desired temperature setpoints with (a) showing the temperature dynamics for hot and cold sides over the entire measurement period. (b) The corresponding input currents i_{cI}; (c) and (d) detailed views of the cold and hot side temperatures are depicted.

Performance of dynamic temperature setpoint

Following the proposal that the desired temperature setpoints can be described either as constants or in a dynamic way (see Eqs. (3.6) and (3.7)), further measurements were performed to examine the effects of both approaches. In Fig. 3.8(a) and in the detailed view of Fig. 3.8(c) and (d), the results for the cold and hot side temperatures are shown. Especially for the cold side, a dynamic approach leads to a smoother and therefore faster reaching performance of the desired value of 263.15 K ($-10°$C). In both cases (cold and hot) the overshoot characteristics of temperatures are noticeably damped. Fig. 3.8(b) shows the corresponding current results for both approaches. As a conclusion, a dynamic expression can reduce the initial energy of the controller without reducing the accuracy of the tracking performance. Moreover, this approach leads to a better behavior of the controller action.

3.4 Controller extension

The objective of a distributed control leads to the question whether the ideal model representation is appropriate for the control task of multiple TECs or whether the model has to be extended to better map the real system. For this purpose, the mathematical model representation is modified for the case of two interacting TECs. The thermal interaction is assumed to be represented by thermal conductive heat flows inside the clamping plate and inside the heat sink. The schematic concept is shown in the enlarged physical model in Fig. 3.9.

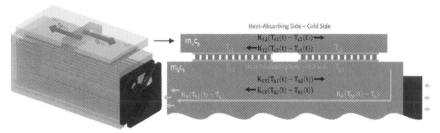

FIGURE 3.9 Extended ice-clamping device actuated by two TECs with interacting thermal heat flows.

According to Fig. 3.9 and with consideration of Eq. (3.1), the cold side temperatures of the first and second TEC change as follows:

$$m_a c_a \frac{dT_{c1}(t)}{dt} =$$
$$- Si_1(t)T_{c1}(t) + \frac{1}{2}Ri_1^2(t) + K(T_{h1}(t) - T_{c1}(t)) - K_{12}(T_{c1}(t) - T_{c2}(t)) + d_c(t),$$
(3.46)

and

$$m_a c_a \frac{dT_{c2}(t)}{dt} =$$
$$- Si_2(t)T_{c2}(t) + \frac{1}{2}Ri_2^2(t) + K(T_{h2}(t) - T_{c2}(t)) - K_{12}(T_{c2}(t) - T_{c1}(t)) + d_c(t),$$
(3.47)

with the new *interacting* conductive term $K_{12}\Delta T$ describing the heat flow inside the material of the cold clamping plate, where K_{12} is the coefficient of the (spatial) thermal conductance. The same applies to the hot side temperature change. The reformulation of Eq. (3.2) yields to

$$m_k c_k \frac{dT_{h1}(t)}{dt} = Si_1(t)T_{h1}(t) + \frac{1}{2}Ri_1^2(t) - K(T_{h1}(t) - T_{c1}(t))$$
$$- K_k(T_{h1}(t) - T_a) - K_{13}(T_{h1}(t) - T_{h2}(t)) \quad (3.48)$$

and

$$m_k c_k \frac{dT_{h2}(t)}{dt} = Si_2(t)T_{h2}(t) + \frac{1}{2}Ri_2^2(t) - K(T_{h2}(t) - T_{c2}(t))$$

$$- K_k\left(T_{h2}(t) - T_a\right) - K_{13}\left(T_{h2}(t) - T_{h1}(t)\right). \quad (3.49)$$

The new term $K_{13}\Delta T$ is the conductive heat flow inside the heat sink with K_{13} the (spatial) thermal conductance coefficient of the heat sink. Similarly to Eq. (3.4), the sliding surfaces are defined as

$$s_1(t) = T_{cd}(t) - T_{c1}(t) + k_{ec}\int_0^t \left(T_{cd}(\tau) - T_{c1}(\tau)\right)d\tau \quad (3.50)$$

and

$$s_2(t) = T_{cd}(t) - T_{c2}(t) + k_{ec}\int_0^t \left(T_{cd}(\tau) - T_{c2}(\tau)\right)d\tau. \quad (3.51)$$

The desired setpoint temperature remains identical for both TECs, since a homogeneous temperature distribution of the clamping plate is required. Nevertheless, for another case of application, this can be easily adjusted.

Following the same control law derivation as described in detail in Section 3.2, the new controlling input currents for the first and second TECs can are formulated as follows:

$$i_{cl_1}(t) = \frac{m_a c_a}{S T_{c1}(t)}\left[- K_{12}\left(T_{c1}(t) \quad T_{c2}(t)\right) \mid a\left(T_a \quad T_{cd}\right)e^{-at}\right.$$
$$\left. - k_{ec}\left(T_{cd}(t) - T_{c1}(t)\right) - \lambda_c\left(s_1(t)\right) + \beta_c \operatorname{sgn}\left(s_1(t)\right)\right] \quad (3.52)$$

and

$$i_{cl_2}(t) = \frac{m_a c_a}{S T_{c2}(t)}\left[- K_{12}\left(T_{c2}(t) - T_{c1}(t)\right) + a\left(T_a - T_{cd}\right)e^{-at}\right.$$
$$\left. - k_{ec}\left(T_{cd}(t) - T_{c2}(t)\right) - \lambda_c\left(s_2(t)\right) + \beta_c \operatorname{sgn}\left(s_2(t)\right)\right]. \quad (3.53)$$

The existence and positivity of currents $i_{cl_1}(t)$ and $i_{cl_2}(t)$ are conditioned by the choice of parameters β and λ, which have *not necessarily* the same values for $i_{cl_1}(t)$ and $i_{cl_2}(t)$.

A block diagram of the controller scheme is shown in Fig. 3.10. Note that the controller is organized in a *central* structure, although the controller encompasses two coupled control laws. These control laws provide two outputs. They are coupled through internal effects as in the thermal model. Moreover, the controller has just one reference for both TECs. Therefore the controller can be seen as a central one.

The measurement setup is depicted in Fig. 3.11. Fig. 3.11(a) shows the structure, and Fig. 3.11(b) shows the photo of the real setup. In Fig. 3.11(c) a detailed view is shown of how the (simulated) external thermal disturbance is applied to the cold side.

The temperature measurement results are shown in Fig. 3.12(a) for the cold sides of the system, and the corresponding controlling input currents are shown

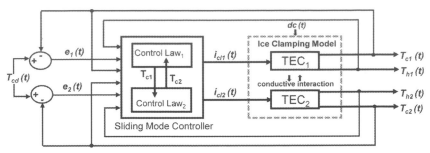

FIGURE 3.10 Block diagram of the proposed control law based on the extended ice-clamping device version activated by two TECs.

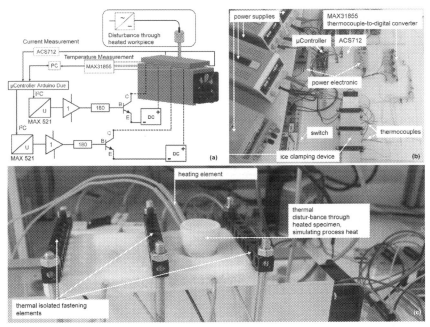

FIGURE 3.11 Experimental setup of the extended ice-clamping system actuated by two TECs; (a) a simplified circuit diagram of the control and measurement strategy; (b) and (c) the corresponding real test configuration.

in Fig. 3.12(b). The subscripts *extended* identify the results of the new control law with additional conductive terms.

The performance of both controllers shows fast responses, and both work effectively and robustly in the presence of thermal disturbances, attenuating the noise to the desired set point temperatures. From the measured results we can conclude that the modeled interaction on the cold sides of the clamping plate between the two TECs is not explicitly necessary, since with and without the additional term, both results show accurate and coherent thermal (delay) behav-

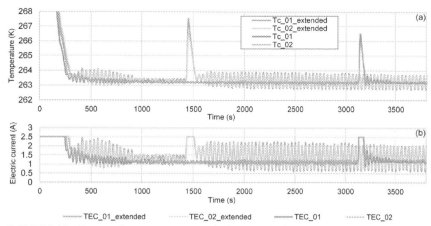

FIGURE 3.12 Comparative measurement results between the first control law, derived from Section 3.2 and the extended version with an additional conductive term to better represent thermal interacting of the cold side dynamics (a), and the corresponding controlling currents (b) in the presence of external thermal disturbance applied on the cold side.

ior of the cold side temperatures. No difference between the two control laws could be identified. Moreover, it is observable that with the extended control law version, the chattering behavior of the SMC is enhanced. Chattering is a typical drawback of the SMC approach. To deal with this, several concepts exist, which may be used to reduce the chattering. To conclude, for the practical use, the SMC shows high robustness in the presence of not-modeled dynamics, such that it was decided to use the simplified control law without taking into account the interacting conductive part. In particular, considering an array of a number of TECs, a central TEC, being enclosed on all four sides by neighboring TECs, will include four (or eight when taking into account the edges) additional conductive terms. The advantage is that not only the complexity is reduced by treating each TEC element individually, but regarding the practical aspect, this facilitates the programming and computation aspects of the Controller.

3.5   Controller comparison

To check whether the choice of SMC is an appropriate one, a controller performance comparison was performed. It was decided to use the classic approach of the PID controller, which has the advantage of being model-free and thus easy to derive and to implement. However, as was stated at the beginning of this chapter, it is known that the classic model-free PID controller approach suffers from multiple drawbacks. In [39] a detailed description of these drawbacks and comparative measurements was given, in which it was demonstrated that it is strictly advisable to include a feedforward regulator to the control law to obtain a good controller performance for the ice-clamping system instead of using only the feedback control. However, introducing a feedforward regulation means that again a model has to be present to derive the feedforward law. Since in [39] only the derivation of the control and regulator law is shown, the proof of stability is missing. To address this, in Section 3.5.1, we introduce a proof of stability and comparative measurements between the PI and SM controls.

Note that it was decided to neglect the derivative part D of the PID control, since the present thermal system is governed by long time constants due to thermal inertia of the components of the system, which can be seen as a system of multiple thermal capacitors. Since the derivative part of the PID controller is appropriate to speed up dynamic response, it, in this sense, will not bring any advantageous effect to the present system. Also, it is a well-known fact that the derivative part leads to an amplification of noise, which leads to the need of an additional filter to mitigate this, introducing an extra effort and complexity to the system [40].

3.5.1 PI controller and feedforward regulator

To obtain a feedforward regulator, if $T_c(t) = T_{cd}(t)$ equals the desired control current input, then rearranging Eq. (3.1) with consideration of Eq. (3.6) leads to

$$m_a c_a \frac{dT_{cd}(t)}{dt} = -S T_{cd}(t) i(t) + \frac{1}{2} R i^2(t) + K\big(T_h(t) - T_{cd}(t)\big). \qquad (3.54)$$

Then, if

$$\dot{T}_h(t) - \dot{T}_{cd}(t) = 0, \qquad (3.55)$$

then by subtracting Eq. (3.1) from Eq. (3.2) we obtain the following relation:

$$i^2(t)\left[\frac{1}{2}R\left(\frac{1}{m_k c_k} - \frac{1}{m_a c_a}\right)\right] + i(t)S\left(\frac{T_{cd}(t)}{m_a c_a} + \frac{T_h(t)}{m_k c_k}\right)$$

$$+ K\big(T_{cd}(t) - T_h(t)\big)\left[-\frac{1}{m_k c_k} - \frac{1}{m_a c_a}\right] + \hat{K}_k\big(T_h(t) - T_a\big)\big(-m_k c_k^{-1}\big)$$

$$+ \dot{T}_{cd}(t) - \dot{T}_h(t) = 0. \qquad (3.56)$$

Hence the *input current* $i_{FF}(t)$ from the feedforward regulator can be designed as

$$i_{FF}(t) = -\left(\frac{R}{m_k c_k} - \frac{R}{m_a c_a}\right)^{-1}\left(\frac{S T_h(t)}{m_k c_k} + \frac{S T_{cd}(t)}{m_a c_a}\right) + \left(\frac{R}{m_k c_k} - \frac{R}{m_a c_a}\right)^{-1}$$

$$\sqrt{\left(\frac{S T_h(t)}{m_k c_k} + \frac{S T_{cd}(t)}{m_a c_a}\right)^2 - 4\left[\left(\frac{R}{2 m_k c_k} - \frac{R}{2 m_a c_a}\right)\big(K\big(T_h(t) - T_{cd}(t)\big)\right.}$$

$$\times\left(- m_k c_k^{-1} - m_a c_a^{-1}\right) + \dot{T}_{cd}(t) - \dot{T}_h(t) - \hat{K}_k\big(T_h(t) - T_a\big)\big(m_k c_k^{-1}\big)\Big]. \qquad (3.57)$$

Now considering the PI controller with Γ_P and Γ_I as the coefficients of the proportional and integral terms, we design the *input current* $i_{PI}(t)$ from the controller as

$$i_{PI}(t) = \Gamma_P\big(T_{cd}(t) - T_c(t)\big) + \Gamma_I \int_0^t \big((T_{cd}(\tau) - T_c(\tau))\mathrm{d}\tau + A_w(t)\big). \quad (3.58)$$

The complete controller can be designed by adding Eq. (3.58) to Eq. (3.57) and, for the new *input current* i_{PIFF}, it follows that

$$i_{\mathrm{PIFF}}(t) = i_{\mathrm{FF}}(t) + i_{PI}(t). \quad (3.59)$$

The term $A_w(t)$ in Eq. (3.58) represents the Anti-windup signal with K_b being a constant to be set to the signal; see Eq. (3.60). It is an additional term, which can be used to prevent an integration windup, since the input to the system has to be saturated. In fact, the input current to a TEC must be saturated towards the positive boundary of the maximum current of a TEC to prevent overcharging. Whereby *sat* describes the saturating level as a constant value corresponding to the physical boundaries of a TEC. Furthermore, a negative current has also to be avoided, since a polarity reversal leads to a reverse of thermal flow. Thus the lower bound is set to zero. In [39] the advantage of the introduction of an Anti-windup action was discussed and proved by measurements of an ice-clamping system. The Anti-windup is defined as

$$A_w(t) = K_b\Big(i_{\mathrm{PIFF}}^*(t) - i_{\mathrm{PIFF}}(t)\Big), \quad (3.60)$$

where

$$i_{\mathrm{PIFF}}^*(t) = \begin{cases} i_{\mathrm{PIFF}}(t) & \text{if } i_{\mathrm{PIFF}}(t) < sat, \\ sat & \text{if } i_{\mathrm{PIFF}}(t) \geq sat. \end{cases} \quad (3.61)$$

It is known that if the system is controlled using a feedforward action that drives the system around desired trajectories, it is possible to analyze the dynamics of it using a linearized model around these trajectories. If $i(t) = \sqrt{u(t)}$, then we obtain the following structure of the system described in Eqs. (3.1) and (3.2):

$$\begin{bmatrix} \dot{T}_c(t) \\ \dot{T}_h(t) \end{bmatrix} = \underbrace{\begin{bmatrix} -\dfrac{K}{m_a c_a} & \dfrac{K}{m_a c_a} \\ \dfrac{K}{m_k c_k} & -\dfrac{K+K_k}{m_k c_k} \end{bmatrix}}_{\mathbf{A}_0} \begin{bmatrix} T_c(t) \\ T_h(t) \end{bmatrix} + \underbrace{\begin{bmatrix} \dfrac{R}{2m_a c_a} \\ \dfrac{R}{2m_k c_k} \end{bmatrix}}_{\mathbf{B}_0} u(t)$$

$$+ \begin{bmatrix} -\dfrac{ST_c(t)}{m_a c_a} \\ \dfrac{ST_h(t)}{m_k c_k} \end{bmatrix} \sqrt{u(t)} + \begin{bmatrix} 0 \\ \dfrac{1}{2m_k c_k} \end{bmatrix} T_a. \quad (3.62)$$

The dynamic behavior of the system can be analyzed around the desired trajectory. In general, if a nonlinear system

$$\dot{x}(t) = f(x(t), u(t)) \tag{3.63}$$

is given, then its linearization around a desired trajectory $T_{cd}(t)$, $T_{hd}(t)$, and $u_d(t)$ is

$$\dot{x}(t) = f(T_{cd}(t), T_{hd}(t), u_d(t)) + \nabla f(T_{cd}(t), T_{hd}(t), u_d(t)) \begin{bmatrix} T_c(t) - T_{cd}(t) \\ T_h(t) - T_{hd}(t) \\ u(t) - u_d(t) \end{bmatrix}, \tag{3.64}$$

where ∇f represents the Jacobian matrix.

In the present case, linearizing the system described in Eq. (3.62) around the trajectories $T_{cd}(t)$, $T_{hd}(t)$, and $u_d(t)$, we obtain the following structure:

$$\begin{bmatrix} \dot{T}_c(t) \\ \dot{T}_h(t) \end{bmatrix} = \begin{bmatrix} -\frac{K}{m_a c_a} & \frac{K}{m_a c_a} \\ \frac{K}{m_k c_k} & -\frac{K+K_k}{m_k c_k} \end{bmatrix} \begin{bmatrix} T_c(t) \\ T_h(t) \end{bmatrix} + \begin{bmatrix} \frac{R}{2m_a c_a} \\ \frac{R}{2m_k c_k} \end{bmatrix} u(t)$$

$$+ \begin{bmatrix} -\frac{S T_{cd}(t)}{m_a c_a} \\ \frac{S T_{hd}(t)}{m_k c_k} \end{bmatrix} \sqrt{u_d(t)} + \underbrace{\begin{bmatrix} -\frac{S\sqrt{u_d(t)}}{m_a c_a} & 0 & -\frac{T_{cd}(t)}{2m_a c_a \sqrt{u_d(t)}} \\ 0 & \frac{S\sqrt{u_d(t)}}{m_k c_k} & \frac{T_{hd}(t)}{2m_k c_k \sqrt{u_d(t)}} \end{bmatrix}}_{\text{Jacobian matrix of the linearization}} \begin{bmatrix} \tilde{T}_c(t) \\ \tilde{T}_h(t) \\ \tilde{u}(t) \end{bmatrix}$$

$$+ \begin{bmatrix} 0 \\ \frac{1}{2m_k c_k} \end{bmatrix} T_a, \tag{3.65}$$

where $\tilde{T}_c(t) = T_c(t) - T_{cd}(t)$, $\tilde{T}_h(t) = T_h(t) - T_{hd}(t)$, $\tilde{u}(t) = u(t) - u_d(t)$, and $u_d(t) = i_{FF}(t)$.

System (3.65) can be rewritten as

$$\begin{bmatrix} \dot{T}_c(t) \\ \dot{T}_h(t) \end{bmatrix} = \underbrace{\begin{bmatrix} -\frac{K}{m_a c_a} & \frac{K}{m_a c_a} \\ \frac{K}{m_k c_k} & -\frac{K+K_k}{m_k c_k} \end{bmatrix}}_{\mathbf{A}_0} \begin{bmatrix} T_c(t) \\ T_h(t) \end{bmatrix}$$

$$+ \underbrace{\begin{bmatrix} -\frac{S\sqrt{u_d(t)}}{m_a c_a} & 0 \\ 0 & \frac{S\sqrt{u_d(t)}}{m_k c_k} \end{bmatrix}}_{\mathbf{A}(t)} \begin{bmatrix} T_c(t) \\ T_h(t) \end{bmatrix} + \underbrace{\begin{bmatrix} \frac{R}{2m_a c_a} \\ \frac{R}{2m_k c_k} \end{bmatrix}}_{\mathbf{B}_0} u(t)$$

$$+ \underbrace{\begin{bmatrix} -\dfrac{T_{cd}(t)}{2m_a c_a \sqrt{u_d(t)}} \\ \dfrac{T_{hd}(t)}{2m_k c_k \sqrt{u_d(t)}} \end{bmatrix}}_{\mathbf{B}(t)} \tilde{u}(t) + \begin{bmatrix} -\dfrac{T_{cd}(t)}{2m_a c_a u_d(t)} - \dfrac{ST_{cd}(t)}{m_a c_a} \\ \dfrac{T_{hd}(t)}{2m_k c_k u_d(t)} + \dfrac{ST_{hd}(t)}{m_k c_k} \end{bmatrix} \sqrt{u_d(t)} + \begin{bmatrix} 0 \\ \dfrac{1}{2m_k c_k} \end{bmatrix} T_a .$$

$$(3.66)$$

The structure of the controller is represented in Fig. 3.13, where we can see a combination of a feedforward action with a PI controller. In particular, the PI part of the controller is as follows:

$$u(t) = \Gamma_P (T_{cd}(t) - T_c(t)) + \Gamma_I \int_0^t \Big((T_{cd}(\tau) - T_c(\tau)) d\tau + A_w(t) \Big) + u_d(t),$$

$$(3.67)$$

where $T_{cd}(t) - T_c(t)$ is the representation of the error $e(t)$ defined by Eq. (3.8). Let

$$x_I(t) = \int_0^t \Big(T_{cd}(\tau) - T_c(\tau) \Big) d\tau \tag{3.68}$$

be the state of the integral controller.

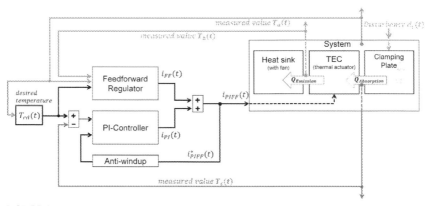

FIGURE 3.13 Block diagram of the control scheme consisting of a feedforward action and a PI regulator.

Then substituting Eqs. (3.67) and (3.68) into Eq. (3.66), we obtain the following structure:

$$\begin{bmatrix} \dot{T}_c(t) \\ \dot{T}_h(t) \end{bmatrix} = \underbrace{\begin{bmatrix} -\dfrac{K}{m_a c_a} & \dfrac{K}{m_a c_a} \\ \dfrac{K}{m_k c_k} & -\dfrac{K+K_k}{m_k c_k} \end{bmatrix}}_{\mathbf{A}_0} \begin{bmatrix} T_c(t) \\ T_h(t) \end{bmatrix}$$

$$+ \underbrace{\begin{bmatrix} -\dfrac{S\sqrt{u_d(t)}}{m_a c_a} & 0 \\[2ex] 0 & \dfrac{S\sqrt{u_d(t)}}{m_k c_k} \end{bmatrix}}_{\mathbf{A}(t)} \begin{bmatrix} T_c(t) \\[1ex] T_h(t) \end{bmatrix}$$

$$+ \underbrace{\begin{bmatrix} \dfrac{R}{2m_a c_a} \\[2ex] \dfrac{R}{2m_k c_k} \end{bmatrix}}_{\mathbf{B}_0} \left(\Gamma_P e(t) + \Gamma_I x_I(t) + u_d(t) \right)$$

$$+ \underbrace{\begin{bmatrix} -\dfrac{T_{cd}(t)}{2m_a c_a \sqrt{u_d(t)}} \\[2ex] \dfrac{T_{hd}(t)}{2m_k c_k \sqrt{u_d(t)}} \end{bmatrix}}_{\mathbf{B}(t)} \left(\Gamma_P e(t) + \Gamma_I x_I(t) + u_d(t) - u_d(t) \right)$$

$$+ \begin{bmatrix} -\dfrac{T_{cd}(t)}{2m_a c_a u_d(t)} - \dfrac{S T_{cd}(t)}{m_a c_a} \\[2ex] \dfrac{T_{hd}(t)}{2m_k c_k u_d(t)} + \dfrac{S T_{hd}(t)}{m_k c_k} \end{bmatrix} \sqrt{u_d(t)} + \begin{bmatrix} 0 \\[2ex] \dfrac{1}{2m_k c_k} \end{bmatrix} T_a. \quad (3.69)$$

The *whole control system*, which consists of the PI controller together with the feedforward regulator, can be represented as the state-varying system

$$\begin{bmatrix} \dot{T}_c(t) \\[1ex] \dot{T}_h(t) \\[1ex] \dot{x}_I(t) \end{bmatrix} = \underbrace{\begin{bmatrix} -\dfrac{K}{m_a c_a} - \dfrac{R\Gamma_P}{2m_a c_a} & \dfrac{K}{m_a c_a} & \dfrac{R\Gamma_I}{2m_a c_a} \\[2ex] \dfrac{K}{m_k c_k} - \dfrac{R\Gamma_P}{2m_k c_k} & -\dfrac{K+K_k}{m_k c_k} & \dfrac{R\Gamma_I}{2m_a c_a} \\[2ex] -1 & 0 & 0 \end{bmatrix}}_{\mathbf{A}_0} \underbrace{\begin{bmatrix} T_c(t) \\[1ex] T_h(t) \\[1ex] x_I(t) \end{bmatrix}}_{\mathbf{X}(t)}$$

$$+ \underbrace{\begin{bmatrix} -\dfrac{S\sqrt{u_d(t)}}{m_a c_a} & 0 & 0 \\[2ex] 0 & \dfrac{S\sqrt{u_d(t)}}{m_k c_k} & 0 \\[2ex] 0 & 0 & 0 \end{bmatrix}}_{\mathbf{A}(t)} \begin{bmatrix} T_c(t) \\[1ex] T_h(t) \\[1ex] x_I(t) \end{bmatrix} + \underbrace{\begin{bmatrix} \dfrac{R}{2m_a c_a} \\[2ex] \dfrac{R}{2m_k c_k} \\[1ex] 1 \end{bmatrix}}_{\mathbf{B}_0} T_{cd}(t)$$

$$+ \begin{bmatrix} \dfrac{R}{2m_a c_a} \\[2ex] \dfrac{R}{2m_k c_k} \\[1ex] 0 \end{bmatrix} u_d(t) + \underbrace{\begin{bmatrix} -\dfrac{T_{cd}(t)}{2m_a c_a \sqrt{u_d(t)}} \\[2ex] \dfrac{T_{hd}(t)}{2m_k c_k \sqrt{u_d(t)}} \\[1ex] 0 \end{bmatrix}}_{\mathbf{B}(t)} \left(\Gamma_P e(t) + \Gamma_I x_I(t) \right)$$

$$+ \begin{bmatrix} -\dfrac{T_{cd}(t)}{m_a c_a} - \dfrac{S T_{cd}(t)}{m_a c_a} \\[2ex] \dfrac{T_{hd}(t)}{m_k c_k} + \dfrac{S T_{hd}(t)}{m_k c_k} \\[1ex] 0 \end{bmatrix} \sqrt{u_d(t)} + \begin{bmatrix} 0 \\[2ex] \dfrac{1}{2m_k c_k} \\[1ex] 0 \end{bmatrix} T_a, \quad (3.70)$$

where $u_d(t)$ represents the forward action defined above.

If system (3.70) is calculated at the desired states $(T_c(t) = T_{cd}(t)$, $T_h(t) = T_{hd}(t)$, and $x_I(t) = x_{Id}(t))$, then

$$
\begin{bmatrix} \dot{T}_{cd}(t) \\ \dot{T}_{hd}(t) \\ \dot{x}_{Id}(t) \end{bmatrix} = \underbrace{\begin{bmatrix} -\frac{K}{m_a c_a} - \frac{R\Gamma_P}{2m_a c_a} & \frac{K}{m_a c_a} & \frac{R\Gamma_I}{2m_a c_a} \\ \frac{K}{m_k c_k} - \frac{R\Gamma_P}{2m_k c_k} & -\frac{K+K_k}{m_k c_k} & \frac{R\Gamma_I}{2m_a c_a} \\ -1 & 0 & 0 \end{bmatrix}}_{\mathbf{A}_0} \underbrace{\begin{bmatrix} T_{cd}(t) \\ T_{hd}(t) \\ x_{Id}(t) \end{bmatrix}}_{\mathbf{X}(t)}
$$

$$
+ \underbrace{\begin{bmatrix} -\frac{S\sqrt{u_d(t)}}{m_a c_a} & 0 & 0 \\ 0 & \frac{S\sqrt{u_d(t)}}{m_k c_k} & 0 \\ 0 & 0 & 0 \end{bmatrix}}_{\mathbf{A}(t)} \begin{bmatrix} T_{cd}(t) \\ T_{hd}(t) \\ x_{Id}(t) \end{bmatrix} + \underbrace{\begin{bmatrix} \frac{R}{2m_a c_a} \\ \frac{R}{2m_k c_k} \\ 1 \end{bmatrix}}_{\mathbf{B}_0} T_{cd}(t)
$$

$$
+ \begin{bmatrix} \frac{R}{2m_a c_a} \\ \frac{R}{2m_k c_k} \\ 0 \end{bmatrix} u_d(t) + \begin{bmatrix} \frac{T_{cd}(t)}{m_a c_a} & \frac{ST_{cd}(t)}{m_a c_a} \\ \frac{T_{hd}(t)}{m_k c_k} + \frac{ST_{hd}(t)}{m_k c_k} \\ 0 \end{bmatrix} \sqrt{u_d(t)} + \begin{bmatrix} 0 \\ \frac{1}{2m_k c_k} \\ 0 \end{bmatrix} T_a.
$$

$$(3.71)$$

By subtracting Eq. (3.71) from Eq. (3.70) we obtain the following expression:

$$
\dot{\mathbf{e}}(t) = \mathbf{A}_0 \mathbf{e}(t) + \Delta \mathbf{f}(t) + \mathbf{B}(t)\big(\Gamma_P e(t) + \Gamma_I x_I(t)\big) \tag{3.72}
$$

with

$$
\mathbf{e}(t) = \begin{bmatrix} T_{cd}(t) - T_c(t) \\ T_{hd}(t) - T_h(t) \\ x_{Id}(t) - x_I(t) \end{bmatrix}. \tag{3.73}
$$

Note that $T_{hd}(t)$ is referred to as the consequence of $T_{cd}(t)$. This means that only $T_{cd}(t)$ represents a desired (setpoint) value such that $T_{hd}(t)$ is the *indirect* result of it when reaching $T_{cd}(t)$. The same applies to $x_{Id}(t)$, represented as the state error result at $T_{cd}(t)$, since $x_I(t)$ is the integral of $e(t)$; see Eq. (3.68).

As the result of the subtraction, we obtain the term $\Delta \mathbf{f}(t)$ of Eq. (3.72) as

$$
\Delta \mathbf{f}(X) = \mathbf{f}(X_d) - \mathbf{f}(X), \tag{3.74}
$$

where X and X_d are the state and the desired state of the system, respectively. It follows that

$$\Delta \mathbf{f}(t) = \underbrace{\begin{bmatrix} -\frac{S\sqrt{u_d(t)}}{m_a c_a} & 0 & 0 \\ 0 & \frac{S\sqrt{u_d(t)}}{m_k c_k} & 0 \\ 0 & 0 & 0 \end{bmatrix}}_{\mathbf{A}(t)} \underbrace{\begin{bmatrix} T_{cd}(t) - T_c(t) \\ T_{hd}(t) - T_h(t) \\ x_{Id}(t) - x_I(t) \end{bmatrix}}_{\mathbf{e}(t)}. \tag{3.75}$$

Proposition 1. *Consider the error of the controlled system around the desired trajectory described by Eq. (3.70) as follows:*

$$\dot{\mathbf{e}}(t) = \mathbf{A}_0 \mathbf{e}(t) + \Delta \mathbf{f}(t) + \mathbf{B}(t)\big(\Gamma_P e(t) + \Gamma_I x_I(t)\big). \tag{3.76}$$

If the nonlinear function field $\Delta \mathbf{f}(t)$ of Eq. (3.76) is a Lipschitz one,[4] then there is a suitable choice of $\Gamma_P > 0$ such that system (3.76) is asymptotically stable around the desired trajectory if and only if

$$\Gamma_I > 0. \tag{3.77}$$

In other words,

$$\lim_{t \to \infty} \mathbf{e}(t) = 0 \quad \forall \Delta \mathbf{f}(t). \tag{3.78}$$

Proof. According to Lyapunov theory, $(\mathbf{A}_0, \mathbf{B}_0)$ is a controllable pair, and for a suitable choice of the controller gains Γ_P, $\mathbf{A}_0$ is a Hurwitz matrix if $\Gamma_I > 0$. This means that there are symmetric positive matrices $\mathbf{P}_0$ and $\mathbf{Q}_0$ that satisfy the *Lyapunov equation*

$$\mathbf{A}_0^T \mathbf{P}_0 + \mathbf{P}_0 \mathbf{A}_0 = -\mathbf{Q}_0. \tag{3.79}$$

Introduce the following Lyapunov function:

$$V(\mathbf{e}(t)) = \frac{1}{2} \mathbf{e}^T(t) \mathbf{P}_0 \mathbf{e}(t). \tag{3.80}$$

Then the time derivative is

$$\dot{V}(\mathbf{e}(t)) = \dot{\mathbf{e}}^T(t) \mathbf{P}_0 \mathbf{e}(t) + \mathbf{e}^T(t) \mathbf{P}_0 \dot{\mathbf{e}}(t). \tag{3.81}$$

From Eq. (3.76) it follows that

$$\dot{V}(\mathbf{e}(t)) = \big(\mathbf{A}_0 \mathbf{e}(t) + \Delta \mathbf{f}(t) + \mathbf{B}(t)\big(\Gamma_P e(t) + \Gamma_I x_I(t)\big)\big)^T \mathbf{P}_0 \mathbf{e}(t)$$

---

[4] Functions with bounded first derivative [41] to avoid becoming infinitely steep. Furthermore, using the Lipschitz condition in combination with the Lyapunov approach is a common practice in engineering for nonlinear systems and control; see [42].

$$+ \mathbf{e}^T(t)\mathbf{P}_0\big(\mathbf{A}_0\mathbf{e}(t) + \Delta\mathbf{f}(t) + \mathbf{B}(t)\big(\Gamma_P e(t) + \Gamma_I x_I(t)\big)\big), \quad (3.82)$$

where the term $\Delta\mathbf{f}(t)$ states the time-varying part of system Eq. (3.70). This yields

$$\dot{V}(\mathbf{e}(t)) = \Big(\mathbf{e}^T(t)\mathbf{A}_0^T + (\Delta\mathbf{f}(t))^T + \big(\mathbf{B}(t)\Gamma_P e(t)\big)^T + \big(\mathbf{B}(t)\Gamma_I x_I(t)\big)^T\Big)\mathbf{P}_0\mathbf{e}(t)$$
$$+ \mathbf{e}^T(t)\mathbf{P}_0\Big(\mathbf{A}_0\mathbf{e}(t) + \Delta\mathbf{f}(t) + \mathbf{B}(t)\big(\Gamma_P e(t) + \Gamma_I x_I(t)\big)\Big). \quad (3.83)$$

Finally, from Eq. (3.79) it follows that

$$\dot{V}(\mathbf{e}(t)) = -\mathbf{e}^T(t)\mathbf{Q}_0\mathbf{e}(t) + (\Delta\mathbf{f}(t))^T\mathbf{P}_0\mathbf{e}(t) + \mathbf{e}^T(t)\mathbf{P}_0(\Delta\mathbf{f}(t))$$
$$+ \big(\mathbf{e}^T(t)\Gamma_P^T + \Gamma_I^T x_I^T(t)\big)\mathbf{B}^T(t)\mathbf{P}_0\mathbf{e}(t) + \mathbf{e}^T(t)\mathbf{P}_0\mathbf{B}(t)\big(\Gamma_P e(t) + \Gamma_I x_I(t)\big). \tag{3.84}$$

Since $\Delta\mathbf{f}(t)$ are Lipschitz functions, there is a set of positive constants L such that for each function $f(t)$,

$$\|f(X_d) - f(X)\| \le L \underbrace{\|X_d - X\|}_{e(t)}. \tag{3.85}$$

The scope of the controller is the reaching of the desired cold side temperature $T_{cd}(t)$, and thus error $e(t) = T_{cd}(t) - T_c(t)$ is considered; therefore $T_h(t)$ and $x_I(t)$ reach their final values, which are limited, because $T_c(t)$ reaches a limited $T_{cd}(t)$. Let $\lambda_{Q_{0m}}$ be the smallest eigenvalue of the matrix Q_0, and let $\lambda_{P_{0M}}$ be the largest eigenvalue of the matrix P_0. Suppose that the following condition is satisfied for $e(t)$ small enough (see [43]):

$$\lambda_{Q_{0m}} \ge L\lambda_{P_{0M}} + M\Gamma_P \lambda_{P_{0M}} + M\Gamma_I \lambda_{P_{0M}}. \tag{3.86}$$

Then

$$\dot{V}(e(t)) \le -(\lambda_{Q_{0m}} - L\lambda_{P_{0M}} - M\Gamma_P \lambda_{P_{0M}} - M\Gamma_I \lambda_{P_{0M}})e^2(t), \tag{3.87}$$

where M is the maximal value of the elements of the matrix $\mathbf{B}(t)$. Once suitable matrices $\mathbf{Q}_0$ and $\mathbf{P}_0$ are chosen, we can also choose Γ_P and Γ_I such that $\mathbf{A}_0$ has negative real eigenvalues and in the meantime satisfies Eq. (3.79).

To guarantee that $\mathbf{A}_0$ is a Hurwitz matrix, the following calculations concerning the characteristic polynomial of matrix $\mathbf{A}_0$ are needed, which verify that all its eigenvalues are strictly negative:

$$\lambda^3 + \lambda^2 + a_2\lambda + a_1\lambda + a_0 = 0, \tag{3.88}$$

where

$$a_2 = \frac{K}{c_k m_k} + \frac{K_k}{c_k m_k} + \frac{K}{c_a m_a} + \frac{\Gamma_P R}{c_a c_k m_k m_a}, \tag{3.89}$$

$$a_1 = \frac{K K_k}{c_a m_a c_k m_k} + \frac{\Gamma_P K R}{c_a m_a c_k m_k} + \frac{\Gamma_P K_k R}{c_a m_a c_k m_k} + \frac{\Gamma_I R}{c_a m_a c_k m_k}, \qquad (3.90)$$

$$a_0 = \Gamma_I R \frac{K c_a m_a + K c_k m_k + K_k c_a m_a}{2 c_a^2 c_k m_k m_a^2}. \qquad (3.91)$$

Now the character of eigenvalues (of the Jacobian matrix) can be used to state a criterion of stability for the nonlinear system. Thanks to the *Routh–Hurwitz criterion*, it is not necessary to know the explicit values. According to this criterion, it is sufficient to examine *"whether the real part is positive, negative, or zero"* [44], to determine the *region* in which the system is stable. The Routh–Hurwitz condition states that the roots of Eq. (3.88) are negative if and only if

$$a_2 > 0, \quad a_0 > 0, \quad \text{and} \quad a_2 a_1 > a_0. \qquad (3.92)$$

□

3.5.2   Conclusions and comparative measurements between PI and SM controllers

Comparative measurements were performed using the setup presented in Fig. 3.4 in Section 3.3. Three different controllers were investigated, the PI controller, the PI with feedforward regulator (referred to as PIFF), and the SMC. The controlled temperature results are shown in Fig. 3.14(a), and in Fig. 3.14(b) the corresponding electric input current is shown. In Fig. 3.15 a detailed view of the cold side dynamics is shown. Since the worst controller performance was recorded for the PI control, the measurements for the disturbance attenuation were only performed for the PIFF and SM controllers, which are depicted in Fig. 3.16.

Due to the switching nature of the SMC, a quantitative comparison between the PI and SM controllers is not directly applicable. In fact, in terms of SM control, they are two basic phases, the reaching phase and the sliding phase, so that the main step response characteristics of interest are the reaching time and the chattering behavior of the SMC. Due to the switching behavior and the magnitude of the chattering, a direct comparison in terms of the rise time, settling time, and over- and undershoots is difficult to obtain with classic step response tools and therefore has to be redefined for the present case.

The results of the performance analysis are depicted in Table 3.2. The rise time t_{rise} is defined as the first crossing point of $T_c(t)$ with the desired value $-10°C$ (or 263.15 K). The settling time t_{set} is defined as the time needed for the state $T_c(t)$ to settle within an error band defined as $\pm 0.2\%$ of the desired value (see the dashed lines in Fig. 3.15(a)). The root mean square (RMS) error is defined as follows:

$$\text{RMS}_{\text{Error}} = \sqrt{\frac{\sum_{t=1*}^{n} (x_{ec,t})^2}{n}}, \qquad (3.93)$$

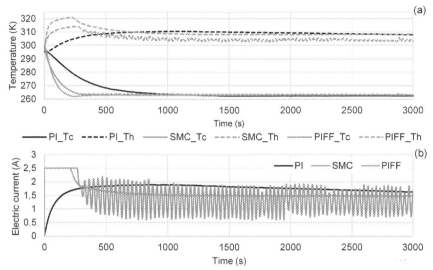

FIGURE 3.14 Comparison of PI controller, PI with feedforward action, and SM controller: (a) measured temperature results and (b) the corresponding input currents.

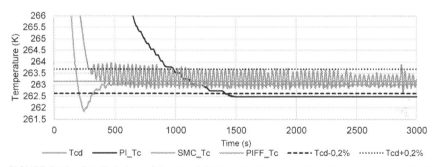

FIGURE 3.15 Detailed view of the Fig. 3.14(a): cold side temperature dynamics as a result of different controllers.

where the error x_{ec} is the deviation of the state $T_c(t)$ from the desired setpoint value; see Eq. (3.8). It was decided to identify the RMS error value at two different starting times. Therefore the subscript $t = 1^*$ in Eq. (3.93) first corresponds to the beginning starting at time $t = 0$ and second, to $t = t_{set_0}$, which is defined as the fastest settling time of the examined controller performances. The best settling time behavior is obtained by the SMC; see Table 3.2. The latter is chosen to better evaluate the performance at steady state without taking into account the transient reaching phase. The steady-state behavior is of greater interest, since this phase represents the operating condition of the clamping device. Furthermore, it was decided to refer to the discrete type of RMS_{Error} (in contrast to the integral type for a continuous distribution), since the recorded measurement points represent discrete time steps due to the digital µController.

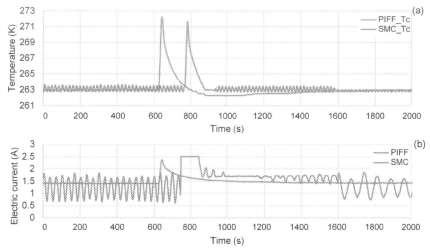

FIGURE 3.16 Controller performance in the presence of a thermal disturbance: (a) measured results for the cold side temperature and (b) the corresponding current inputs.

TABLE 3.2 Comparison of performance of PI Controller with and without feedforward regulator and SMC.

	PI	PIFF	SMC
Rise time (s)	1.123	193	381
Settling time (s)	–	320	301
$RMS_{Error}\ t = 0$ (K)	8.19	4.58	4.72
$RMS_{Error}\ t = t_{set_0}$ (K)	2.56	0.17	0.32

The quantitative comparison shows that the rise time performs best for the proposed open-loop regulation paired with the PI controller. On the other hand, the fast rise time results in an undershoot of the state. Although the SMC control shows slightly slower rising dynamics, its settling time performance almost equals the rise time with no undershoots, so that the overall transient performance shows the best behavior. Note that these rise time results represent the launch phase, so that the system starts at room temperature to cool down to the desired value, which initially takes a while to achieve the steady-state condition. Once the system reaches operation conditions, the unclamping procedure (by reversing the current polarity) will heat up the plate only to 0 or 1 degrees to thaw the ice, so that further cooling phases will be faster, since the temperature difference between starting point and desired setpoint value is decreased.

The worst controller performance is recorded for the pure feedback PI control, having the slowest dynamics with a permanent steady-state error higher than the defined error tolerance (see dashed lines in Fig. 3.15(a)), such that no settling time can be obtained. The PIFF control has the best error performance with the lowest RMS errors. Especially for the steady state, the PIFF error is over 15 times smaller compared to the PI. In comparison with the SM controller, the error difference gets much smaller (two times). However, since the sliding nature of the SMC leads to chatter around the desired trajectory, this, in the classical sense, does not represent a real steady-state error.

In the presence of the same external disturbance, applied to the cold sides, the SMC shows a faster reaction and a significantly better settling time characteristic compared to the PIFF, depicted in Fig. 3.16(a), and the corresponding input current depicted in Fig. 3.16(b). Moreover, in contrast to the PIFF, the SMC has no undershoot.

In terms of a homogeneous temperature distribution on the plate, the chattering of the SMC has to be minimized.

References

[1] A. Mironova, Industrial Application Study and Development of a Control Strategy for an Ice Clamping Device, PhD thesis, Leuphana University of Lueneburg, Institute for Production Tehnology and Systems, ISBN 13: 978-3-96548-007-0 (Print), Sierke Verlag, 2018.

[2] W.-H. Chen, C.-Y. Liao, C.-I. Hung, A numerical study on the performance of miniature thermoelectric cooler affected by Thomson effect, Applied Energy 89 (1) (2012) 464–473. Special issue on Thermal Energy Management in the Process Industries.

[3] N.A. Abukhshim, P.T. Mativenga, M.A. Sheikh, Heat generation and temperature prediction in metal cutting: a review and implications for high speed machining, International Journal of Machine Tools and Manufacture 46 (7) (2006) 782–800.

[4] J.P. Kaushish, Manufacturing Processes, vol. 2, PHI Learning Private Limited, New Delhi, 2010.

[5] M.D. Thakor, S.K. Hadia, A. Kumar, Precise temperature control through thermoelectric cooler with PID controller, in: 2015 International Conference on Communications and Signal Processing (ICCSP), April 2015, pp. 1118–1122.

[6] G. Engelmann, M. Laumen, K. Oberdieck, R.W. De Doncker, Peltier module based temperature control system for power semiconductor characterization, in: 2016 IEEE International Power Electronics and Motion Control Conference (PEMC), Sept 2016, pp. 957–962.

[7] J. Li, X. Zhang, C. Zhou, J. Zheng, D. Ge, W. Zhu, New applications of an automated system for high-power LEDs, IEEE/ASME Transactions on Mechatronics 21 (2) (April 2016) 1035–1042.

[8] A. Mironova, P. Mercorelli, A. Zedler, A multi input sliding mode control for Peltier cells using a cold–hot sliding surface, Journal of the Franklin Institute (2017).

[9] Y.Z. Li, K.M. Lee, J. Wang, Analysis and control of equivalent physical simulator for nanosatellite space radiator, IEEE/ASME Transactions on Mechatronics 15 (1) (Feb 2010) 79–87.

[10] M.A. Capcha, W. Ipanaqué, R. De Keyser, Comparison of model-based and non-model-based strategies for nonlinear control of a three-tank system, in: 2017 22nd IEEE International Conference on Emerging Technologies and Factory Automation (ETFA), Sept 2017, pp. 1–4.

[11] A.T. Ansari, H. Kala, S. Abirami, K. Thivakaran, R.A.R. Zepherin, Model identification and comparison of different controller for the air-temperature process, in: 2015 International Conference on Circuits, Power and Computing Technologies [ICCPCT-2015], March 2015, pp. 1–6.

[12] S. Song, J. Wang, Dynamic model of thermoelectric cooler and temperature control based on adaptive Fuzzy-PID, Applied Mechanics and Materials 130–134 (2012) 1919–1924.

[13] A.A. Aly, A.S.A. El-Lail, Fuzzy temperature control of thermoelectric cooler, in: IEEE International Conference on Industrial Technology, 2006, pp. 1580–1585.

[14] A.M. Yusop, R. Mohamed, A. Ayob, Model building of thermoelectric generator exposed to dynamic transient sources, in: IOP Conference Series: Materials Science and Engineering, vol. 53, 2013, p. 012015.

[15] H.S. Choi, S. Yun, K. Whang, Development of a temperature-controlled car-seat system utilizing thermoelectric device, Applied Thermal Engineering 27 (17–18) (2007) 2841–2849.

[16] J. Wang, K. Zou, J. Friend, Minimum power loss control – thermoelectric technology in power electronics cooling, in: 2009 IEEE Energy Conversion Congress and Exposition, San Jose, CA, 2009.

[17] H. Aschemann, A. Rauh, An integro-differential approach to control-oriented modelling and multivariable norm-optimal iterative learning control for a heated rod, in: 2015 20th International Conference on Methods and Models in Automation and Robotics (MMAR), Aug 2015, pp. 447–452.

[18] K.D. Young, V.I. Utkin, U. Ozguner, A control engineer's guide to sliding mode control, IEEE Transactions on Control Systems Technology 7 (3) (May 1999) 328–342.

[19] J.Y. Hung, W. Gao, J.C. Hung, Variable structure control: a survey, IEEE Transactions on Industrial Electronics 40 (1) (Feb 1993) 2–22.

[20] A. Levant, L. Alelishvili, Discontinuous homogeneous control, in: Modern Sliding Mode Control Theory, 2008, pp. 71–95.

[21] T. Xiao, H.-X. Li, Sliding mode control design for a rapid thermal processing system, Chemical Engineering Science 143 (Supplement C) (2016) 76–85.

[22] T. Hatano, M. Deng, S. Wakitani, Operator and sliding mode based nonlinear control for cooling and heat-retention system actuated by Peltier devices, in: 2015 International Conference on Advanced Mechatronic Systems (ICAMechS), Aug 2015, pp. 343–348.

[23] P. Princes Sindhuja, B. Senthil Kumar, K. Suresh Manic, Fast terminal sliding mode controller for temperature control of an unstable chemical reactor, in: 2016 IEEE International Conference on Computational Intelligence and Computing Research, ICCIC 2016, 2017.

[24] C. Vecchio, Sliding Mode Control: theoretical developments and applications to uncertain mechanical systems, PhD thesis, Department of Industrial and Information Engineering, University of Pavia, Italy, 2008.

[25] Q. Zhang, P. Friedberg, K. Poolla, C.J. Spanos, Enhanced spatial PEB uniformity through a novel bake plate design, in: Proceedings AEC/APC XVII Symposium, Austin, Texas, 2005, pp. 1–5.

[26] M. Guiatni, A. Drif, A. Kheddar, Thermoelectric modules: recursive non-linear ARMA modeling, identification and robust control, in: Industrial Electronics Society. 33rd Annual Conference of the IEEE, 2007, pp. 568–573.

[27] S.A. Zahiripour, A.A. Jalali, Systematic approach of extracting sliding manifold in robust stabilizing of stochastic multi-input systems, Journal of the Franklin Institute 353 (2) (2016) 378–397.

[28] Y. Feng, F. Han, X. Yu, Chattering free full-order sliding-mode control, Automatica 50 (4) (2014) 1310–1314.

[29] A. Nasiri, S.K. Nguang, A. Swain, Adaptive sliding mode control for a class of MIMO nonlinear systems with uncertainties, Journal of the Franklin Institute 351 (4) (2014) 2048–2061. Special Issue on 2010-2012 Advances in Variable Structure Systems and Sliding Mode Algorithms.

[30] A. Polyakov, L. Fridman, Stability notions and Lyapunov functions for sliding mode control systems, Journal of the Franklin Institute 351 (4) (2014) 1831–1865. Special Issue on 2010–2012 Advances in Variable Structure Systems and Sliding Mode Algorithms.

[31] A.M. Lyapunov, The General Problem of the Stability of Motion (in Russian). PhD thesis, Mathematical Society of Kharkov, University of Kharkov, Ukraine, 1892.

[32] A.M. Lyapunov, The General Problem of the Stability of Motion, 1 edition, Taylor and Francis, London, Washington DC, 1992.

[33] Review on computational methods for Lyapunov functions.

[34] P. Kachroo, M. Tomizuka, Integral action for chattering reduction and error convergence in sliding mode control, in: 1992 American Control Conference, June 1992, pp. 867–870.

[35] J.-J.E. Slotine, W. Li, Applied Nonlinear Control, Pearson, Upper Saddle River, NJ, 1991. The book can be consulted by contacting: BE-ABP-CC3: Pfingstner, Juergen.

[36] W.L. Staats, J.G. Brisson, Active heat transfer enhancement in air cooled heat sinks using integrated centrifugal fans, International Journal of Heat and Mass Transfer 82 (2015) 189–205.

[37] S.L. Ma, J.W. Chen, H.Y. Li, J.T. Yang, Mechanism of enhancement of heat transfer for plate-fin heat sinks with dual piezoelectric fans, International Journal of Heat and Mass Transfer 90 (2015) 454–465.

[38] V. Egan, J. Stafford, P. Walsh, E. Walsh, An experimental study on the design of miniature heat sinks for forced convection air cooling, Journal of Heat Transfer 131 (7) (2009).

[39] A. Mironova, P. Mercorelli, A. Zedler, E. Karaman, A model based feedforward regulator improving PI control of an ice-clamping device activated by thermoelectric cooler, Number 8014064, 2017, pp. 484–489.

[40] K.K. Tan, Q.G. Wang, C.C. Hang, Advances in PID Control, 1 edition, Springer-Verlag, London, 1999.

[41] N. Weaver, Lipschitz Algebras, World Scientific Publishing Co. Pte. Ltd., Singapore, 1999.

[42] P. Mercorelli, A motion-sensorless control for intake valves in combustion engines, IEEE Transactions on Industrial Electronics 64 (4) (April 2017) 3402–3412.

[43] Q. Zhu, A. Kaddouri, L.A. Dessaint, O. Akhrif, A nonlinear state observer for the sensorless control of a permanent-magnet AC machine, IEEE Transactions on Industrial Electronics 48 (6) (Dec 2001) 1098–1108.

[44] E.X. DeJesus, C. Kaufman, Routh–Hurwitz criterion in the examination of eigenvalues of a system of nonlinear ordinary differential equations, Physical Review A 35 (Jun 1987) 5288–5290.

Chapter 4

A Digital Twin for part quality prediction and control in plastic injection molding

Alexander Rehmer[a], Marco Klute[b], Hans-Peter Heim[b], and Andreas Kroll[a]

[a]*Department of Measurement and Control, University of Kassel, Kassel, Germany,* [b]*Institute of Material Engineering - Polymer Engineering, University of Kassel, Kassel, Germany*

Funding

The research project "Digital Twin of Injection Molding (DIM)" (FKZ: 0107/20007409) was funded by the State of Hesse and the European Regional Development Fund (ERDF 2014-2020).

4.1 Introduction

After the first injection molding (IM) machine was patented in 1872 [3], the large-scale integration of this processing method into series production started towards the end of the first quarter of the 20th century with the introduction of new polymer materials, which allowed more complex part geometries and shorter processing cycle times. With the invention of screw IM machines in 1946 [4], process parameters such as the injection velocity could be controlled more precisely, resulting in increased qualities of the produced parts. This is one of the reasons why plastic IM has been established as the most widespread manufacturing process in the plastic processing industry. It is employed by almost 70 % of all plastic processing companies [1,2]. Although today piston IM machines are still used for special processes, e.g., micro-IM of heat-sensitive plastics, almost all plastic IM machines use plasticizing screws. The design and operating principle of such processing machines are described in detail in Section 4.2.

In particular, the very short cycle times and the use of multiple cavities within an injection mold, make quality control of each manufactured component enormously difficult, which is why the plastics processing industry has usually resorted to line test inspections. In addition, the monitoring of machine values with an initial analysis of their correlation to the crucial quality variables has established itself as a proven method (e.g., [5–9]). Rapid developments in the field of automation and digitalization of production processes have opened up new opportunities for the IM process to implement the methods created in the context of Industry 4.0 and the Internet of Things (IoT) for linking production chains through communication interfaces. In particular, the Unified Architecture (UA) of the Open Platform Communications (OPC) Foundation standardizes the communication protocol between different machines and peripherals, thus enabling all accruing machine and process data to be collected centrally. Due to the availability of such high-resolution and comprehensive data, machine learning methods can be effectively applied in corresponding production lines. Most of the research in this area is focused on improving and providing a good prediction of the part quality during the IM process, thereby aiding the machine operator in choosing the optimal machine setpoints (e.g., [10–14]). A Digital Twin (DT), for example, can represent a combination of the monitoring of process values and prediction of component quality derived from this. However, its implementation is associated with some challenges, especially in the provision of all machine, process, and quality values in sufficient resolution and within the process cycle time.

4.1.1 Digital Twin

A Digital Twin can be defined as a high-fidelity representation of the operational dynamics of its physical counterpart, enabled by near real-time synchronization between the cyberspace and the physical space [15]. It can, amongst others, be used for simulation, monitoring, and control of the physical asset. Depending on the characteristics of the physical system and also the intended purpose of the DT, different digital representations are appropriate, i.e., a DT of a physical system is not a unique entity. A DT intended for monitoring may be significantly different from a DT intended for control of the physical system. The DT and the Physical Production System (PPS) it represents constitute a Cyber Physical Production System (CPPS). The DT is a prerequisite for developing a CPPS and is also essential for achieving smart manufacturing: By using a DT in conjunction with optimization procedures the decision-making process of the operator can be aided. Via direct feedback control from the DT to the physical system even autonomous smart manufacturing can be enabled [15]. The benefits for producers are a significant reduction in the manual effort for supervising and controlling CPPSs, which in turn frees up skilled workers.

In companies that use the IM process in series production, one machine operator is usually responsible for monitoring several injection molding machines. This employee usually only intervenes if individual machines indicate a malfunction or if random checks of the manufactured components reveal a quality deviation outside the tolerance limits. Based on experience, the machine operator then adjusts the machine setpoints to compensate for the quality deviation. By using DTs of the individual machines these adjustments are made on the basis of all correlations that influence quality, so that the number of production waste can be significantly reduced by preventing bad part production. In addition, downtime is reduced, and start-up processes are accelerated.

4.1.2   Challenges

Both the realization of a DT and its integration into the CPPS is a challenging task [16]. To mitigate the efforts associated with realizing a DT, reference architectures have been proposed for injection molding [15–17]. Whereas these works mainly focus on software and data integration, the aspect of how to obtain an accurate DT, i.e., a dynamical model, of the plastic IM process has not been at the center of attention. As the results of the model-based optimization procedures will heavily depend on the accuracy of the model, this aspect is crucial. In addition, estimating a dynamic model that maps machine and process values to part quality is especially challenging in IM: As opposed to other batch processes, the measurements of part qualities can only be performed at the end of each batch. The DT must therefore be able to map trajectories of machine and process values to a single quantity datum.

A prerequisite for realizing and continuously updating a DT is the measuring of all quantities relevant for predicting part quality. A standard IM machine exclusively measures so-called machine values, i.e., quantities that are measured in the injection molding machine, rather than in the mold cavity where the part is formed. The process values that describe the state of the cavity or even quality values that describe the relevant properties of the produced part and are the ultimate quantity of interest are not measured. The implementation of the infrastructure (hard- and software-wise) necessary to measure all relevant quantities poses another challenge. Especially the realization of in-line quality measurements is difficult. In-line quality measuring systems are laborious to implement, and some quality variables cannot be measured nondestructively.

Another challenge for quality prediction are external influences that affect the IM process and can have unpredictable effects on part quality. Due to these influences, the same machine settings at different times can lead to deviating qualities. Significant influencing factors here are environmental influences, such as temperature and humidity within the production halls, fluctuations within the cooling water circuits, wear of machine and mold components, and material-related influences, such as batch fluctuations, varying residual moisture content, or deviating mixing ratios. Particularly due to the increased use of recycled materials, the fluctuating material properties have an increasing influence on part quality, making a quality prediction very difficult without continuous monitoring of the process variables.

4.1.3   Solution approach

An industrial plastic injection molding machine (ALLROUNDER 470 S, Arburg GmbH + Co KG, Loßburg, Germany), shown in Fig. 4.1, is equipped with sensors and an in-line quality measuring system. Thereby all relevant quantities to build a DT that maps the physical process from machine setpoints s to the resulting part quality Q are made available. Different modeling approaches and model structures for obtaining an accurate quality model, which together with the process model constitute the DT, are investigated, and their performances are evaluated in a case study. In particular, a novel switched system internal dynamics model approach for building a dynamic quality model is presented. Although not part of this work, the Digital Twin can then be continuously updated via the retrieved machine, process, and quality data. Via model-based numerical optimal control the optimal machine setpoints for a desired quality Q_{ref} can be calculated from batch to batch, thereby essentially implementing a quality control loop while leaving the machines internal control loops unchanged. The CPPS is depicted in Fig. 4.2: The given physical assets, the injection molding machine and the produced part (green – the lightest gray in print version), are augmented via a quality measuring system (orange – light gray in print version), and via various data interfaces (blue – mid gray in print version) data is transferred in real-time to the DT, i.e., the models and their corresponding functions (red – dark gray in print version).

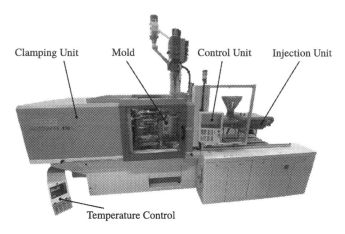

FIGURE 4.1   Injection molding machine ALLROUNDER 470S.

4.2 Plastic injection molding

The basic design of modern injection molding machines is very similar among different manufacturers and consists of five functional units depicted in Fig. 4.1. To better understand the impact of the machine parameters and the resulting process values, a typical injection molding cycle as that used in the case study will be described before the values themselves, and the current process control strategies will be explained.

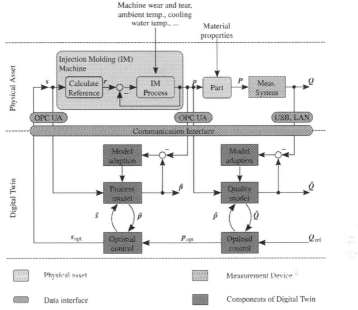

FIGURE 4.2 The devices, models and functions constituting the CPPS of the plastic injection molding process.

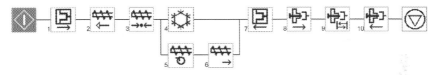

FIGURE 4.3 Injection molding process cycle used in the case study.

4.2.1 Process cycle and machine components

While the mold specifies the geometry of the parts to be produced, the clamping unit performs the mold and ejector movements required during the injection molding cycle and provides the clamping force needed to prevent the mold from opening due to high injection pressures. As depicted in Fig. 4.4, the injection unit consists of a hopper, through which the base material is fed into the process in granular form, and a barrel tempered by heating bands, which contains the screw responsible for material conveying, melting, and dosing through a rotational movement. In addition, a translatory movement of the screw injects the melted material into the cavity of the mold.

Within the control unit of the machine, the operator can define the process sequence. An exemplary sequence, as also used in the case study, is shown in Fig. 4.3. First, the mold is closed (1) so that the previously dosed quantity of melt can be injected into the cavity under high pressure (2). To prevent shrinkage of the part during cooling in the mold, a packing pressure is maintained for a defined time after injection (3), allowing the cavity to be kept filled volumetrically. During the subsequent residual cooling time (4), the melt volume required for the next cycle is dosed against a defined back pressure (5) and then relieved by decompression (6). Following the residual cooling time, the mold opens (7), and the ejector movement (8–10) demolds the finished part from the cavity.

4.2.2 Machine setpoints and measured process variables

The production process of an IM machine can be divided into three phases depending on the controlled machine values, the screw velocity and hydraulic pressure. To clarify the screw positions resulting from the three phases, they are depicted in Fig. 4.4.

1) Injection: At the beginning of the injection phase, the screw is in the dosing position. A defined melt volume, located in front of the screw, is then injected with a defined injection velocity until a set switching point is reached, which is defined by the screw position. Since this phase is velocity-controlled, only a hydraulic pressure limit is set on the machine to avoid damage.

2) Packing: To compensate for material shrinkage resulting from cooling, a defined pressure curve over a specified time is needed. Therefore the packing phase is pressure-controlled, whereas the pressure reference signal is composed of multiple consecutive pressure ramps. Three pressure levels were set in the case study presented in this chapter, with only height and duration of the second level being varied as part of the experimental design. The packing phase starts immediately after the screw reaches the switching point and ends after the set time of the pressure levels is over.

3) Cooling: Cooling of the melt begins immediately upon entry into the mold cavity. However, since the time between the start of injection and the end of the packing phase is not sufficient to cool the plastic part to a temperature that ensures deformation-free demolding, an additional cooling time is required. Since this cooling phase is pressureless, the injection unit can simultaneously meter the melt volume for the next cycle against a set back pressure so that the screw is again in the dosing position.

Table 4.1 lists the important machine parameters that were set at the control unit of the IM machine during the case study and the resulting process values that were monitored. The process variables measured in the cavity, the cavity pressure p_{cav} and cavity wall temperature T_{cav}, of a few cycles with different realizations of the setpoint s are shown in Fig. 4.5. In addition to the listed variables, for each set target value, the corresponding actual values were also recorded. Typically, an IM machine records only the trajectories of the clamping and injection unit, i.e., the hydraulic pressures, positions, and velocities. For monitoring the resulting process values inside the mold cavity, the machine was additionally equipped with a combined pressure and temperature sensor (Type 6190C, Kistler Instrumente GmbH, Sindelfingen, Germany) positioned in the rigid side of the mold. The monitoring and storing of the data will be described in Section 4.3.

TABLE 4.1 Machine setpoints s and measured process variables $p(t)$.

Machine setpoint s_i	Unit	Process Variable p_i	Unit
Nozzle temperature	°C	Injection pressure	bar*
Mold temperature	°C	Screw position	cm^3
Injection flow rate	cm^3/s	Injection flow rate	cm^3/s
Switching point	cm^3	Cavity pressure	bar*
Packing pressure	bar*	Cavity wall temperature	°C
Packing time	s		
Back pressure	bar*		
Cooling time	s		

*In IM the pressure is usually measured and stated in bar, and therefore no conversion to SI units has been made here.

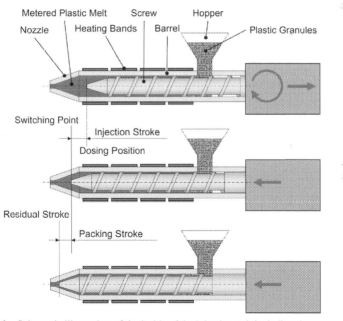

FIGURE 4.4   Schematic illustration of the inside of the injection unit including the movement and important positions of the screw.

4.2.3   State of the art: process control in injection molding

The injection molding process is affected by various disturbances (e.g., environmental influences, material viscosity, mold temperature, closing behavior, wear and tear of non-return valve), which make a stable production of high-precision parts difficult. To compensate these disturbances, process control systems are

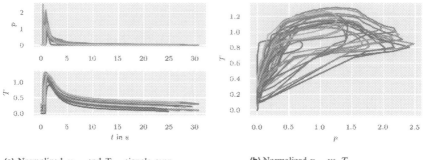

(a) Normalized p_{cav} and T_{cav} signals over time t

(b) Normalized p_{cav} vs. T_{cav}

FIGURE 4.5 Subset of measured normalized cavity pressure p_{cav} and normalized cavity wall temperature T_{cav} signals recorded during execution of the experimental design described in Sec. 4.5.

developed. Depending on which variables are controlled, three different control strategies can be distinguished [18]:

- Control of machine values, i.e., process variables that are measured machine-side and therefore reflect the state of the injection molding machine, such as the screw velocity and the hydraulic pressure.
- Control of process values, i.e., process variables that are measured cavity-side and therefore reflect the state of the process at the location of part formation, i.e., the cavity, e.g., the cavity pressure and temperature.
- Control of quality variables, i.e., the properties of the manufactured parts that contribute to their functioning

In industrial applications the control of machine values is prevalent. However, aforementioned disturbances affect the process values and hence the part quality, i.e., the part quality may vary even if the machine values are reproduced perfectly each cycle [19]. Therefore the efforts of the scientific community in recent decades have been directed towards the control of in-cavity process-variables, especially the cavity pressure, which are more closely related to the quality variables. However, due to the nonlinear behavior of the material, PID controllers whose parameters are adapted to the current system state are employed for this task. These have to be parameterized for every mold, which is very time-consuming and therefore not suitable for industrial production [19]. To overcome these obstructions to an introduction of cavity pressure control on a wide industrial scale, the concepts of model-based optimization (MO) and iterative learning control (ILC) have been applied [19,20]: An optimal cavity pressure trajectory is calculated based on the pvT-model (material-specific relationship between pressure p, temperature T, and specific volume v), i.e., the quality model. Via an ILC, which can be model-free or model-based, the controller output is then adapted from cycle to cycle to minimize the tracking error. The control of quality variables has not yet been achieved. The main obstacles are the necessity for in-line quality measurements and a quality model to predict the quality characteristics of the final part.

4.3 Data acquisition and management

In the past, different approaches have been developed for obtaining process data from injection molding processes. Here, too, the approaches differ in the type of data to be obtained. For example, process data describing the behavior inside the cavity of the injection mold can be acquired by additional sensor technology and read out via external monitoring systems [21–23]. In this way, Ke et al. [24] were able to integrate seven pressure sensors into a mold cavity to analyze the filling behavior as precisely as possible.

Zhao et al. [25] also used external sensors for process data acquisition, but by doing so they did not record process data in the cavity, but from the injection unit. For this, they used three different types of sensors and data acquisition cards to record pressures, positions, temperatures, and times.

To meet the requirements associated with the smart factory concept and Industry 4.0 in terms of communication capability and the degree of inter-connectivity along the process chain, machine manufacturers have developed communication interfaces enabling them to integrate their machines into fully networked environments. Most injection molding machines therefore not only allow visualization of the process variables in the machine display, but also provide various export options for the process data generated during the manufacturing process such as USB interfaces [26] or network servers such as the aforementioned Open Platform Communication Unified Architecture (OPC UA) [27].

OPC UA is a platform-independent service-oriented architecture developed by the OPC Foundation as a standard for data exchange between machines and systems. Based on OPC UA, companion specifications such as EUROMAP 83 [28] and EUROMAP 77 [29] have been developed specifically for plastics processing machines. The companion specification EUROMAP 77 defines, among other things, an OPC UA ObjectType, which is used for the root object representing an injection molding machine with all its subcomponents. Martins et al. [27] have used this standard to record process parameters such as the injection velocity, cycle time, and maximum injection pressure.

4.3.1 Machine and process values acquisition via OPC UA

Although nearly all machine and process variables can be read out from the injection molding machine via the described standards, this method is only suitable to a limited extent for the formation of DTs, since the data transfer rate is not fast enough for a real-time transfer, especially when several variables are considered. However, in addition to exporting one-time values such as the injection time or the maximum injection pressure, some injection molding machines also allow the export of so-called measurement and monitoring charts (in tabular form) that contain trajectories of set process variables over the cycle time. This can be, for example, the hydraulic pressure or the injection flow trajectories. External sensors that are read out via a measuring amplifier connected to the machine can also be mapped in these charts.

In the case study described in this chapter, a manufacturer-independent Python script was developed, which exports one-time machine values such as switching point, nozzle, and heating band temperatures (including their target and actual values), resulting one-time process variables such as injection time and residual melt volume after the packaging phase, as well as time-series of target and actual process variables such as hydraulic pressure and injection velocity trajectories. Process values such as the cavity pressure and temperature time-series were also recorded in this way via the in-mold sensor connected to the injection molding machine.

Within the machine control system, each setpoint and actual value as well as the aforementioned measurement and monitoring charts are assigned an internal manufacturer-dependent label. Via the OPC UA specifications, each of these labels is linked to a unique manufacturer-independent NodeID. The developed script continuously monitors a trigger signal and queries all parameters listed in Fig. 4.6 after the signal occurs. The trigger signal had to be selected in such a way that at the time of triggering, all cycle-related process variables relevant for the formation of the DT were fully mapped in the corresponding charts and the one-time actual values were available. Additionally, after triggering, there needs to be enough time to query the values before they were overwritten by the values of the next cycle. Therefore the opening of the injection mold (7 in Fig. 4.3) was chosen.

Although the monitoring charts only monitor one signal, the measurement charts can include up to four different signals. By dividing the process variables time-series into multiple equally distributed consecutive time-series, which are all mapped in different charts, the frequency of the trajectories can be adjusted. The set recording duration of a chart is thereby divided into 512 equally distributed measuring points due to the programming of the IM machine. By dividing the cycle time over multiple charts the measuring points add up as a multiple of 512 depending on the number of charts. The maximum frequency at which the sensor values can be recorded is 500 Hz for the machine used. However, since there is only a limited number of available charts (4 measurement charts, 8 monitoring charts, and 4 extended monitoring charts), a compromise must be found between achievable frequency and mapping of the relevant time section of the cycle time.

To determine an appropriate sampling frequency, the most dynamic signal, i.e., the measured injection pressure, was sampled at the highest sampling rate possible, i.e., 500 Hz. The amplitude and power spectrum are shown in Fig. 4.7. Before applying the Fourier transform, the signal was centered, and a Hann window was applied to reduce the effect of the offset between the first and last signal values on the spectrum. We can observe that the frequencies below 25 Hz account for more than 90 % of the total sum of Fourier coefficients and more

```
1   SIGNALS = {
2       'cycle_counter':              signal_struct('ns=2;i=238982'),
3       'monitoring_chart_1':         signal_struct('ns=2;i=178052'),
4       'monitoring_chart_2':         signal_struct('ns=2;i=179612'),
5       'monitoring_chart_3':         signal_struct('ns=2;i=181172'),
6       'monitoring_chart_4':         signal_struct('ns=2;i=182732'),
7       'monitoring_chart_5':         signal_struct('ns=2;i=184292'),
8       'monitoring_chart_6':         signal_struct('ns=2;i=185852'),
9       'measurement_chart_1':        signal_struct('ns=2;i=142812', num_signals=4),
10      'measurement_chart_2':        signal_struct('ns=2;i=144482', num_signals=4),
11      'measurement_chart_3':        signal_struct('ns=2;i=573962', num_signals=4),
12      'cycle_time':                 signal_struct('ns=2;i=160842'),
13      'heating_zone_1_actual':      signal_struct('ns=2;i=207272'),
14      'heating_zone_2_actual':      signal_struct('ns=2;i=207422'),
15      'heating_zone_3_actual':      signal_struct('ns=2;i=207572'),
16      'heating_zone_4_actual':      signal_struct('ns=2;i=207722'),
17      'heating_zone_5_actual':      signal_struct('ns=2;i=207872'),
18      'heating_zone_1_target':      signal_struct('ns=2;i=207282'),
19      'heating_zone_2_target':      signal_struct('ns=2;i=207412'),
20      'heating_zone_3_target':      signal_struct('ns=2;i=207562'),
21      'heating_zone_4_target':      signal_struct('ns=2;i=207712'),
22      'heating_zone_5_target':      signal_struct('ns=2;i=207862'),
23      'injection_velocity_target':  signal_struct('ns=2;i=201802'),
24      'packing_velocity_1':         signal_struct('ns=2;i=201172'),
25      'packing_pressure_1_target':  signal_struct('ns=2;i=201292'),
26      'packing_velocity_2':         signal_struct('ns=2;i=416782'),
27      'packing_pressure_2_target':  signal_struct('ns=2;i=201332'),
28      'packing_time_2_target':      signal_struct('ns=2;i=201322'),
29      'packing_velocity_3':         signal_struct('ns=2;i=416792'),
30      'packing_pressure_3_target':  signal_struct('ns=2;i=201372'),
31      'packing_time_3_target':      signal_struct('ns=2;i=201362'),
32      'dosing_volume_target':       signal_struct('ns=2;i=201972'),
33      'dosing_volume_actual':       signal_struct('ns=2;i=201762'),
34      'switching_point_target':     signal_struct('ns=2;i=201112'),
35      'switching_point_actual':     signal_struct('ns=2;i=202432'),
36      'residual_volume':            signal_struct('ns=2;i=202672'),
37      'backpressure_target':        signal_struct('ns=2;i=201962'),
38      'max_pressure':               signal_struct('ns=2;i=202472'),
39      'switching_pressure':         signal_struct('ns=2;i=202522'),
40      'dosing_time':                signal_struct('ns=2;i=202732'),
41      'injection_time':             signal_struct('ns=2;i=202582'),
42      'timestamp':                  signal_struct('ns=2;i=117522'),
43  }
```

FIGURE 4.6 Recorded charts and values including their corresponding NameSpaceIndex (ns) and NodeID (i).

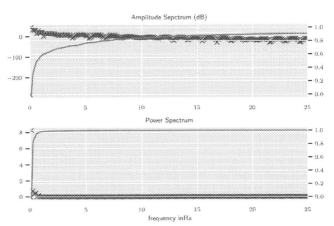

FIGURE 4.7 First part of amplitude spectrum (above) and power spectrum (below) of the injection pressure signal sampled at 500 Hz. The red (mid gray in print version) curve represents the share of cumulative sum up to the respective frequency relative to the total sum.

than 99.9 % of the total signal power. Therefore we concluded that a sampling rate of 50 Hz is sufficient for data acquisition. To achieve this sampling rate in the case study, the cycle time had to be divided among three consecutive charts, each with a recording duration of 10.16 s. The process variables were divided among the various charts as follows:

- Monitoring charts 1–3:   injection velocity
- Monitoring charts 4–6:   screw position
- Measurement charts 1–3:   target and actual injection pressure, pressure and temperature in the mold cavity

The segmentation of the cycle time among the three charts is depicted in Fig. 4.8. The recording of the displayed trace starts directly at the beginning of the process cycle and ends when the trigger is reached. In the case study, this trigger was chosen to be the opening of the injection mold (7 in Fig. 4.3). Data during the time span Δt, which occurs between the trigger and the start of the next cycle, is not recorded, as it is not relevant for model building. Within this time span, only the injection mold moves, which is not relevant for the part quality.

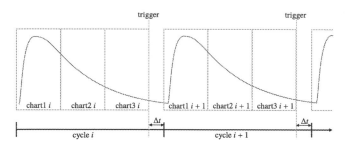

FIGURE 4.8 Visualization of splitting each cycle time into three charts.

All recorded machine and process variables are stored in a Hierarchical Data Format 5 File (HDF5) in a cycle-related manner. Fig. 4.9 shows the chosen structure of the files. For simplified visualization, each parameter is denoted with the manufacturer-dependent label and unambiguously assigned to the corresponding cycle. For each parameter, the corresponding recorded value (block0_values) is stored in addition to the description (block0_items) shown in Fig. 4.6. The charts, e.g., f3103I, are stored in tabular form, as shown in Fig. 4.9, where column 0 represents the time, 1 the value of the injection velocity at the corresponding time, and 2 the machine state (not considered in this study).

4.3.2 In-line part quality data acquisition

In IM series production with very short cycle times and, possibly, multi-cavity injection molds, in-line quality control is very costly, and in some cases the costs exceed the benefits. However, especially in the case of complex technical parts,

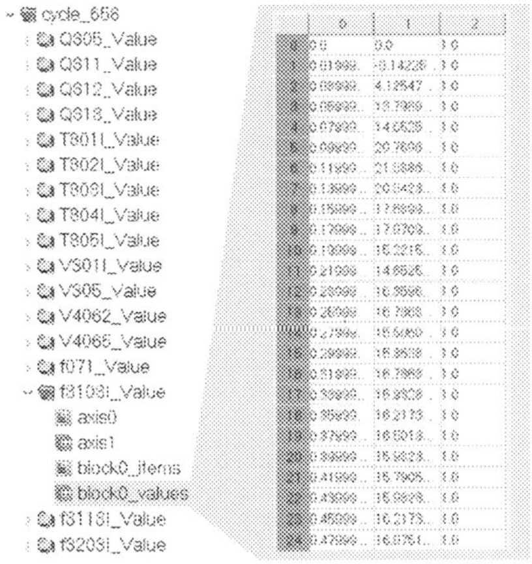

FIGURE 4.9 Structure of the HDF5 file.

which are subject to high requirements in terms of load-bearing capacity, durability, and optical properties, undetected quality deviations can result in costly consequences. Quality control of each manufactured part is desirable, but it is not possible to record all relevant quality characteristics within the process cycle and to measure non-destructively. Quality parameters that can be detected in the process cycle include weight and geometric tolerances such as dimensions, shrinkage, and warpage and also optically detectable defects such as sink marks, color differences, streaks, and weld lines. Since quantifiable values are recorded by measuring the component weight and its dimensions, and conclusions can be drawn about the process behavior with the aid of these quality variables, these characteristics were taken into account in the case study. For this purpose, both a scale (Entris II, Sartorius AG, Göttingen, Germany) and a digital measurement projector (IM-7020, Keyence Corporation, Osaka, Japan) were integrated into the process (Fig. 4.10). These measurements are carried out manually. Since the quality variables are recorded within the process cycle time by this measurement setup, they can be assigned to the cycle-based process data. However, the quality characteristics of the part produced in cycle i can only be measured during cycle $i + 1$ (Fig. 4.8).

Fig. 4.11 shows the tamper-evident closure, which was produced in the case study, and its dimensions are measured with a measurement projector. Whereas only the diameter of the perforated circle, denoted as the inner diameter D_i in the case study, was measured on the upper side of the closure, both the diameter D_o and the roundness were determined for the outer diameter on the lower side. As further quality characteristics, the web widths W_j and W_t were measured on both sides of the perforated circle.

FIGURE 4.10 Quality measuring setup including a scale to measure the parts' weight and a digital measurement projector to measure the parts' dimensions.

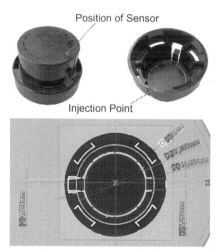

FIGURE 4.11 Tamper-evident closure and its geometrical dimensions (screenshot from measuring projector) measured in the case study ([1] inner diameter D_i, [2] outer diameter D_o, [3] width at joint W_j, [4] width at tab W_t, [5] roundness).

4.4 Control-oriented modeling of final part quality

A data-driven internal dynamics approach for dynamic modeling of the part quality characteristics is developed. The proposed approach is in principal a nonlinear state-space model capable of mapping process value trajectories to final part quality. This quality model can be employed to calculate reference process value trajectories via numerical optimal control. In contrast to pvT-optimization (Section 4.2.3), the data-driven quality model does not rely on any assumptions and is able to map arbitrary process values to the part quality. This means in turn that based on this model, reference trajectories can be generated not only for cavity pressure, but also for all process variables deemed relevant. Also, the process values' trajectories are optimized directly with respect to the part quality as opposed to the minimal shrinkage, as is the case in pvT-optimization. The downsides of the proposed approach are that the resulting parameter estimation problem is rather complex and time-intensive to solve, and the need for training data is rather high.

4.4.1 Final part quality prediction

To obtain a consistent and desirable batch end-product quality, precise quality prediction models are necessary that can be employed for in-batch or batch-to-batch optimization. Many batch processes, such as the plastic injection molding process, do not allow for continuous in-process quality measurements. Quality variables of interest, e.g., the weight, density, or geometrical features, can only be quantified once the part is ejected from the machine. The task of predicting the quality characteristics of the emerging part at the end of the batch is known as final part quality prediction [30]. Although the formation of quality characteristics is a dynamic process, i.e., its future evolution is a function of its current state, it has almost exclusively been treated as a static modeling task [13,24,30–33]. One of the reasons is that the estimation of a dynamic model that maps process value trajectories to a single batch-end quality measurement is an unusual and difficult modeling task. In plastic injection molding, the final part quality is usually directly predicted from process setpoints [16,32]. This approach works reasonably well; as long as the process is operating at steady state, process dynamics do not change, and no disturbances act on the process. In plastic injection molding processes with cavity pressure control, the cavity pressure reference trajectory can be derived from the pvT-diagram of the material. This is known as pvT-optimization [34]. The pressure reference trajectory is derived from the pvT-diagram with the intent to obtain a part with a certain specific volume v and minimize shrinkage in the process. pvT-optimization does not provide a link to any other quantity of interest, such as surface characteristics or mechanical properties of the produced part. In particular, Hopmann et al. [23] documented that in an injection compression molding application, the cavity pressure trajectory that yielded the best quality differed from the pressure trajectory that led to the least shrinkage. This suggests that apart from cavity pressure, other process values have to be taken into account and that minimizing the shrinkage does not ensure an optimal part quality in every regard.

In most applications, besides plastic injection molding, multivariate statistical process control (MSPC) methods, such as multi-way principal component analysis (MWPCA) and multi-way projection to latent structures (MWPLS), are employed to correlate process variable trajectories with final product quality [35]. Although these methods exploit all the information in the data, as static models, they are subject to certain restrictions, e.g., all batches need to have the exactly same duration [33]. Hence methods must be employed that somehow scale all trajectories to the same length, e.g., dynamic time warping. By doing so the original information contained in the measured data is affected to an unknown extent.

4.4.2 Preliminaries

Dynamic models can amongst other criteria be differentiated into *external* and *internal dynamics* approaches; see Fig. 4.12. In the far more widespread external dynamics approach, a static model $f(\cdot)$ is provided with past inputs $u_{k-j} \in \mathbb{R}^{n_u}$ $j = 1, \dots, n$, and outputs $y_{k-i} \in \mathbb{R}^{n_y}$ $i = 1, \dots, m$, to predict the current output y_k; n_u and n_y denote the dimensions of the input and output signal, whereas n and m are the orders of the model, i.e., the maximal shifts of the input and output signals. Hence the dynamics are realized via external filters, which in their most simple form correspond to time delays. The model equation for a time-invariant external dynamics model without dead time is

$$y_k = f\left(y_{k-1}, \dots, y_{k-m}, u_{k-1}, \dots, u_{k-n}\right), \tag{4.1}$$

where q^{-1} is the time-shift operator, i.e., $q^{-1}u_k = u_{k-1}$. The internal dynamics approach realizes process dynamics not via external filters, but via the internal model states $x_k \in \mathbb{R}^{n_x}$ and therefore results in a state space representation of the identified system. The states usually do not have any physical meaning and are merely a means of realizing dynamic behavior. An internal dynamics model is provided with the current input and the previous state to predict the next output:

$$\begin{aligned} x_{k+1} &= h\left(x_k, u_k\right), \\ y_k &= g\left(x_k\right), \end{aligned} \tag{4.2}$$

where h and g are the right-hand sides of the state and output equations, respectively. Both the external and internal dynamics approaches are viable for the problem of final part quality prediction from process value measurements. Irrespectively of the chosen dynamics, models can only be trained in *parallel* configuration, i.e., as simulation or output-error (OE) model. Training in *series-parallel* configuration, i.e., as one-step ahead predictor or equation-error model, is not possible since the model output $\hat{y}_k$ is only available at the very last time instance $k = T$ of each batch. This also means that only the prediction error of the last time step $e_T = \hat{y}_T - y_T$ is available for calculating the parameter updates during optimization, i.e., in case of a quadratic loss function

$$\mathcal{L} = \frac{1}{2}\left(\hat{y}_T - y_T\right)^T\left(\hat{y}_T - y_T\right). \tag{4.3}$$

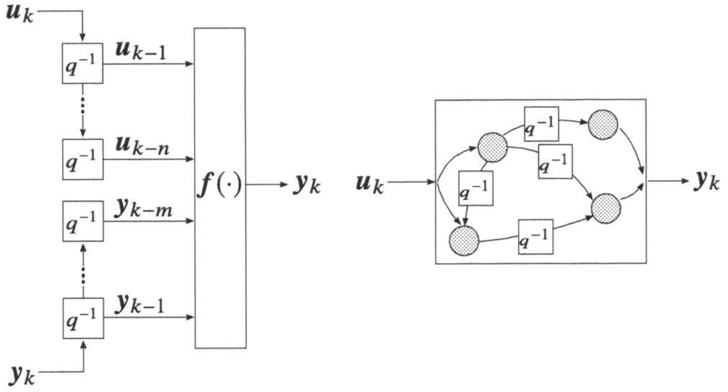

FIGURE 4.12   External (left) and internal dynamics approach (right).

Major advantages of the internal dynamics approach are the lower-dimensional input space, especially for higher-order multiple-input multiple-output (MIMO) systems, and that most advanced controller synthesis methods require a state-space representation. Additionally, determining the dynamic order can be cumbersome in case of an external dynamics approach, especially in the MIMO-case and if input and output are allowed to have different delays. In an internal dynamics approach, determining the order of the models equates to choosing the dimension of the internal state vector. However, training recurrent model structures poses a complex optimization problem: The evolution of the internal states must reflect the true process dynamics but can only be deduced from input–output data. In the external dynamics approach on the other hand, the process dynamics are fixed by the filters specified by the user.

In the last decade, recurrent model structures that alleviate these problems have been developed, most notably, the long short-term memory (LSTM) [36] and the gated recurrent unit (GRU), which will both be subsumed under the term Gated Units in the following. Although a subject of ongoing research, it appears that Gated Units restrict the dynamics that can be represented in comparison to traditional Recurrent Neural Network approaches. This results in fewer bifurcation boundaries and hence fewer regions with very large gradients in the model parameter space, which makes them less sensitive to initialization and easier to train [37]. In the following, modeling approaches for predicting the final part quality with external and internal dynamics models will be presented. External dynamics models without output feedback, e.g., Finite Impulse Response (FIR) and its nonlinear variants, will be excluded from consideration. These models treat each time instance as a distinct input. This would result in thousands of inputs (and model parameters) depending on the length and number of measured process variables. The amount of experimental data needed to estimate these parameters makes this approach unfeasible.

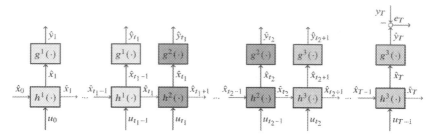

FIGURE 4.13 Switched internal dynamics model for final part quality prediction.

4.4.3 Internal dynamics quality model

The machine-value-controlled IM process is a time-varying process due to the switching between different controllers at defined switching time instances between the injection, packing, and cooling phases. The formation of the part quality characteristics can also be assumed to be a time-varying process, although for different reasons: The part changes its aggregate state from fluid to solid and hence its mechanical properties. A fluid will respond differently, e.g., to a certain applied pressure, than a solid body. Accounting for this time-varying behavior could be done by assuming that the model parameters are functions of the internal model states, but this would further exacerbate the complexity of the optimization problem. The quality model was therefore assumed to be a time-varying switched system comprised of three time-invariant subsystems representing the injection, packing, and cooling phase ($i = 1, 2, 3$, respectively). The parameter vectors of the state and output equations of each subsystem are denoted Θ^i and Φ^i, respectively:

$$x_{k+1} = h^i \left(x_k, u_k; \Theta^i \right), \quad x_0 = 0, \quad k \in \mathbb{Z}, \quad i = \begin{cases} 1, & k \leq t_1, \\ 2, & t_1 < k \leq t_2, \\ 3, & k > t_2, \end{cases} \quad (4.4)$$

$$y_k = g^i \left(x_k; \Phi^i \right).$$

The switching time instances t_i coincide with the end of the respective phase and are known in advance. Since the internal state is thought to be an, albeit very abstract, representation of the emerging part, the model is assumed to be continuous in the states, i.e., all subsystems are of the same order, and $x_{t_i}^{i+1} = x_{t_i}^i$ at the switching instances. The unfolded switched system is depicted in Fig. 4.13. For the aforementioned reasons, the recurrent part h^i of each subsystem will be modeled using a GRU. The GRU is depicted in Fig. 4.14. Its state equation is

$$x_{k+1} = f_z \odot x_k + (1 - f_z) \odot f_c. \quad (4.5)$$

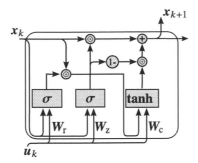

FIGURE 4.14   Gated Recurrent Unit architecture.

The operator $\odot$ denotes the Hadamard product. The activation of the so-called reset gate $\boldsymbol{f}_r$, update gate $\boldsymbol{f}_z$, and the output gate $\boldsymbol{f}_c$ are given by

$$\boldsymbol{f}_r = \sigma \left(\mathbf{W}_r \cdot [\boldsymbol{x}_k, \boldsymbol{u}_k]^T + \boldsymbol{b}_r \right),$$
$$\boldsymbol{f}_z = \sigma \left(\mathbf{W}_z \cdot [\boldsymbol{x}_k, \boldsymbol{u}_k]^T + \boldsymbol{b}_z \right), \qquad (4.6)$$
$$\boldsymbol{f}_c = \tanh \left(\mathbf{W}_c \cdot [\tilde{\boldsymbol{x}}_k, \boldsymbol{u}_k]^T + \boldsymbol{b}_c \right)$$

with $\tilde{\boldsymbol{x}}_k = \boldsymbol{f}_r \odot \boldsymbol{x}_k$, $\mathbf{W}_r, \mathbf{W}_z, \mathbf{W}_c \in \mathbb{R}^{n_x \times n_x + n_u}$, $\boldsymbol{b}_r, \boldsymbol{b}_z, \boldsymbol{b}_c \in \mathbb{R}^{n_x}$, and $\boldsymbol{f}_r, \boldsymbol{f}_z, \boldsymbol{f}_c :$ $\mathbb{R}^{n_x} \to \mathbb{R}^{n_x}$, where σ denotes the logistic function, and **tanh** is the hyperbolic tangent function (both are applied element-wise).

The feedforward-part $\boldsymbol{g}^i$ maps the internal model state $\boldsymbol{x}_k$ to the model output $\boldsymbol{y}_k$. However, only $\boldsymbol{y}_T$, i.e., the part quality measurement at the very last time instance T of each batch, is measured. Therefore $\boldsymbol{\Phi}^i$, $i = 1, 2$, will not enter the loss function as it has no effect on $\boldsymbol{y}_T$ and will not be adjusted during training. Usually, a feed forward neural network with a nonlinear hidden layer and a linear output layer is employed:

$$\boldsymbol{h}_k = \tanh \left(\mathbf{W}_h \cdot \boldsymbol{x}_k + \boldsymbol{b}_h \right),$$
$$\boldsymbol{y}_k = \mathbf{W}_o \cdot \boldsymbol{h}_k + \boldsymbol{b}_o, \qquad (4.7)$$

where $\boldsymbol{h}_k$ is the activation of the neurons in the hidden layer. Parameter initialization was found to be crucial to obtain useful models. As in any (stable) dynamical system, the contribution of the GRU's state $\boldsymbol{x}_k$ at a given time instance k to the output $\boldsymbol{y}_{k+\Delta T}$ at a later time instance decreases exponentially with ΔT. Since the model to be optimized is recurrent in the states, the same is true for its parameters. This phenomenon is known as the *vanishing gradient* problem. If care is not taken during initialization, especially the contribution of the first two subsystems ($i = 1, 2$) will be negligibly small. The solution to this dilemma is initializing the bias of the update gate $\boldsymbol{b}_z^i$, $i = 1, 2$, with large positive values. By doing so the state equation (4.5) becomes $\boldsymbol{x}_{k+1} \approx \boldsymbol{x}_k$, i.e., the GRU just passes on the state. This ensures the maximal possible gradient, at least at the very beginning of the optimization procedure. If necessary, the optimizer will then reduce the bias to an appropriate value, such that the GRU dynamics approximates the dynamics of the true process. For this reason, the bias must also not be chosen too large; otherwise, it would take an excessive amount of optimization steps to reduce it to an appropriate value. For this case study, the best results were obtained by drawing $\boldsymbol{b}_z$ from a random uniform distribution $\mathcal{U}$ with support [4, 10], i.e., $\mathcal{U}_{[4,10]}$. Without this initialization, the estimated models were merely able to reproduce the mean of the training data.

4.4.4 External dynamics quality model

Although no continuous output measurements are available, the external dynamics approach is still applicable: The model is trained in parallel configuration, i.e., as a simulation model, with output measurements not required when evaluating the model. It should be noted that a recommended initialization strategy for output-error (OE) models is to train the model as a one-step-ahead predictor and use the result to initialize the OE optimization. This procedure can potentially yield an initial parameterization that is close to a good local minimum of the NOE loss function. Due to lacking continuous output measurements, this procedure cannot be applied, and we have to resort to random initialization. As discussed in Section 4.4.3, a switched system approach with three subsystems is chosen to model the time dependency of the formation process of the part quality characteristics. For convenience of notation, the maximum time delays of the input and output are assumed to be equal, and each subsystem is assumed to have the same maximal delay n:

$$\hat{\boldsymbol{y}}_k = \boldsymbol{f}^i \left(\left[\hat{\boldsymbol{y}}_{k-1} \quad \cdots \quad \hat{\boldsymbol{y}}_{k-n} \quad \boldsymbol{u}_{k-1} \quad \cdots \quad \boldsymbol{u}_{k-n} \right]^T ; \boldsymbol{\theta}^i \right)$$

$$\hat{\boldsymbol{y}}_i = \boldsymbol{0} \; \forall \, j = 1, \ldots, n-1 \qquad k \in \mathbb{Z}, \quad j = \begin{cases} 1, & k \leq t_1, \\ 2, & t_1 < k \leq t_2, \\ 3, & k > t_2. \end{cases} \tag{4.8}$$

Since the external dynamics approach requires the (unknown) first n output measurements as initialization, they are set to zero in accordance to the internal dynamics model (4.4). The unfolded external dynamics switched system is depicted in Fig. 4.15. As the external model is to be trained in parallel configuration, care has to be taken that the initial model is not unstable. Finding a stable parameterization for a nonlinear model is difficult. It is easier to choose a model structure whose output cannot blow up by design. A multilayer perceptron (MLP) with bounded activation functions satisfies this assumption, as opposed, e.g., to a polynomial model, and is therefore an appropriate candidate structure.

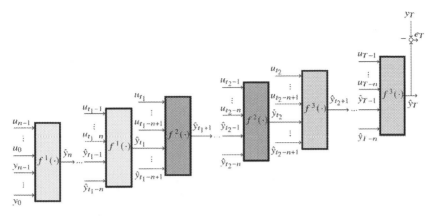

FIGURE 4.15   Switched external dynamics model for final part quality prediction.

4.4.5   Static quality model

Estimating a static model that maps process setpoints, i.e., $u = s$, to the final part quality, i.e., $y = Q$, can be considered as the current industrial state-of-the-art. This approach will therefore be considered as a baseline in the subsequently presented case study. Employing a static model entails the implicit assumption that the process is at steady state. This implies that data collected for a number of cycles after changing the process setpoint cannot be used for model estimation, increasing the time spent on collecting data for model training. In addition, during process operation, the model cannot be used whenever even minor disturbances occur, such as variations in the time interval between cycles, e.g., because a part got stuck. As these events happen quite frequently, it is proposed to include process value measurements at the beginning of a cycle, i.e., $p_0 = p(t = 0)$ as inputs to the static model, i.e., $u^T = \left[s^T, p_0^T \right]$. By doing so the initial state of the process is explicitly considered. Variations in the time between cycles would become visible in different temperatures of the mold cavity. The model equation of the static approach is simply

$$y = f(u).$$

4.5   Case study: tamper-evident closure quality prediction

To investigate whether dynamic quality models have additional value compared to the less complex and less time-consuming static approach, a case study is conducted. The objective of this case study is the prediction of the mass m and inner diameter D_i of the tamper-evident closure depicted in Fig. 4.11. The material chosen for the production of this closure is an unreinforced black polyamide 6 (PA6) of the type B3S (BASF SE, Ludwigshafen am Rhein, Germany).

 To effectively train models, the training data must originate from as large a process window as possible to include all correlations and nonlinear effects of

TABLE 4.2 Upper and lower limits of the setpoint parameters used for the face centered composite design of experiment.

Setpoint parameter	Unit	Lower limit	Upper limit
Nozzle temperature	°C	250	260
Mold temperature	°C	40	50
Injection flow rate	cm³/s	16	48
Switching point	cm³	13	14
Packing pressure	bar*	500	600
Packing time	s	3	5
Back pressure	bar*	25	75
Cooling time	s	15	20

*In IM the pressure is usually measured and stated in bar; therefore no conversion to SI units has been made here.

the processing parameters. However, a compromise must be found between a large process window, the achievable quality, and the processability of the material. A large number of Design of Experiments (DoE) have been developed to analyze precisely such correlations between the individual parameters and to describe them by means of regression analysis. Since the injection molding process reacts very sluggishly when certain parameters (e.g., temperatures) are varied and it takes several cycles until a constant process is established again after the change, randomized experimental designs, such as Latin Hypercubes, are sometimes difficult to implement in the real process. Heinisch et al. [38] compared the performance of artificial neural networks trained with data retrieved from different DoEs used in IM simulations. It can be shown that central composite designs (CCDs) outperformed the other DoEs and thus provide the best data basis for training neural networks. CCDs extend a full factorial design (FFD) by so-called star points (SP), which are located exactly on the coordinate axes and have a distance α to the origin, which is the central point (CP) of the FFD. For this case study, a process setpoint $s \in \mathbb{R}^8$ is defined by the eight setpoint parameters defined in Table 4.2. With two factor levels for each setpoint parameter and two repetitions of the central point (for the estimation of process dispersion), the resulting experimental design contains a total of 274 experiments (FFD: $2^8 = 256$, SP: $2 \cdot 8 = 16$, CP: 2) with 10 repetitions each. A face centered CCD ($\alpha = 1$) was chosen.

For the production of the parts, the aforementioned IM machine (Fig. 4.1) was used. The injection mold was equipped with an additional combined temperature and pressure sensor, as described in Section 4.2.2. During the manufacturing of the parts, only the setpoins listed in Table 4.2 were varied according to the face centered CCD, and for each experiment, ten parts were produced.

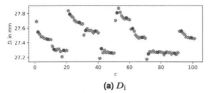
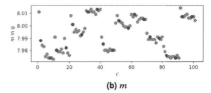

(a) D_i

(b) m

FIGURE 4.16 Measured inner diameter D_i and mass m of the tamper-evident closure over the first ten charges. Gray samples were used for training, red (dark gray in print version) samples for model validation.

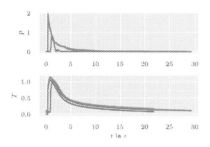
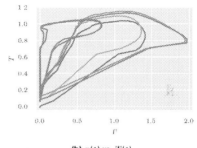

(a) Normalized cavity pressure $p(t)$ and temperature $T(t)$ over time t

(b) $p(t)$ vs. $T(t)$

FIGURE 4.17 Process measurements from three different factor combinations (cycles 1–10, 51–61, and 71–81).

A subset of the acquired data, i.e., the first ten factor combinations, are depicted in Fig. 4.16. From Fig. 4.16 we can observe that the process setpoint has a noticeable effect on the parts' inner diameter D_i and mass m.

We can also observe that different process setpoints also lead to distinctively different process variable trajectories. Fig. 4.17 shows the normalized cavity pressure p_{cav} and normalized cavity temperature T_{cav} of different factor combinations that respectively corresponded to parts with small, medium, or large inner diameter D_i. Normalization was performed with respect to the maximum cavity pressure p_{cav}^{max} and cavity temperature T_{cav}^{max} that occurred during the first cycle. It is also evident from Fig. 4.16 that with respect to the quality variables, a good amount of variation also exists within most factor combinations. Especially, a pronounced transient behavior is noticeable with respect to D_i. For a dynamic quality model to predict these variations, they must be reflected in the process measurements. Therefore the cavity pressure p_{cav} and cavity temperature T_{cav} of factor combinations with and without pronounced transient behavior with respect to D_i were investigated. As an example, the measurements of the first (cycles 1–10) and ninth factor combination (cycles 81–90) are shown in Figs. 4.18 and 4.19, respectively. The process measurements taken during the cycles of the first factor combination are appreciably different from another, whereas there is barely any difference between the process measurements of the ninth factor combination. Most noticeable is a significant increase in the ini-

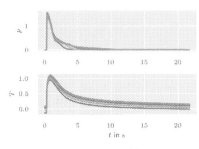

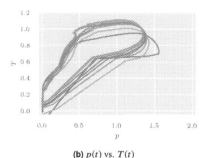

(a) Normalized cavity pressure $p(t)$ and temperature $T(t)$ over time t

(b) $p(t)$ vs. $T(t)$

FIGURE 4.18   Process measurements from first factor combination (cycles 1–10).

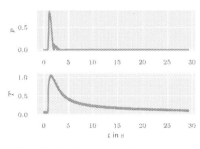

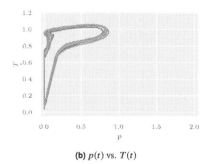

(a) Normalized cavity pressure $p(t)$ and temperature $T(t)$ over time t

(b) $p(t)$ vs. $T(t)$

FIGURE 4.19   Process measurements from ninth factor combination (cycle 71–81).

tial and maximal cavity temperature T_{cav} over the first ten cycles, which is not the case during cycles 81–90. This effect is due to the fact that some setpoint changes require more time than others, which allows the mold more or less cool-down time. In addition, the cavity pressure trajectories vary mostly around $t \approx 2.5$ s. Closer inspection revealed that the cause was an increase in cavity pressure, which took place at different points at time, but precisely when the measured cavity pressure was 104 °C. Most likely, the pressure spike was due to changes on the level of the microstructure, which occurred at different points in time as a result of different initial cavity temperatures. Based on these observations, the cavity temperature is expected to be a decisive quantity in predicting the part quality.

Of each factor combination, the second and the third last repetitions were chosen as validation data $\mathcal{D}_{val}$, whereas the remaining data $\mathcal{D}_{train}$ were used to train the model. By doing so it was made sure that the model ability to explain the transient behavior is validated. The model structures considered for this task are listed in Table 4.3. Implementation and optimization of all nonlinear models was performed using Casadi [39] and Ipopt [40]. 20 multi-starts were performed for each candidate structure. The performance of each model config-

TABLE 4.3 Considered candidate model structures for quality prediction.

Label	Dynamics	Definition
PR_s^n	static	nth-degree polynomial regression (PR) model with interactions mapping process setpoints s and possibly initial process value measurements p_0 to product quality Q.
MLP_s^n	static	MLP with n neurons in single hidden layer mapping process setpoints s and possibly initial process value measurements p_0 to product quality Q.
ED^n	ext. dynamics	External dynamics (ED) model consisting of 3 subsystems mapping process measurements p to product quality Q, as described in Sec. 4.4.4.
ID^n	int. dynamics	Internal dynamics (ID) model consisting of 3 subsystems mapping process measurements p to product quality Q. The recurrent model part is realized via GRUs with n internal states. The mapping from internal state to product quality is realized via an MLP with 10 tanh neurons in the hidden layer.

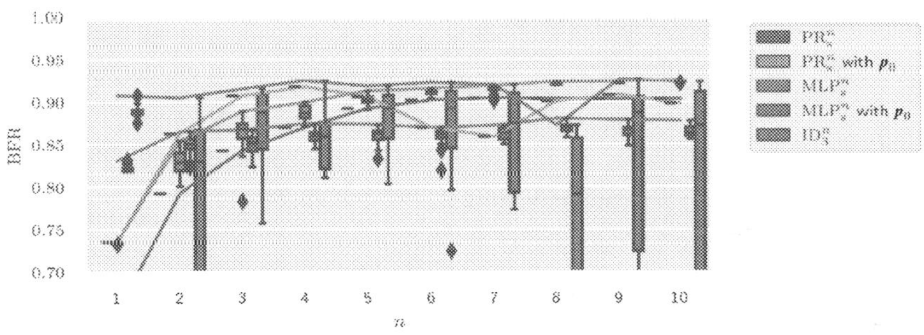

FIGURE 4.20   BFR of the model structures in Table 4.3 for the quality prediction task tamper-evident closure.

uration is measured in terms of the Best Fit Rate (BFR), which corresponds to the coefficient of determination restricted to positive values:

$$BFR = \max\left\{ 0, 1 - \left(\frac{\sum_c (y_c - \hat{y}_c)^2}{\sum_c (y_c - \bar{y}_c)^2} \right)^{\frac{1}{2}} \right\}. \quad (4.9)$$

The performance in terms of the best fit rate (BFR) (4.9) on $\mathcal{D}_{val}$ of each model structure is depicted in Fig. 4.20. For model structures that are nonlinear in the parameters, a boxplot is given to visualize the effect of initialization dependence. The lines connect the best identified models. The first thing standing out from Fig. 4.20 is that the proposed switched internal dynamics approach ID consistently yields the model with the best fit, with the exception of $n = 8$

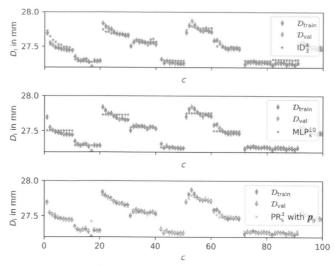

FIGURE 4.21 Prediction of ID_3^4 (top), MLP_s^{10} (middle), and PR_s^4 with p_0 (bottom) on $\mathcal{D}_{train}$ and $\mathcal{D}_{val}$.

internal states. For this configuration, the optimizer was apparently not able to find a good local optimum. The superior performance of the internal dynamics models is remarkable, since the task of predicting batch end quality from raw sensor measurements is considerably harder than a static mapping from setpoints to the part quality. The results regarding the polynomial models PR and the MLP seem somewhat conflicting. Comparing the polynomial models with and without initial process measurements p_0 suggests that the additional information helps predicting the part quality, as PR_s^4 with p_0 is the best polynomial model. The decreasing performance for PR_s^n with p_0 for $n > 4$ is most likely due to the tendency of higher-degree polynomials for oscillatory interpolation behavior. The best MLP and second best model overall on the other hand is MLP_s^{10}, a purely static MLP, which does not receive initial process measurements p_0 as input. As discussed previously, variations in quality measurements are larger between factor combinations than within a factor combination and are therefore mostly governed by the machine setpoint. A good fit of the static model was therefore expected. It is however surprising that the MLPs without initial process measurements p_0 performed as well as the internal dynamics model in terms of the Best Fit Rate. To investigate this outcome, the predictions of the best internal dynamic model ID_3^4, the best static model with initial process measurements PR_s^4 with p_0, and the best static model without initial conditions MLP_s^{10} on $\mathcal{D}_{train}$ and $\mathcal{D}_{val}$ are compared; see Fig. 4.21. Fig. 4.21 shows that the models using process measurements to predict the part quality are able to explain variations within the same setpoint (factor combination) reasonably well. The reason why MLP_s^{10} has a slightly higher BFR than PR_s^4 with p_0 despite not being able to explain any variation within a cer-

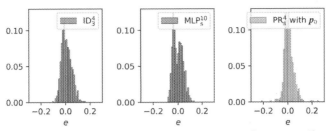

FIGURE 4.22 Frequency distribution of the prediction error e of ID_3^4 (left), MLP_s^{10} (middle), and PR_s^4 with p_0 (right) on $\mathcal{D}_{val}$.

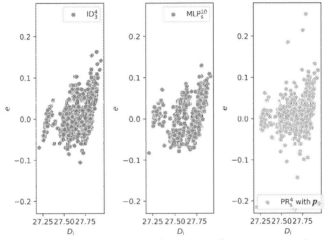

FIGURE 4.23 Prediction error e vs. D_i of ID_3^4 (left), MLP_s^{10} (middle), and PR_s^4 with p_0 (right) on $\mathcal{D}_{val}$.

tain setpoint becomes apparent from the histograms of the prediction errors in Fig. 4.22. PR_s^4 with p_0 has more prediction errors that exceed ± 0.1 mm compared to MLP_s^{10}. These larger errors have a disproportionate effect since the BFR is quadratic in e. It should also be mentioned that the prediction errors of all three models are not normally distributed according to the Shapiro–Wilk test. Fig. 4.23 clearly shows that within three distinguishable clusters, the prediction error slightly increases with increasing D_i, i.e., smaller D_i tend to be overestimated, whereas larger D_i tend to be underestimated within these clusters. Hence there exists some, albeit minor, deterministic relation none of the models was able to capture, either because the model is not complex enough or some model inputs are missing. A phenomenon that might be difficult to model even for the most complex considered candidate structures is the change in monotonicity that can be observed for neighboring setpoints, e.g., cycles 21–30 and 51–60 in Fig. 4.16a. During both experiments, the cavity temperature increases steadily with every cycle. During cycles 21–30, D_i decreases monotonously, as is expected since thermal expansion causes the cavity to shrink. During cycles 51–60, however, D_i increases sharply during the first few cycles before it decreases and approaches a constant value. These changes in monotonicity can be observed for setpoints directly adjacent in the setpoint space, a nonlinear behavior that is difficult to capture. Considered additional model inputs were the ambient temperature, ambient humidity, and the number of completed production cycles during that day (as a measure for the duration of continuous production of the machine). None of these inputs improved predictive performance.

4.6   Conclusions and outlook

4.6.1   Conclusions

The purpose of this work was to provide and implement a framework for the development of a Digital Twin of an industrial plastic injection molding machine. Therefore a system for process data acquisition was developed, which records the process data (single values and trajectories) cycle-related via the OPC-UA interface of the IM machine during series production. An additional temperature and pressure sensor was installed in the cavity of the injection mold for the analysis of quality-relevant state variables. It was shown that the sampling frequency of the recorded process variables is high enough to allow access to all the information contained in the trajectories. For the development of a Digital Twin, quality variables must be recorded in-line in addition to the process variables. For this purpose, the quality data of the manufactured tamper-evident closure were recorded during the following process cycle within the cycle time. The weight was recorded with the aid of a laboratory scale, and the dimensional accuracy (diameter and distances) was measured by a digital measuring projector. By applying a design of experiment (face centered central composite design) the described experimental setup was used to sample a process window as large as possible for training different models. Subsequently, the crucial aspect of finding an appropriate model representation of the plastic injection molding process for part quality prediction was investigated.

Various static and dynamic modeling approaches were compared in a case study. The case study revealed that there is a substantial amount of variation in the part quality not only between but also within a certain process setpoint. A proposed internal dynamics approach for part quality prediction from raw process variable measurements was able to explain more than 90 % of the variance in the dataset. However, a static polynomial model considering the process setpoint and process variable measurements at the beginning of a cycle performed almost as well. Since the effort of training the latter is considerably less, they have been found to provide the best trade-off in this specific case study. From the point of view of the practitioner, it is certainly most welcome that the easiest to implement and fastest to estimate model approach yields highly competitive prediction models.

4.6.2   Outlook

Future work will cover the estimation of quality models for more involved quantities, such as mechanical properties of the produced part. It is suspected that the quantities considered in this case study, i.e., part weight and geometric properties, are influenced by the actual course of the process to little extent. As long as the cavity is filled properly and its volume remains constant (due to thermal expansion), these quantities will vary little. The proposed internal dynamics approach is expected to have an advantage over static approaches if the course of the process influences the considered quality variables considerably. This is expected to be the case with mechanical properties, such as a parts tensile strength, where especially the evolution of the temperature is a significant factor.

In the future, the Digital Twin will also be employed for quality optimization. The static models estimated in this work are in principal already fit for this purpose. To take into account the effect of creeping disturbances like machine wear and tear, as well as changes in ambient temperature, the parameters should be re-estimated or estimated recursively once the prediction error increases. As mentioned above, a dynamic process model, from which in conjuncture with the dynamic quality model the optimal process setpoints can then be inferred, needs to be estimated to complete the dynamical Digital Twin. This aspect will be covered in future work as well. Based on the experience gained from the case study conducted in this work, an incremental increase in predictive performance is expected at best.

Finally, to truly close the quality optimization loop, the optimal machine setpoints need to be applied automatically to the machine. Up to now, producers only have reading access to the nodes in the OPC-UA address space of the injection molding machine. So to achieve autonomous smart manufacturing, the permission to write to the variables of the machine has to be enabled by injection molding machine manufacturers.

References

[1] VDI-Statusreport Industrie 4.0 in Spritzgießunternehmen, https://www.vdi.de/ueber-uns/presse/publikationen/details/vdi-statusreport-industrie-40-in-spritzgiessunternehmen, February 2019. (Accessed 29 July 2020).
[2] H.-P. Heim, Specialized Injection Molding Techniques, William Andrew, 2015.
[3] V. Goodship, Injection Moulding: A Practical Guide, De Gruyter, 2020.
[4] A.M. Merrill, Plastics Technology, vol. 1, Rubber/Automotive Division of Hartman Communications, Incorporated, 1955.
[5] K.-M. Tsai, C.-Y. Hsieh, W.-C. Lo, A study of the effects of process parameters for injection molding on surface quality of optical lenses, Journal of Materials Processing Technology 209 (7) (2009) 3469–3477.
[6] M. Rohde, A. Ebel, F. Wolff-Fabris, V. Altstädt, Influence of processing parameters on the fiber length and impact properties of injection molded long glass fiber reinforced polypropylene, International Polymer Processing 26 (3) (2011) 292–303.
[7] M. Mohan, M.N.M. Ansari, R.A. Shanks, Review on the effects of process parameters on strength, shrinkage, and warpage of injection molding plastic component, Polymer-Plastics Technology and Engineering 56 (1) (2017) 1–12.

[8] E. Farotti, M. Natalini, Injection molding. Influence of process parameters on mechanical properties of polypropylene polymer. A first study, Procedia Structural Integrity 8 (2018) 256–264.

[9] P. Sälzer, M. Feldmann, H.-P. Heim, Wood-polypropylene composites: influence of processing on the particle shape and size in correlation with the mechanical properties using dynamic image analysis, International Polymer Processing 33 (5) (2018) 677–687.

[10] A. Tellaeche, R. Arana, Machine learning algorithms for quality control in plastic molding industry, in: 2013 IEEE 18th Conference on Emerging Technologies & Factory Automation (ETFA), 2013, pp. 1–4.

[11] O. Ogorodnyk, O.V. Lyngstad, M. Larsen, K. Wang, K. Martinsen, Application of machine learning methods for prediction of parts quality in thermoplastics injection molding, in: Kesheng Wang, Yi Wang, Jan Ola Strandhagen, Tao Yu (Eds.), Advanced Manufacturing and Automation VIII, Springer Singapore, Singapore, 2019, pp. 237–244.

[12] F. Finkeldey, J. Volke, J.-C. Zarges, H.-P. Heim, P. Wiederkehr, Learning quality characteristics for plastic injection molding processes using a combination of simulated and measured data, Journal of Manufacturing Processes 60 (2020) 134–143.

[13] H. Jung, J. Jeon, D. Choi, J.-Y. Park, Application of machine learning techniques in injection molding quality prediction: implications on sustainable manufacturing industry, Sustainability 13 (8) (2021) 4120.

[14] S.K. Selvaraj, A. Raj, R. Mahadevan, U. Chadha, V. Paramasivam, A review on machine learning models in injection molding machines, Advances in Materials Science and Engineering (2022) 1–28.

[15] Y. Lu, C. Liu, I. Kevin, K. Wang, H. Huang, X. Xu, Digital twin-driven smart manufacturing: connotation, reference model, applications and research issues, Robotics and Computer-Integrated Manufacturing 61 (2020) 101837.

[16] P. Bibow, M. Dalibor, C. Hopmann, B. Mainz, B. Rumpe, D. Schmalzing, M. Schmitz, A. Wortmann, Model-driven development of a digital twin for injection molding, in: International Conference on Advanced Information Systems Engineering, Springer, 2020, pp. 85–100.

[17] F. Tao, M. Zhang, Digital twin shop-floor: a new shop-floor paradigm towards smart manufacturing, IEEE Access 5 (2017) 20418–20427.

[18] S. Stemmler, M. Vukovic, M. Ay, J. Heinisch, D. Abel, C. Hopmann, Cross-phase model-based predictive cavity pressure control in injection molding, in: 2019 IEEE Conference on Control Technology and Applications (CCTA), IEEE, 2019, pp. 360–367.

[19] C. Hopmann, D. Abel, J. Heinisch, S. Stemmler, Self-optimizing injection molding based on iterative learning cavity pressure control, Production Engineering 11 (2) (2017) 97–106.

[20] S. Stemmler, M. Vukovic, M. Ay, J. Heinisch, Y. Lockner, D. Abel, C. Hopmann, Quality control in injection molding based on norm-optimal iterative learning cavity pressure control, IFAC-PapersOnLine 53 (2) (2020) 10380–10387.

[21] H. Karbasi, H. Reiser, Smart mold: real-time in-cavity data acquisition, in: First Annual Technical Showcase & Third Annual Workshop, Canada, 2006.

[22] H.S. Park, D.X. Phuong, S. Kumar, AI based injection molding process for consistent product quality, Procedia Manufacturing 28 (2019) 102–106.

[23] C. Hopmann, A. Reßmann, J. Heinisch, Influence on product quality by PVT-optimised processing in injection compression molding, International Polymer Processing 31 (2) (2016) 156–165.

[24] K.-C. Ke, M.-S. Huang, Quality prediction for injection molding by using a multilayer perceptron neural network, Polymers 12 (8) (2020) 1812.

[25] P. Zhao, H. Zhou, Y. He, K. Cai, J. Fu, A nondestructive online method for monitoring the injection molding process by collecting and analyzing machine running data, The International Journal of Advanced Manufacturing Technology 72 (5) (2014) 765–777.

[26] B. Silva, J. Sousa, G. Alenya, Data acquisition and monitoring system for legacy injection machines, in: 2021 IEEE International Conference on Computational Intelligence and Virtual Environments for Measurement Systems and Applications (CIVEMSA), IEEE, 2021, pp. 1–6.

[27] A. Martins, B.M. Lopes e Silva, H. Costelha, C. Neves, J. Lyons, J. Cosgrove, An approach to integrating manufacturing data from legacy injection moulding machines using OPC UA, in: 37th International Manufacturing Conference (IMC37), 2021.

[28] EUROMAP, Euromap 83 – OPC UA for Plastics and Rubber Machinery – General Type Definitions, June 2021.

[29] EUROMAP, Euromap 77 – OPC UA interfaces for plastics and rubber machinery – Data exchange between injection moulding machines and MES, June 2020.

[30] X. Tang, Y. Li, J. Guo, Z. Xie, Final quality prediction for multi-phase batch process based on phase cumulative product quality model, Transactions of the Institute of Measurement and Control 36 (5) (2014) 696–708.

[31] W.-C. Chen, P.-H. Tai, M.-W. Wang, W.-J. Deng, C.-T. Chen, A neural network-based approach for dynamic quality prediction in a plastic injection molding process, Expert Systems with Applications 35 (3) (2008) 843–849.

[32] W.-C. Chen, D. Kurniawan, Process parameters optimization for multiple quality characteristics in plastic injection molding using Taguchi method, BPNN, GA, and hybrid PSO-GA, IJPEM 15 (8) (2014) 1583–1593.

[33] J. Wan, O. Marjanovic, B. Lennox, Uneven batch data alignment with application to the control of batch end-product quality, ISA Transactions 53 (2) (2014) 584–590.

[34] W. Michaeli, J. Gruber, Prozessführung beim Spritzgießen – direkte Regelung des Werkzeuginnendrucks steigert die Reproduzierbarkeit, Zeitschrift Kunststofftechnik 6 (2005) 1–12.

[35] Y. Yao, F. Gao, A survey on multistage/multiphase statistical modeling methods for batch processes, Annual Reviews in Control 33 (2) (2009) 172–183.

[36] F.A. Gers, J. Schmidhuber, F. Cummins, Learning to forget: continual prediction with LSTM, Neural Computation 12 (10) (2000) 2451–2471.

[37] A. Rehmer, A. Kroll, The effect of the forget gate on bifurcation boundaries and dynamics in recurrent neural networks and its implications for gradient-based optimization, in: Preprints of the International Joint Conference on Neural Networks (IJCNN 2022), Padua, Italy, 18.–23. July, 2022.

[38] J. Heinisch, Y. Lockner, C. Hopmann, Comparison of design of experiment methods for modeling injection molding experiments using artificial neural networks, Journal of Manufacturing Processes 61 (2021) 357–368.

[39] J.A.E. Andersson, J. Gillis, G. Horn, J.B. Rawlings, M. Diehl, CasADi – a software framework for nonlinear optimization and optimal control, Mathematical Programming Computation 11 (2018) 1–36.

[40] A. Wächter, L.T. Biegler, On the implementation of an interior-point filter line-search algorithm for large-scale nonlinear programming, Mathematical Programming 106 (1) (2006) 25–57.

Part II

Motion control and autonomous robots as a challenge in Industry 4.0 process

Paolo Mercorelli[a], Hamidreza Nemati[b], and Quanmin Zhu[b]

[a]*Institute for Production Technology and Systems, Leuphana University of Lueneburg, Lueneburg, Germany,* [b]*Department of Engineering Design and Mathematics, University of the West of England, Bristol, United Kingdom*

Retina Implant, a government-sponsored German company, is developing a retinal prosthesis that may allow blind people to gain partial vision. It involves implanting a small microchip inside the patient's eye. Photocells absorb light falling on the eye and transform it into electrical energy, which stimulates the intact nerve cells in the retina. The nervous impulses from these cells stimulate the brain, where they are translated into sight. At present the technology allows patients to orientate themselves in space, identify letters, and even recognise faces . . .

—Yuval Noah Harari
Sapiens: A Brief History of Humankind

Introduction and challenges with autonomous robots and motion control

With the rapid development of autonomous mobile robots and driverless vehicles, a pioneering idea suggested in [1], known as Simultaneous Localization And Mapping (SLAM), has become a hot topic in the research field recently. Due to the limitations of the indoor operating environment, GPS cannot be used to restrict positioning errors, and SLAM opens another door for the development of the indoor robot positioning and guidance. Among different technologies, LiDAR-based SLAM has already become a relatively mature scheme, but the cost-effective problem is still prominent. As a practical solution, the low-cost visual SLAM (VSLAM) has become a research hot spot in recent years. In fact, sensor selection is highly dependent on computational cost, measuring range, environment (including complex geographical features and obstacles), cost-effectiveness, and accuracy [2]. Improving the autonomous navigation capacity of mobile robots yields concrete economic advantages for the firms that use them. However, developing a self-sufficient navigation system from scratch is no simple feat. Savioke CEO Steve Cousins recently told The Robot Report that the company would rather devote resources to regions where it has a competitive edge than to those where it does not. I think what Brain Corp. is doing has the potential to completely change the cost landscape. They have the purchasing power to negotiate lower per-sensor costs, and they design everything to work together smoothly. With the help of SLAM technology, the mobile robot can perceive the surrounding environment according to its own sensors during the movement so as to perform autonomous positioning and incrementally construct an environmental map [4,5].

Many researchers and engineers in the field of robotics remain intent on creating their very own versions of autonomous navigation. That is something that Brain Corp. gets. Brain Corp's Director of Innovation Paul Behnke and Vice President of Innovation Phil Duffy (who spoke at The Robot Report's first Robotics Summit & Showcase event) presented a webinar titled Robotics: The Decathlon of Startups to educate robotics developers on the difficulties of creating an autonomous navigation system. This part focuses on different aspects on motion and control of autonomous robots starting from different problems on SLAM and also with advanced problems on control.

Getting the software right

Behnke said, "That's the first issue that comes to mind when you think about designing robots for different kinds of surroundings." Autonomous robots will need complex algorithms to successfully navigate busy public locations like airports, shopping centers, warehouses, and more. The close quarters and ever-shifting impediments of such places make planning out the best possible pathways difficult. The difficulty is in creating software that addresses these concerns from the perspective of the end user.

Gathering enough real-world data

Physical obstructions in a robot working area are just one of several factors that might hinder its ability to navigate autonomously. Autonomous navigation is made more challenging by locations with few or no landmarks, as well as by factors such as time of day. For instance, in [3], visual

SLAM dataset including various sequences has been acquired to visualize the impact of a sensor on the precision of SLAM algorithms with different modalities such as RGB, IR, and depth images in both passive and active stereo modes. A lot of these problems are edge situations that do not become apparent until after the software is written and the robots are tested in the real world. Cases on the edge are the sucker punches you never saw coming [6].

Creating precision motion control

According to Duffy, developing a system with precise and accurate motion control is the true key to producing a robot with autonomous navigation. In many cases, transporting a robot simply involves moving it from one location to another. Industrial floor care machines were Brain Corporation's first product to hit the market, and they had to go as near to a wall or other impediment as possible to clean the most surface area.

Reducing false positives

The ability to identify and interact with humans is critical for developing new services for consumers. Your company is doomed before it even begins if your robot cannot distinguish between a human and a package on the floor. Having a system that can distinguish people differently from barriers is vital whether you are designing your algorithms or seeking for navigation solutions to employ in your robotics project. Solving the human factor is essential unless you want to remove all humans from the robot working environment.

Installation must be simple

Nontechnical workers just do not have the necessary expertise to complete the procedure. This comprehensive and technically complicated launch has the potential to stall this crucial early stage; sending an engineer to each new customer's location is neither scalable nor feasible. Customer identification of personnel capable of taking on installation as a new project is one approach to overcoming this obstacle. Preventive design maintenance is another option. Design your robot to seem like other things your customer's workers are already acquainted with. The interface must be both simple and easy to use.

References

[1] R. Smith, M. Self, P. Cheeseman, Estimating uncertain spatial relationships in robotics, Machine Intelligence and Pattern Recognition 5 (1988) 435–461.

[2] M.S.A. Khan, D. Hussain, K. Naveed, U.S. Khan, I.Q. Mundial, A.B. Aqeel, Investigation of widely used SLAM sensors using analytical hierarchy process, Journal of Sensors 2022 (2022) 1–15.

[3] I.E. Bouazzaoui, S. Rodriguez, B. Vincke, A.E. Ouardi, Indoor visual SLAM dataset with various acquisition modalities, Data in Brief 39 (2021) 1–11.

[4] W.D. Chen, F. Zhang, Review on the achievements in simultaneous localization and map building for mobile robot, Control Theory and Applications 22 (3) (2005) 455–457.

[5] M. Csorba, Simultaneous Localization and Mapping, University of Oxford, 1997, pp. 56–89.

[6] B. Liu, P. De Giovanni, Green process innovation through Industry 4.0 technologies and supply chain coordination, Annals of Operations Research (2019) 1–36.

Chapter 5

SLAM algorithms for autonomous mobile robots

Sufang Wang[a] and Weicun Zhang[b]

[a]*Second Academy of China Aerospace Science and Industry Corporation, Institute 706, Beijing, China,* [b]*University of Science and Technology Beijing, School of Automation and Electrical Engineering, Beijing, China*

5.1 Introduction

For autonomous mobile robots, there are three aspects involved in navigation and positioning [1,5]:

1. where am I?
2. where am I going?
3. How should I get there?

These three issues correspond to three technologies respectively:

1. Positioning, to get the current position of a robot;
2. Perception, to get the target location for the robot;
3. Path planning, to decide the optimal trajectory for the robot to move to the target location.

With the help of SLAM technology, the mobile robot can perceive the surrounding environment according to its own sensors during the movement so as to perform autonomous positioning and incrementally construct an environmental map [2,3].

Although the research on autonomous mobile robots has become a hot topic in the field of high-tech field, we have seen and applied very little in our daily life. Most of them are based on theoretical research and laboratory experiments. According to published papers, most of them are based on a single sensor, such as liDAR SLAM and visual SLAM. SLAM using liDAR has mature algorithms and solutions. With the follow-up of hardware devices, visual SLAM (VSLAM) has also been developed. To meet the requirements as much as possible, different products adopt different sensors.

The SLAM technology of mobile robots has very important theoretical significance and application values. For driverless vehicles, SLAM helps to build 3D environment models and position navigation by liDAR. In the military field, SLAM facilitates mobile robots to reach many harsh environments that humans cannot reach. It is also essential to realize intelligent reconnaissance and combat of robots. SLAM can also be used to locate and remove hazardous explosives [4]. In our daily life, SLAM can be used to support household robots to walk autonomously and successfully avoid obstacles to complete high-quality tasks.

In this chapter, we try to review related algorithms and technologies widely used in SLAM research, especially in VSLAM research, including the classification of SLAM, the Kalman filtering algorithms in SLAM and VSLAM, and comparisons of different kinds of SLAM systems and algorithms.

5.2   SLAM classification

Intelligent robots are gradually entering our daily life. Their make our daily life easier and more convenient. However, simple robots cannot walk autonomously; to address such an issue, the auxiliary role of the liDAR is needed to realize the robot intelligence. It can acquire the information of the robot environment in real time. However, a robot itself cannot understand the information scanned by liDAR. Thus the powerful SLAM navigation algorithm is very important for the robot to achieve an intelligent walking. Widely used liDAR products, such as SICK liDAR (Sick AG, based in Waldkirch, Germany, is a global manufacturer of sensors and sensor solutions for industrial applications), Velodyne liDAR (Velodyne is a US-based company focused on R&D and production of liDAR sensors for autonomous driving) and SLAMTEC RPliDAR (SLAMTEC company, based in ShangHai, China, was founded in 2013 and focused on the development and application of robot autonomous localization and navigation solution as well as related core sensors), have been validated successfully in many applications.

As a practical solution, the low-cost visual SLAM (VSLAM) has become a research hot spot in recent years. VSLAM is usually implemented together with cameras. According to different working approaches, VSLAM is classified into three types: Monocular, Stereo Vision, and RGB-D, respectively. Here by RGB-D we mean Red, Green, Blue plus Depth. Red, green, and blue can be combined in various proportions to obtain any color in the visible spectrum.

Using only one camera in VSLAM is called monocular SLAM, whereas using multiple cameras as sensors is called stereo VSLAM, which is the most widely used VSLAM. The combination of a monocular camera and an infrared sensor is RGB-D SLAM.

5.3   liDAR-SLAM

The key issue in SLAM is positioning. The methods to solve positioning problems are divided into two categories, probabilistic and non-probabilistic methods. The probability-based method is the mainstream of positioning technology in liDAR-SLAM, and the Bayesian estimation is its basis. There are mainly two algorithms of the probability-based method, the Kalman filter (KF) and particle filter (PF) algorithms. Next, we will introduce these algorithms with focus on the multiple model Kalman filtering, also known as multiple model adaptive estimation (MMAE).

5.3.1   Kalman Filter

Considering that there are system and measurement noises with Gaussian distributions, we need to process sensor data in SLAM system according to some kinds of filtering algorithms. Among others, the Kalman Filter (KF) provides the optimal estimates of the system states in SLAM systems [6–9]. The premise of using KF is that the system under consideration is a linear system with Gaussian noises. However, the actual system is often described by nonlinear state space equations. Thus the standard KF is not suitable. The extended KF (EKF) linearizes the nonlinear system through first-order Taylor expansion [10]. The unscented KF (UKF) is a method of approximating nonlinear distribution using a sampling strategy [11]. It does not need to calculate the Jacobian matrix and has a higher linearization accuracy, and its performance is better than that of EKF.

KF-SLAM occupies a dominant position in many solutions, because it has some advantages in convergence and affordable implementation complexity. But it has the problem of lack of self-closed loop capability and associated fragility. When an error occurs in the data association, it will eventually be brought into the entire SLAM state estimation, sometimes even causing the entire prediction process to diverge. Therefore a robust data association method is very important.

Practical implementation of the Kalman filter is often difficult due to the uncertainties and nonlinearities of the modeling of dynamic systems. Extensive research has been done to address the modeling uncertainty and nonlinearity problems in state estimation, filtering, and control. Among others, the multiple model Kalman filter, also known as the MMAE scheme has received much attention. The multiple model concept coincides with the famous engineering logic of "divide and conquer". The thought of multiple models for adaptive estimation originated from Magill [42]. Later on, many scholars such as Lainiotis [43], Athans et al. [44,45], Anderson and Moore [46], and Li and Bar-Shalom [47] studied MMAE for different purposes of applications.

There are mainly three aspects in designing an MMAE system: first, to select a "local" model set to include the parameter uncertainties or nonlinearity of the system; second, to design "local" KF set corresponding to the "local" model set, in which each "local" KF is designed based on each "local" model; third, to design or select a weighting algorithm to calculate weights for each "local" KF. After that, each "local" KF generates its own state estimates and a corresponding output error (residual) to feed the weighting algorithm. The "global" MMAE state estimate is then a weighted sum of all "local" KF's estimates. We use the block diagram in Fig. 5.1 to describe the principle of an MMAE system. In Fig. 5.1, $x(k) \in \mathbb{R}^n$ is the state of the system, $u(k) \in \mathbb{R}^m$ is the control input, $y(k) \in \mathbb{R}^q$ is the system output, $\omega(k) \in \mathbb{R}^q$ is the measurement

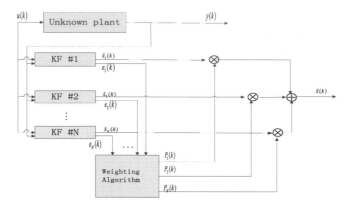

FIGURE 5.1 Block diagram of MMAE.

noise that cannot be measured with uncertain covariance $E[\omega(k)][\omega(k)]^T = R(k) = \{R_1, \dots, R_N\}$, and $v(k) \in \mathbb{R}^n$ is the system noise that cannot be measured with uncertain covariance $E[v(k)][v(k)]^T = Q(k) = \{Q_1, \dots, Q_N\}$.

Here we introduce two improved weighting algorithms to simplify the classical weighting algorithm, which was developed from a dynamic hypothesis test and Bayesian formula. With the improved weighting algorithm, the convergence conditions have been relaxed than those of the classical weighting algorithm.

Consider an uncertain discrete-time system $P(\vartheta)$ described by the following state-space equation:

$$\begin{aligned} x(k+1) &= A(k, \vartheta)x(k) + B(k, \vartheta)u(k) + v(k), \\ y(k) &= C(k, \vartheta)x(k) + \omega(k). \end{aligned} \tag{5.1}$$

The matrices $A(k, \vartheta)$, $B(k, \vartheta)$, $C(k, \vartheta)$, $R(k, \vartheta)$, and $Q(k, \vartheta)$ are assumed to be piecewise continuous, uniformly bounded in time, and containing unknown constant parameters denoted by a vector $\vartheta \in \mathbb{R}^l$. The initial condition $x(0)$ is assumed to be deterministic but unknown. Consider a finite set of candidate parameter values $\Theta := \vartheta_1, \vartheta_2, \dots, \vartheta_N$ indexed by $i \in 1, \dots, N$. The MMAE algorithm can be described as follows:

$$\hat{x}(k) = \sum_{i=1}^{N} p_i(k)\hat{x}_i(k), \tag{5.2}$$

where $\hat{x}(k)$ is the estimate of the state $x(k)$ at time k, and $p_i(k)$, $i = 1, \dots, N$ are time-varying weights generated by the weighting algorithm, which will be given in the next section. In (5.2), each "local" state estimate $\hat{x}_i(k)$, $i = 1, \dots, N$

is generated by the corresponding "local" KF described as follows.

$$\hat{x}_i(k|k-1) = A_i \hat{x}(k-1) + B_i u(k-1),$$
$$P_i(k|k-1) = A_i P_i(k-1)A_i^T + Q_i,$$
$$K_i(k) = P_i(k|k-1)C_i^T (C_i P_i(k|k-1)C_i^T + R_i)^{-1}, \qquad (5.3)$$
$$\hat{x}_i(k|k) = \hat{x}_i(k|k-1) + K_i(k)(y(k) - C_i \hat{x}_i(k|k-1)),$$
$$P(k) = (I - K_i(k)C_i)P(k|k-1),$$

where $A_i := A(k, \vartheta_i)$, $B_i := B(k, \vartheta_i)$, $C_i := C(k, \vartheta_i)$, $Q_i := Q(k, \vartheta_i)$, $R_i := R(k, \vartheta_i)$, and $\hat{x}_i(k/k)$ is $\hat{x}_i(k)$ in Eq. (5.2).

We expect that if the jth model M_j in the model set is (or close to) the real plant model, then the corresponding jth KF will generate the optimal state estimation $\hat{x}_j(k)$. In addition, if the jth weight $p_j(k)$ converge to 1 and others to 0, hen the state estimates of the MMAE will converge to $\hat{x}_j(k)$.

Thus the key problem for an MMAE system is to design an effective weighting algorithm with insurance of correct convergence, as well as an effective model set that includes the real model or the model closest to the system.

5.3.1.1 Weighting algorithm

First of all, we need to define the residual/error signal of each "local" KF:

$$r_i(k) = y(k) - y_i(k) = y(k) - C_i \hat{x}_i(k). \qquad (5.4)$$

The classical weighting algorithm can be described as follows:

$$\begin{cases} p_i(k+1) = \dfrac{\beta_i(e^{-\frac{1}{2}r_i'(k+1)S_i^{-1}r_i(k+1)})}{\sum_{j=1}^{N} \beta_j(e^{-\frac{1}{2}r_j'(k+1)S_j^{-1}r_j(k+1)})p_j(k)} p_i(k), \\ p_i(0) = \frac{1}{N}, i = 1, \ldots, N, \end{cases} \qquad (5.5)$$

where $r_i(k), i = 1, \ldots, N$ is the residual of the ith Kalman filter, $S_i, i = 1, \ldots, N$, is the steady-state constant residual covariance matrix of $r_i(k)$, $\beta_i = \frac{1}{(2\pi)^{m/2}\sqrt{(det(S_i))}}$ is a constant scaling factor, and m is the number of measurements. For more details of design and convergence analysis of the classical weighting algorithm, see [48,49].

To further relax the convergence condition of the classical weighting algorithm, two improved weighting algorithms have been put forward by Zhang [50,51] to replace the classical weighting algorithm.

Algorithm 1:

$$l_i(0) = \frac{1}{N}; \ p_i(0) = l_i(0), \qquad (5.6)$$

$$l_i'(k) = 1 + \frac{1}{k} \sum_{q=1}^{k} \|r_i(q)\|^2, \tag{5.7}$$

$$l_{\min}(k) = \min_i l_i'(k), \tag{5.8}$$

$$l_i(k) = l_i(k-1) \frac{l_{\min}(k)}{l_i'(k)}, \tag{5.9}$$

$$p_i(k) = \frac{l_i(k)}{\sum_{i=1}^{N} l_i(k)}, \tag{5.10}$$

where $\| \cdot \|$ denotes the Euclidean norm.

According to [50,51], we have the following convergence result of weighting algorithm (5.6)–(5.10). For more details of the proof of the theorem, see Lemma 1 in Appendix.

Theorem 3. *Let $M_j \in \mathbb{M}$ be the model closest to the true plant in the following sense with probability one:*

$$\begin{cases} \sum_{q=1}^{k} \|r_j(q)\|^2 < \sum_{q=1}^{k} \|r_i(q)\|^2 \ \forall k \geq 1, \\ \frac{1}{k} \sum_{q=1}^{k} \|r_j(q)\|^2 \to \sigma_j, \\ \frac{1}{k} \sum_{q=1}^{k} \|r_i(q)\|^2 \to \sigma_i, \\ \sigma_j < \sigma_i, \ i \neq j, \end{cases} \tag{5.11}$$

where σ_j is a constant, and σ_i may be constant or infinity.
Then the weighting algorithm (5.6)–(5.10) leads to

$$p_j(k) \to 1; \ p_i(k) \to 0, \ i = 1, \dots, N, \ i \neq j. \tag{5.12}$$

It is worth pointing out that the convergence condition for the weighting algorithm (5.6)–(5.10) is weaker than that for the classical weighting algorithm. To be specific, the convergence condition (5.11) means the discriminability of $r_i(k)$, whereas the convergence conditions for classical weighting algorithm include ergodicity, stationarity, and discriminability of $r_i(k)$; for more detail, we refer to [48].

To get a sharper convergence rate, we may use the following weighting algorithm.

Algorithm 2:

$$l_i(0) = \frac{1}{m}; \ p_i(0) = l_i(0), \tag{5.13}$$

$$l_i'(k) = 1 + \frac{1}{k} \sum_{q=1}^{k} \|r_i(q)\|^2, \tag{5.14}$$

$$l_{\min}(k) = \min_i l_i'(k), \tag{5.15}$$

$$\beta_i(k) = \frac{l_{\min}(k)}{l_i'(k)}, \tag{5.16}$$

$$l_i(k) = \begin{cases} l_i(k-1) & \text{if } \beta_i(k) = 1, \\ l_i(k-1)[\beta(k)]^{\text{ceil}(\frac{1}{1-\beta(k)})} & \text{if } \beta_i(k) < 1, \end{cases} \tag{5.17}$$

$$p_i(k) = \frac{l_i(k)}{\sum_{i=1}^m l_i(k)}, \tag{5.18}$$

where $\text{ceil}(x)$ is the ceiling function that generates the smallest integer not less than x, i.e.,

$$\text{ceil}(x) = \min\{n \in \mathbb{Z} | x \leq n\}.$$

According to [51], we have the following convergence result of weighting algorithm 2. For more details of the proof of the theorem, see Lemma 2 in Appendix.

Theorem 4. *Let $M_j \in \mathbb{M}$ be the model closest to the true plant in the following sense with probability one:*

$$\sum_{q=1}^k \|r_j(q)\|^2 < \sum_{q=1}^k \|r_i(q)\|^2 \; \forall k \geq 1, \; i \neq j. \tag{5.19}$$

Then the weighting algorithm (5.13)–(5.18) leads to

$$p_j(k) \to 1; \; p_i(k) \to 0, \; i = 1, \dots, N, \; i \neq j. \tag{5.20}$$

Remarks. These two algorithms, i.e., Eqs. (5.6)–(5.10) and Eqs. (5.13)–(5.18), both can be used in MMAE; it should be chosen according to specific engineering conditions, such as software and hardware configurations.

We further have the following results on the multiple model Kalman filters.

Let us consider the situation that the model set $\mathbb{M}$ includes the unique real model of system (5.1). Other complicated situations will be considered in the future research work. We have the following results on the convergence of the proposed MMAE system.

Theorem 5. *Let the following conditions be satisfied:*

1. *$M_j \in \mathbb{M}$ is the only real model of system (5.1) in the following sense with probability one:*

$$\begin{cases} \sum_{q=1}^k \|r_j(q)\|^2 < \sum_{q=1}^k \|r_i(q)\|^2 \; \forall k \geq 1, \\ \frac{1}{k} \sum_{q=1}^k \|r_j(q)\|^2 \to \sigma_j, \\ \frac{1}{k} \sum_{q=1}^k \|r_i(q)\|^2 \to \sigma_i, \\ \sigma_j < \sigma_i, \; i \neq j, \end{cases} \tag{5.21}$$

where σ_j is a constant, and σ_i may be a constant or infinity.

2. *Each Kalman filter is designed to ensure the stability, i.e., the state estimates of each Kalman filter are bounded.*

Then the state estimates of MMAE with weighting algorithm (5.6)–(5.10) converge to the optimal estimates given by the jth KF corresponding to M_j, i.e.,

$$\hat{x}(k) \to \hat{x}_j(k). \tag{5.22}$$

Proof. According to Theorem 3, condition 1 of Theorem 5, i.e., Eq. (5.21), leads to

$$p_j(k) \to 1; \ p_i(k) \to 0, \ i = 1, \ldots, N, \ i \neq j. \tag{5.23}$$

Further, condition 2 of Theorem 5 guarantees that

$$\hat{x}_i(k) < \infty, \ i = 1, \ldots, N. \tag{5.24}$$

Then by (5.23) and (5.24) we have

$$
\begin{aligned}
\hat{x}(k) &= \sum_{i=1}^{N} p_i(k)\hat{x}_i(k) \\
&= p_j(k)\hat{x}_j(k) + \sum_{i=1,i\neq j}^{N} p_i(k)\hat{x}_i(k) \\
&\to 1 \cdot \hat{x}_j(k) + \sum_{i=1,i\neq j}^{N} 0 \cdot \hat{x}_i(k) \\
&\to \hat{x}_j(k),
\end{aligned}
\tag{5.25}
$$

which completes the proof. $\square$

Theorem 6. *Let the following conditions be satisfied:*

1. *$M_j \in \mathbb{M}$ is the only real model of system (5.1) in the following sense with probability one:*

$$\sum_{q=1}^{k} \|r_j(q)\|^2 < \sum_{q=1}^{k} \|r_i(q)\|^2, \ \forall k \geq 1, \ i \neq j. \tag{5.26}$$

2. *Each Kalman filter is designed to ensure the stability, i.e., the state estimates of each Kalman filter are bounded.*

Then the state estimates of MMAE with weighting algorithm (5.13)–(5.18) converge to the optimal estimates given by the jth KF corresponding to M_j, i.e.,

$$\hat{x}(k) \to \hat{x}_j(k). \tag{5.27}$$

Proof. According to Theorem 4, condition 1 of Theorem 6, i.e., Eq. (5.26), leads to

$$p_j(k) \to 1; \ p_i(k) \to 0, i = 1, \dots N, i \neq j \tag{5.28}$$

Further, condition 2 of Theorem 6 guarantees that

$$\hat{x}_i(k) < \infty, \ i = 1, \dots, N. \tag{5.29}$$

Then by (5.28) and (5.29) we have

$$
\begin{aligned}
\hat{x}(k) &= \sum_{i=1}^{N} p_i(k)\hat{x}_i(k) \\
&= p_j(k)\hat{x}_j(k) + \sum_{i=1,i\neq j}^{N} p_i(k)\hat{x}_i(k) \\
&\to 1 \cdot \hat{x}_j(k) + \sum_{i-1,i\neq j}^{N} 0 \cdot \hat{x}_i(k) \\
&\to \hat{x}_j(k),
\end{aligned}
\tag{5.30}
$$

which completes the proof. ☐

5.3.2 Particle filter

As a new type of filter, the particle filter gets rid of the assumption of the system linearity and the constraints of the Gaussian assumptions on the system and measurement noises. The particle filter can approximate any probability distribution, and the calculation is simple and convenient. It can also effectively solve the robot positioning problem [12]. Murphy et al. [13,14] found that if a robot motion trajectory is known, then the probability between landmark positions is conditionally independent. Therefore the Rao–Blackwellized decomposition is proposed and implemented, which provides a theoretical basis for the particle filter to solve the SLAM problem. Based on this, Montemerlo et al. [15] demonstrated the feasibility of using Rao–Blackwellized Particle Filter (RBPF) to solve the SLAM problem and proposed the FastSLAM algorithm.

The FastSLAM uses an improved particle filter to estimate the posterior distribution of the robot path. In Fig. 5.2 [16], each particle maintains a state estimate and a set of individual feature location information. Each particle represents a path traveled by a robot. Each feature is individually estimated using an EKF. Each particle maintains a number of *M* EKFs, and there are a total of *N* particles to predict the robot state [17]. Compared with other methods, the Fast-SLAM reduces the sampling space, thus greatly reducing the complexity and improving the calculation speed; it has a high precision and better robustness; it can be applied to a non-Gaussian, nonlinear, and unknown posterior density

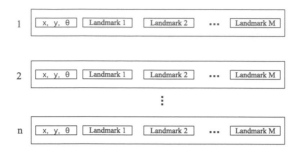

FIGURE 5.2 RBPF-SLAM algorithm.

function. However, it was found in the experiments that it requires more particles to avoid estimation divergence and that divergence is closely related to data association. In addition, as a new algorithm for solving SLAM problem, particle filters also have many space for improvement and optimization.

5.3.3 Graph-based optimization

Lu and Milios [41] first proposed graph-based optimization. In their paper, the two-dimensional map created by liDAR is taken as an example, and the influence of all frames is taken into consideration to form a spatial constraint relationship. Compared with the filter method that only considers the influence of the previous frame, if at a random moment between x_0 and x_1 an error occurs, then the later posture is difficult to correct. However, as the surroundings size increases, the posture error rate increases. Therefore the filter method is not suitable to estimate posture in big surroundings. But graph-based optimization can optimize all frames to calculate correct postures. Graph-based optimization estimates the trajectory of the entire robot and continuously optimizes the linear points to minimize the errors. Since the precision after optimizing multiple times is significantly higher than the filter method, it is suitable for applying to estimate posture in big scenario, but the efficiency will be reduced accordingly. Graph-based optimization generally optimizes graphs constructed by SLAM front-end estimation and loop closure detection. So graph-based optimization is also called back-end optimization. Considering that most application situations of robots are nonlinear systems, posture graph can be solved by nonlinear least squares algorithm. Basic idea of nonlinear least squares is that given a nonlinear system with nonlinear state equations, we need to find an optimal solution to minimize the error of estimates (observations). Also, to find the target function of nonlinear least squares, we assume that the error obeys Gaussian distribution.

5.4 Visual SLAM algorithm

Visual SLAM is mainly divided into visual front-end and optimized back-end. The front-end is also called visual odometry (VO). It estimates rough camera movement based on the information of adjacent images and provides a good initial value for the back-end. The implementation method of VO is divided into the feature point method and the direct method according to whether or not features need to be extracted. The feature-based method is stable in operation and insensitive to light and dynamic objects. So it is considered as the mainstream method of VO [18].

5.4.1 Feature-based method

Davision [19] first proposed a monocular visual SLAM (MonoSLAM) system that uses EKF as the back-end to track sparse feature points on the front-end. After that, the keyframe-based monocular visual SLAM gradually developed [20–23]. The most representative of these is parallel tracking and mapping (PTAM). Klein et al. [21] proposed a simple and effective method for extracting key frames, parallelizing the tracking and mapping, and for the first time, PTAM used nonlinear optimization as the back-end. Mur-Artal et al. [23] inherited and improved PTAM and innovatively proposed three threads to implement the monocular visual SLAM system based on PTAM of the dual thread. The entire system is implemented around the ORB feature ORB-SLAM. Subsequent studies have shown that the ORB-SLAM system is equally applicable to monocular, stereo, and RGB-D models and has good general-purpose use. The entire SLAM processes can be divided into three threads: tracking thread (real-time tracking feature point), optimizing thread (for partial bundle adjustment), and loop closure detecting and optimizing thread of global pose graph. Also, applying identical ORB feature to the three threads makes them interact with each other, which diminishes the accumulated errors when constructing the graph and guarantees the global consistence between the motion trajectory and map. With the above-mentioned improvements, ORB-SLAM2 system becomes applicable to monocular, binocular, and RGB-D modes and has good versatility. ORB-SLAM2 algorithm framework is shown in Fig. 5.3.

ORB-SLAM2 system is built with four parallel threads: tracking, local mapping, loop closure detecting, and global bundle adjustment (BA) optimization. Among them, the fourth thread executes after having the confirmation of the previous thread. We give the specific explanation for the four threads as follows:

1. tracking: pre-processes the input to get features at the location of significant key point.
2. local mapping: executes partial BA algorithm to achieve local map regulation and optimization.
3. loop closure detecting: this thread is divided into two steps: first, cycle inspection and verification; and second, cyclical correction and optimization of the posture map. Compared with monocular ORB-SLAM camera, which may result in a scale drift, stereo/depth information makes the scale observable. Geometry verification and target optimization are no longer needed while system deals with scale drift issues. Also, based on rigid body transfor-

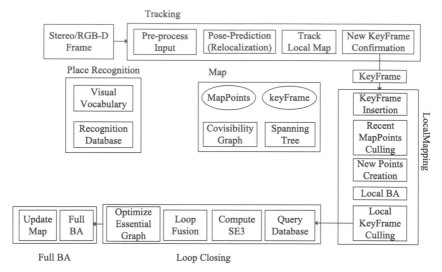

FIGURE 5.3   ORB-SLAM2 algorithm framework.

mation instead of similarity, the global BA optimization of the fourth thread is executed after the posture graph.

4. global BA: after optimizing the posture graphs, the optimized system structure and motion result can be calculated by executing global BA. This system embeds a DBoM2-based position recognition model used for relocation, which can effectively avoid tracking failures, like occlusion. This system applies scenes graph re-initialization and loop closure detection, etc. This system applies ORB feature to localizing and constructing map; it also has good scale invariance and ideal rotation invariance. Besides, it can promptly extract features to match, so that it can meet the requirements of real-time operation.

Because a monocular camera captures a two-dimensional projection of a three-dimensional space object, which is a single image, and there is uncertainty in the motion and trajectory estimated by moving the camera, the true depth of the object cannot be determined. Therefore there are enormous difficulties in the three-dimensional reconstruction work. At this time, stereo cameras and RGB-D cameras appear, but stereo cameras have a major problem in terms of calculations under the existing conditions. However, SLAM based on RGB-D data simplifies the complexity of 3D reconstruction. Henry et al. [24] first proposed a method of 3D reconstruction of indoor environment using RGB-D camera, extracting the SIFT features in the color image and finding the corresponding depth information on the depth image. Then Random Sample Consensus (RANSAC) method is used to match the 3D feature points, and the corresponding rigid motion transformation is calculated. RANSAC is applied to many cases with incorrect data, which can handle data with error matching and use it as the initial value of the iterative closest point (ICP) to find more accurate position and pose. RGB-D SLAM usually uses the ICP algorithm for pose estimation and camera motion transfer matrix optimization. The basic concept of ICP algorithm is to minimize the registration error function between the paired points in two point sets P_1 and P_2. The registration error function is defined as

$$e(R, t) = \min_{R,t} \frac{1}{N} \sum_{i=1}^{N} \|v_{p1}(i) - R v_{p2}(i) - t\|^2,$$

where R is the rotation matrix, t is the translation vector, N is the number of point pairs, and $v_{p1}(i)$ and $v_{p2}(i)$ are the coordinates vectors of the ith point pair. To be specific, we have the following expressions of R and t:

$$R = \begin{bmatrix} 1 & 0 & 0 \\ 0 & \cos\alpha & \sin\alpha \\ 0 & -\sin\alpha & \cos\alpha \end{bmatrix} \begin{bmatrix} \cos\beta & 0 & -\sin\beta \\ 0 & 1 & 0 \\ \sin\beta & 0 & \cos\beta \end{bmatrix} \begin{bmatrix} \cos\gamma & \sin\gamma & 0 \\ -\sin\gamma & \cos\gamma & 0 \\ 0 & 0 & 1 \end{bmatrix}$$

and

$$t = \begin{bmatrix} t_x & t_y & t_z \end{bmatrix}^T,$$

where α, β, γ are the rotation angles around the coordinates axes x, y, z, respectively.

The procedures of ICP algorithm in RGB-D SLAM is as follows [25].

1. Read in the two point sets P_1, P_2;
2. Select the pair of points. Search P_2 for the point closest to P_1 to form a point pair. Find out all the pairs of points in the two point sets;
3. Calculate two barycentric coordinates based on the pairs of points in the two point sets;
4. From the new point set calculate the rotation matrix R and the translation vector t;
5. From the obtained R and t calculate the new point set P_2' after the rigid transformation of the point set P_2;
6. If the absolute value of the difference between the sum of squares of the distances from P_2 to P_2' is less than the threshold value for two consecutive times, then it is converged and the iterations stop. Otherwise, steps 1–6 are repeated until the algorithm convergence is achieved.

Among many matching algorithms for depth images, such as Generic Algorithm, RANSAC, and ICP, the latter is the most widely used. With ICP algorithm, three-dimensional data will be processed directly to depth images and do not need to assume and distribute object features. After selecting the proper initial value, the ICP algorithm has good convergence performance, thus getting the global optimal value and obtaining more accurate matching results. So it quickly becomes a mainstream algorithm for depth image matching [26]. Although feature-based SLAM method has many advantages, the extraction of key points and the calculation of feature points are very time-consuming, and some useful image information may be discarded because only feature points are used, such as a white wall, because there are few feature points. In such a situation, it is difficult to accurately calculate the motion of the camera.

5.4.2 Direct method

According to the classification of the number of pixels used, the direct method can be divided into sparse, dense, and semi-dense. Assume that P is a spatial point of a known location. When P is derived from a sparse key point, it is called the sparse direct method. When P comes from a portion of pixels, it is called the semi-dense direct method. If all pixels are used, then it is called the dense direct method, which can build a complete map. Many people have been devoted to the study of the direct method [27–33]. Irani and Anandan [27] gave a detailed and in-depth description of the direct method. Silveira et al. [28] applied the direct method to the visual SLAM and described main advantages and limitations. Subsequently, a sparse-based semi-direct monocular visual odometry (SVO) was proposed [29]. This algorithm has high accuracy and good robustness and is faster than the most advanced methods currently available. This sparse method eliminates the needs of feature extraction and robust matching techniques for motion estimation and is suitable for estimating the state of microcars in GPS denied environments. Next, Engel et al. [30] proposed a large-scale direct monocular SLAM (LSD-SLAM); LSD-SLAM algorithm framework is shown in Fig. 5.4. Compared to other direct methods, it reconstructs the keyframe's pose map and the environment's semi-dense and highly accurate three-dimensional map in real time. Usenko et al. [33] proposed a novel direct visual-inertial odometry method for stereo cameras that use the complementarity of visual and inertial data to improve the accuracy of 3D reconstruction maps.

Newcombe et al. [34] integrate all the deep data and image information from Kinect into the observation scene and reconstruct the 3D model to get the global map. Particularly, it allows reconstruction of dense maps in real time, making a big step towards augmented reality (AR). Henry et al. [35] adopt a joint optimization algorithm to apply RGB-D cameras to the robot field in indoor environments. Kerl et al. [36] proposed a visual SLAM method based on a direct dense RGB-D camera. The error terms of this method are the photometric and depth errors. The optimal camera pose is solved using the g2o optimization library. Also, an entropy-based method for key frame selection and loop closure detection is proposed. Thus the path error is greatly reduced. Compared to the feature method, the direct method does not need to extract image features. It has a fast speed of execution, high robustness to the photometric error of the image, but a high requirement for the camera internal reference. When there is geometric noise, the algorithm performance decreases quickly. In the event of image motion blur, camera position can still be achieved, but the direct method has poor robustness to large baseline motion.

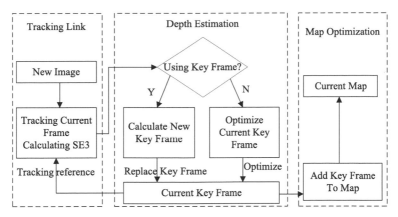

FIGURE 5.4   LSD-SLAM algorithm framework.

5.4.3   Scheme comparison

LiDAR sensors' advantages are a wide visual range and a high precision, which can measure the angle and distance of surrounding obstacles precisely, so that it can help robots to avoid obstacles, but liDAR is relatively expensive. In contrast, camera has some advantages, such as lower cost, lightweight, easy installation, convenience, and flexibility. Also, camera can extract plenty of scenes information; therefore visual SLAM becomes popular for SLAM researchers these years. Visual SLAM and density reconstruction use little CPU. It provides enough capability for application logic and processing of other sensors. By providing the situational awareness of robots or drones the depth camera can help solving SLAM problems better [37].

The performance of visual SLAM depends on the environment in which it is operating. The ideal conditions are summarized as follows:

1. Light is sufficient, without large changes in lighting. The camera must be able to identify features in the scene. Generally, a more complex scene, with lots of objects or geometry, is the best situation. Blank walls, floors, or ceilings are the worst cases. Many reflective surfaces, such as glass or mirrors, can cause problems. Also, direct sunlight can interfere with the depth cameras, which affects the accuracy of occupancy mapping.
2. When the scene is mostly motionless, Visual SLAM works best. If people or objects are moving, then performance will be more or less affected. If the entire scene is moving, like in an elevator, SLAM will not work at all.
3. When the camera motion is primarily translation, not a rotation, Visual SLAM works best. When rotation is necessary, it is best to rotate slowly.
4. When SLAM begins, the camera must be stationary, and there must be sufficient visual features. If the camera is aimed at a blank wall, floor, or ceiling that is enough to block the camera view, SLAM may not initialize properly.

liDAR and camera have their own advantages and disadvantages. It is very easy for the camera to identify the same object, but it is difficult for the liDAR. If the camera tells the liDAR that the two frames are the same object, then it is possible to know what the speed and displacement of this object are between the two frames by the liDAR. It can be seen that recognition and tracking are easy to achieve, resulting in more accurate maps.

5.5 Conclusions

Great progress has been made in SLAM technologies and products after more than thirty years of research and application by predecessors; especially, SLAM based on liDAR has already been a relatively mature scheme in positioning and guidance for indoor robots, but the high cost is still the primary problem to be addressed. Therefore the low-cost visual SLAM has become a research hot spot in recent years. However, no matter which sensor is used alone, there are some defects. The multi-sensor fusion technology based on liDAR, vision sensor, and inertial measurement unit [38,39] not only can realize the cooperative operation among sensors, but also greatly enhance the robustness. It is believed that the research and application of multi-sensor fusion technology will bring wider space to driverless, robotics, augmented reality, and virtual reality. In addition, SLAM is combined with deep learning to perform image processing [40], to generate semantic maps of the environment, and to improve the human–computer interaction techniques, so that intelligence can be better realized.

Appendix

Lemma 1 ([19]). *Consider weighting algorithm (5.6)–(5.10). Suppose M_j is closest in the model set $\mathbb{M} = \{M_i, i = 1, 2, \ldots, N\}$ to the true plant in the following sense with probability one;*

$$\frac{1}{k}\sum_{q=1}^{k} r_j^2(q) < \frac{1}{k}\sum_{q=1}^{k} r_i^2(q) \; \forall k \geq k^*, \; i \neq j, \tag{5.31}$$

$$\lim_{k\to\infty}\frac{1}{k}\sum_{q=1}^{k} r_j^2(q) = \sigma_j, \; \lim_{k\to\infty}\frac{1}{k}\sum_{q=1}^{k} r_i^2(q) = \sigma_i, \; \sigma_j < \sigma_i, \; i \neq j, \tag{5.32}$$

where k^* is an unknown limited time instant, σ_j is a constant, and σ_i may be constant or infinity.
Then we have

$$\lim_{k\to\infty} p_j(k) = 1; \; \lim_{k\to\infty} p_i(k) = 0, \; i \neq j.$$

Proof. It is not difficult to see that algorithms (5.6)–(5.10) together with (5.31) guarantee with probability one that

$$
\begin{cases}
l'_{\min}(k) = l'_j(k), \\
\dfrac{l'_{\min}(k)}{l'_j(k)} = 1, \qquad \forall k \geq k^*, \ i \neq j. \\
\dfrac{l'_{\min}(k)}{l'_i(k)} < 1,
\end{cases}
\tag{5.33}
$$

Further, considering (5.32), we have

$$
\lim_{k \to \infty} \frac{l'_{\min}(k)}{l'_i(k)} = \frac{1 + \sigma_j}{1 + \sigma_i} < 1, \ i \neq j.
\tag{5.34}
$$

Putting (5.33), (5.34), and (5.9) together, we obtain

$$
\lim_{k \to \infty} l_j(k) = l_j(k^*) > 0; \ \lim_{k \to \infty} l_i(k) = 0, \ i \neq j.
\tag{5.35}
$$

Then from (5.10) we have

$$
\lim_{k \to \infty} p_j(k) = 1; \ \lim_{k \to \infty} p_i(k) = 0, \ i \neq j,
\tag{5.36}
$$

which completes the proof. □

Lemma 2 ([25]). *Consider weighting algorithm (5.13)–(5.18). Suppose there is a model, say $M_j \in \mathbb{M}$, which is closest to the true plant in the following sense with probability one:*

$$
\sum_{q=1}^{k} r_j(q)^2 < \sum_{q=1}^{k} r_i(q)^2 \ \forall k \geq d, i \neq j,
\tag{5.37}
$$

where d is an unknown limited time instant.
Then weighting algorithm (5.13)–(5.18) guarantees

$$
p_j(k) \to 1; \ p_i(k) \to 0, \ i = 1, \ldots, N, \ i \neq j.
\tag{5.38}
$$

Proof. It is not difficult to see that algorithms (5.13)–(5.18) together with (5.37) guarantee with probability one that

$$
\begin{cases}
l_{\min}(k) = l'_j(k), \\
\dfrac{l_{\min}(k)}{l'_j(k)} = 1, \qquad \forall k \geq d, \ i \neq j. \\
\dfrac{l_{\min}(k)}{l'_i(k)} < 1,
\end{cases}
\tag{5.39}
$$

Further, we know that if $\beta_i(k) = \frac{l_{min}(k)}{l_i'(k)} < 1$ and

$$\lim_{k\to\infty} \beta_i(k) = \lim_{k\to\infty} \frac{l_{min}(k)}{l_i'(k)} = 1, \; i \neq j, \tag{5.40}$$

then we still have

$$\lim_{k\to\infty} [\beta_i(k)]^{\text{ceil}(\frac{1}{1-\beta_i(k)})} = \frac{1}{e} < 1. \tag{5.41}$$

Putting (5.39), (5.41), and (5.17) together, we obtain

$$l_j(k) \to l_j(d); \; l_i(k) \to 0, \; i \neq j. \tag{5.42}$$

Thus from (5.18) we have

$$p_j(k) \to 1; \; p_i(k) \to 0, \; i \neq j, \tag{5.43}$$

which completes the proof. □

References

[1] J.J. Leonard, H.F. Durrant-Whyte, I.J. Cox, Dynamic map building for an autonomous mobile robot, The International Journal of Robotics Research 11 (4) (1992) 286–298, Sage Publications Inc.

[2] W.D. Chen, F. Zhang, Review on the achievements in simultaneous localization and map building for mobile robot, Control Theory and Applications 22 (3) (2005) 455–457.

[3] M. Csorba, Simultaneous Localization and Mapping, University of Oxford, 1997, pp. 56–89.

[4] F.A. Cheein, N. Lopez, C.M. Soria, F.A. di Sciascio, F.L. Pereira, R. Pereira, SLAM algorithm applied to robotics assistance for navigation in unknown environments, Journal of NeuroEngineering and Rehabilitation 7 (10) (2010) 1–16.

[5] H. Durrant-Whyte, Where am I? A tutorial on mobile vehicle localization, Industrial Robot 21 (2) (1994) 11–16.

[6] S. Yavuz, Z. Kurt, M.S. Bicer, Simultaneous localization and mapping using Extended Kalman Filter, in: 17th IEEE Signal Processing and Communications Applications Conference, Antalya, 2009, pp. 700–703.

[7] Y.W. Wei, Z.Y. Zuo, Improvement of the simultaneous localization and map building algorithm applying scaled unscented transformation, in: International Conference on Industrial Mechatronics and Automation, 2009, IEEE, Chengdu, 2009, pp. 371–374.

[8] J.G. Kang, W.S. Choi, S.Y. An, et al., Augmented EKF based SLAM method for improving the accuracy of the feature map, in: 2010 IEEE/RSJ International Conference on Intelligent Robots and Systems (IROS), Taipei, 2010, pp. 3725–3731.

[9] D.B. Wang, H.W. Liang, T. Mei, Lidar Scan matching EKF-SLAM using the differential model of vehicle motion, in: IEEE Intelligent Vehicles Symposium (IV), 2013, Gold Coast, QLD 36 (1) (2013) 908–912.

[10] S.J. Julier, J.K. Uhlmann, A counter example to the theory of simultaneous localization and map building, in: IEEE International Conference on Robotics and Automation, Seoul, Korea, 2001, pp. 4238–4243.

[11] S.J. Julier, The spherical simplex unscented transformation, in: Proceedings of the American Control Conference, Denver, 2003, pp. 2430–2434.

[12] M.S. Arulampalam, S. Maskell, et al., A tutorial on particle filters for online nonlinear/non-Gaussian Bayesian tracking, IEEE Transactions on Signal Processing 50 (2) (2002) 174–188.

[13] K. Murphy, Bayesian map learning in dynamic environments, Advances in Neural Information Processing Systems 12 (2008) 1015–1021.
[14] K. Murphy, S. Russell, Rao-Blackwellised particle filtering for dynamic Bayesian networks 43 (2) (2001) 499–515, Springer New York.
[15] M. Montemerlo, S. Thrun, D. Koller, Fast SLAM (simultaneous localization and mapping), in: Proceedings of the AAAI National Conference on Artificial Intelligence, AAAI, Menlo Park, CA, USA, 2002, pp. 593–598.
[16] E.Y. Wu, Z.Y. Xiang, M.Y. Shen, J.L. Liu, Robot SLAM algorithm based on laser range finder for large scale environment, Journal of Zhejiang University 41 (12) (2007) 1982–1986.
[17] S. Thrun, D. Fox, W. Burgard, F. Dallaert, Robust Monte Carlo localization for mobile robots, Artificial Intelligence 128 (1–2) (2001) 99–141.
[18] X. Gao, T. Zhang, Y. Liu, Q.R. Yan, Visual SLAM Fourteen Lectures: From Theory to Practice, Publishing House of Electronics Industry, Beijing, 2017, pp. 132–204.
[19] A.J. Davison, I.D. Reid, N.D. Molton, O. Stasse, Monoslam: real-time single camera SLAM, IEEE Transactions on Pattern Analysis and Machine Intelligence 29 (6) (2007) 1052–1067.
[20] M.X. Quan, S.H. Piao, G. Li, An overview of visual SLAM, CAAI Transactions on Intelligent Systems 11 (6) (2016) 768–776.
[21] G. Klein, D. Murray, Parallel tracking and mapping for small AR workspaces, in: IEEE and ACM International Symposium on Mixed and Augmented Reality, Nara, Japan, 2007, pp. 225–234.
[22] R. Mur-Artal, J.D. Tardos, Fast delocalization and loop closing in key frame-based SLAM, in: IEEE International Conference on Robotics and Automation, New Orleans, LA, 2014, pp. 846–853.
[23] R. Mur-Artal, J.M.M. Montiel, J.D. Tardos, ORB-SLAM: a versatile and accurate monocular SLAM system, IEEE Transactions on Robotics 31 (5) (2015) 1147–1163.
[24] P. Henry, M. Krainin, E. Herbst, et al., RGB-D mapping: using depth cameras for dense 3D modeling of indoor environments 31 (5) (2014) 647–663, Springer Berlin Heidelberg.
[25] K. Zhu, H.F. Liu, Q.Y. Xia, Survey on monocular visual SLAM algorithms, Application Research of Computers 35 (1) (2018) 1 6.
[26] S.F. Li, P. Wang, Z.K. Shen, A survey of iterative closest point algorithm, Signal Processing 25 (10) (2009) 1582–1588.
[27] M. Irani, P. Anandan, About direct methods [M], in: International Workshop on Vision Algorithms: Theory and Practice, 1999, pp. 267–277.
[28] G. Silveira, E. Malis, P. Rives, An efficient direct approach to visual SLAM 24 (5) (2008) 969–979, IEEE Press.
[29] C. Forster, M. Pizzoli, D. Scaramuzza, SVO: fast semi-direct monocular visual odometry, in: IEEE International Conference on Robotics and Automation (ICRA), 2014, pp. 15–22.
[30] J. Engel, T. Schops, D. Cremers, LSD-SLAM: Large-Scale Direct Monocular SLAM. Computer Vision-ECCV 2014, Springer International Publishing, 2014, pp. 834–849.
[31] J. Engel, J. Sturm, D. Cremers, Semi-dense visual odometry for a monocular camera, in: Proceedings of the IEEE International Conference on Computer Vision, 2013, pp. 1449–1456.
[32] J. Engel, V. Koltun, D. Cremers, Direct sparse odometry, IEEE Transactions on Pattern Analysis and Machine Intelligence 99 (2017) 1–10.
[33] V. Usenko, J. Engel, J. Stückler, D. Cremers, Direct visual-inertial odometry with stereo cameras, in: IEEE International Conference on Robotics and Automation (ICRA), 2016, pp. 1885–1892.
[34] R.A. Newcombe, S. Izadi, O. Hilliges, et al., KinectFusion: real-time dense surface mapping and tracking, in: IEEE International Symposium on Mixed and Augmented Reality, 2011, pp. 127–136.
[35] P. Henry, M. Krainin, E. Herbst, et al., RGB-D mapping: using kinect-style depth cameras for dense 3D modeling of indoor environments, The International Journal of Robotics Research 31 (5) (2012) 647–663.

[36] C. Kerl, J. Sturm, D. Cremers, Dense visual SLAM for RGB-D cameras, in: IEEE/RSJ International Conference on Intelligent Robots and Systems 8215 (2) (2014) 2100–2106.

[37] M. Wu, J.Y. Sun, Extended Kalman filter based moving object tracking by mobile robot in unknown environment, ROBOT 32 (3) (2010) 334–343.

[38] Wen Zhang, Research on Autonomous Navigation Method for Indoor Robots Based on Multi Sensor Fusion, University of Science and Technology of China, 2017.

[39] Q.L. Lai, S.P. Yu, S.Y. Ding, Mobile robot SLAM and path planning system, Computer Knowledge and Technology 33 (13) (2017) 24–37.

[40] Y. Zhao, G.L. Liu, G.H. Tian, et al., A survey of visual SLAM based on deep learning, ROBOT 39 (6) (2017) 889–896.

[41] F. Lu, E.E. Milios, Globally consistent range scan alignment for environment mapping, Autonomous Robots 4 (1997) 333–349.

[42] D.T. Magill, Optimal adaptive estimation of sampled stochastic processes, IEEE Transactions on Automatic Control 10 (1965) 434–439.

[43] D.G. Lainiotis, Partitioning: a unifying framework for adaptive systems I: estimation II: control, IEEE Transactions on Automatic Control 64 (1976) 1182–1198.

[44] M. Athans, et al., The stochastic control of the F-8c air-craft using a multiple model adaptive control (MMAC) method part I: equilibrium flight, IEEE Transactions on Automatic Control 22 (1977) 768–780.

[45] M. Athans, et al., Investigation of the multiple method adaptive control (MMAC) method for flight control systems, Technical Report NASA-CR-2916, NASA Technical Reports, 1979.

[46] B.D.O. Anderson, J.B. Moore, Optimal Filtering, Prentice Hall, New Jersey, USA, 1979.

[47] X.R. Li, Y. Bar-Shalom, Multiple-model estimation with variable structure, IEEE Transactions on Automatic Control 41 (1996) 478–493.

[48] S. Fekri, M. Athans, A. Pascoal, Issues, progress and new results in robust adaptive control, International Journal of Adaptive Control and Signal Processing 20 (10) (2006) 519–579.

[49] S. Fekri, M. Athans, A. Pascoal, Robust multiple model adaptive control (RMMAC): a case study, International Journal of Adaptive Control and Signal Processing 21 (1) (2007) 1–30.

[50] W. Zhang, Stable weighted multiple model adaptive control: discrete-time stochastic plant, International Journal of Adaptive Control and Signal Processing 27 (7) (2013) 562–581.

[51] W. Zhang, Further results on stable weighted multiple model adaptive control: discrete-time stochastic plant, International Journal of Adaptive Control and Signal Processing 29 (12) (2015) 1497–1514.

Chapter 6

Optimization of motion control smoothness based on Eband algorithm

Sufang Wang[a], Chuanxu An[b], and Weicun Zhang[c]

[a]Second Academy of China Aerospace Science and Industry Corporation, Institute 706, Beijing, China, [b]Institute of Software Chinese Academy of Sciences, Beijing, China, [c]University of Science and Technology Beijing, School of Automation and Electrical Engineering, Beijing, China

6.1 Introduction

ROS autonomous mobile robots [1] mainly need to solve the problems of positioning, navigation, planning, and control. Navigation is the core technology of an autonomous mobile robot, and the navigation function is mainly implemented based on the move base framework of ROS. Navigation [2–4] enables the robot to autonomously conduct the corresponding guidance according to the internal predetermined information or by obtaining the external environment based on sensors, so as to plan a path suitable for the robot to walk in the environment. Therefore the key to successful navigation lies in path planning [5,6], which consists of two parts, global and local path plannings [7]. The function of global path planning is to use path planning algorithms to plan a feasible path in a two-dimensional grid map, whereas local path planning is to make dynamic planning and motion control based on the global path to achieve the effect of real-time obstacle avoidance [8,9].

Common local path planning algorithm include dynamic window approach (DWA) [10], timed elastic band (TEB) algorithm [11], pure pursuit tracking algorithm [12], model predictive control (MPC) [13–15], Eband algorithm [16], etc. Among them, DWA is to obtain the current velocity set that the robot can reach by obtaining the current velocity of the robot and the set maximum acceleration. By calculating all the velocity costs in the velocity set, the velocity with the lower cost is selected and sent to the robot. Due to the limitation of velocity and acceleration, DWA has short sampling time, and only safe trajectories will be considered. Therefore forward simulation is easy to implement. It has low computational complexity and is suitable for differential chassis. However, the forward-looking distance calculated by DWA is relatively short, and the algorithm effect is greatly reduced after lengthening the forward-looking dis-

Modeling, Identification, and Control for Cyber-Physical Systems Towards Industry 4.0
https://doi.org/10.1016/B978-0-32-395207-1.00017-2

tance, so it is impossible to avoid obstacles in advance. In complex scenes the effect of path planning and dynamic obstacle avoidance is poor. TEB [17,18] is an algorithm where the starting point and the target point state are specified by the user/global planner, and N rubber band-shaped control points (robot posture) are inserted in the middle. To display the kinematics information of the trajectory, the motion time is defined between points. TEB [19] can optimize the trajectory ahead and has a better obstacle avoidance effect on dynamic obstacles, but the TEB algorithm has a higher computational complexity. The velocity and angular velocity in the control cycle are calculated by the distance, angle difference, and time difference between the two states, which leads to large fluctuations in the velocity and angle during the control process. Eband is a method similar to the artificial potential field forming gravitational and repulsive forces to generate elastic bands for local planning to improve the velocity of robot walking and the continuity of actions [20].

In view of the lack of consideration of robot motion in some planners, when encountering dynamic obstacles, even if the global planner can re-plan the path, the current position may deviate from the planned path [21,22]. In this chapter, we use the Eband local path planner to implement dynamic obstacle avoidance and the Eband algorithm to optimize the previously planned path, but its motion control changes with the fluctuation of the "bubble", sometimes resulting in an insufficiently smooth path. Therefore this paper mainly optimizes the algorithm for the smoothness motion control of Eband and implements it through code. The effectiveness of this method is proved by comparing the optimized algorithm with the original Eband algorithm.

6.2 Eband algorithm implementation principle

The Eband algorithm uses a method similar to the artificial potential field forming gravitational and repulsive forces to generate elastic bands for local planning. The input of the Eband algorithm is the global path, local cost map, and the current pose of the robot. The output of the Eband algorithm is the control velocity (angular velocity and linear velocity). Eband is represented by bubbles; each bubble is in the free space, and each bubble overlaps with two adjacent bubbles. The radius of each bubble is the shortest distance between the robot and obstacle. The position of the bubble is determined by calculating the resultant force (the gravitational force generated by adjacent points and the repulsive force generated by the obstacle). Through the center position of the next bubble, the linear and angular velocities of the control can be calculated. The specific process is as follows.

Fig. 6.1(a) shows the path planned by the robot global path planner, and Fig. 6.1(b) shows the path after using the Eband local path planning algorithm. To improve the shape of the path, there are two main forces, the tightening force and the repulsive force generated by the obstacle, until the two forces are balanced, making the originally slack path look tight.

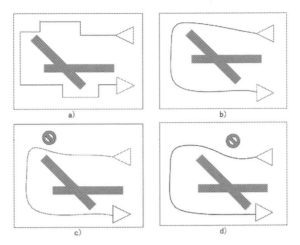

FIGURE 6.1   a), b), c), d) Path planning diagram.

Figs. 6.1(c) and 6.1(d) are the paths after planning by the Eband local path planner and making the corresponding adjustments according to the obstacles in the environment as the robot moves.

A bubble is a circular area that represents a point on the robot path that does not collide with an obstacle. When it is close to the obstacle, the bubble is small, and when it is far away, the bubble increases accordingly. Define the bubble function as follows:

$$B(b) = q : \|b - q\| < \rho(b), \tag{6.1}$$

where b represents the center of the current bubble, q represents any random position in the current bubble, and $\rho(b)$ represents the radius of the bubble with b as the center. After that, the generated bubbles are processed, such as moving the bubbles forward or backward, so that the bubbles generated near the obstacle are dense, and the bubbles generated far away from the obstacle are sparse. The bubble generation along the path and changes with obstacles are shown in Fig. 6.2. As shown in Fig. 6.2, the hollow circle on the upper left represents an obstacle. The proximity of the obstacle causes the nearby bubbles to become smaller and denser. After the obstacle is far away, the bubbles become larger and sparse. When an obstacle suddenly appears near the robot, the bubble jumps; or when the robot moves in a narrow corridor with obstacles such as trash cans, fire hydrants, and debris, and when there is a sudden turn in a narrow section of road, the bubble becomes smaller. If the bubble is too small, then the band may break. At this time the robot re-plans the path and interpolation, and the target point of the local path changes, which causes the robot to freeze.

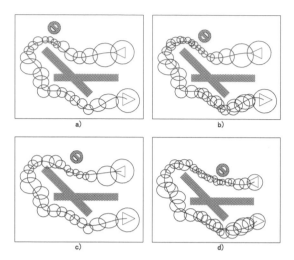

FIGURE 6.2   a), b), c), d) Bubble generation along the path and changes with obstacles.

6.3   Improved Eband algorithm

The jump of the Eband bubble in a specific scene directly leads to the jump of the current target point of the local path and the fluctuation of the motion control velocity, which will result in the robot motion jam. To solve this problem, in this chapter, we present an improved Eband algorithm including three aspects: curvature algorithm optimization, velocity product factor denoising, and velocity interpolation.

6.3.1   Curvature algorithm optimization

Aiming to change the radius of the bubble caused by the sudden obstacle, which leads to the change of the curvature of the motion, the curvature calculation is optimized to smooth the large curvature change that causes the linear velocity of robot to drop sharply.

On the generated local path, three bubbles are intercepted, band[0], band[1], band[2]; band[0] represents the pose of the robot, and the subsequent two bubbles represent the next two poses of the robot. Then band[0], band[1], and band[2] correspond to pose[0], pose[1], and pose[2]. Using the X and Y coordinates in each pose, the angle change between the front and rear poses can be obtained.

Here we use vectors to calculate the angle changes of the two poses of the robot; band[0] and band[1] are set to vector a, and band[1] and band[2] are set to vector b:

$$a = (x_1, y_1), \ b = (x_2, y_2), \tag{6.2}$$

where

$$x_1 = \text{pose}[1].\text{pose.position.x-pose}[0].\text{pose.position.x},$$
$$y_1 = \text{pose}[1].\text{pose.position.y-pose}[0].\text{pose.position.y},$$
$$x_2 = \text{pose}[2].\text{pose.position.x-pose}[1].\text{pose.position.x},$$
$$y_2 = \text{pose}[2].\text{pose.position.y-pose}[1].\text{pose.position.y}.$$

Further we define

$$\text{curvaturefactor} = \cos(ab) = (a \cdot b)/(|a||b|) \qquad (6.3)$$

where $a \cdot b$ means the dot product of vectors a and b. To be specific, we have

$$\cos(ab) = \frac{x_1 x_2 + y_1 y_2}{\sqrt{(x_1^2 + y_1^2)(x_2^2 + y_2^2)}}. \qquad (6.4)$$

The value range of $\cos(ab)$ is [0, 1], which means that when the turing angle is 0°, the curvature factor does not affect the overall linear velocity, and when the turing angle is close to 90°, the linear velocity will directly become 0. In the real scene, it is to brake sharply and turn 90°, and the linear velocity is decelerated to 0.

6.3.2   Velocity product factor denoising

To overcome the problem of robot motion jam caused by noise in the velocity product factor of Eband in narrow sections, the linear denoising of velocity product factor in narrow sections is carried out.

The influence of the velocity product mainly depends on the narrowness of the environment. When the environment is too narrow, the corresponding bubble radius of the Eband rapidly decreases. At this time the linear velocity should also be decelerated to ensure safety. The linear velocity formula output by the algorithm is

$$\text{Max\_el\_lin} = \text{max\_vel\_lin\_} * \text{curvature\_factor} * \text{velocity\_multiplier}, \qquad (6.5)$$

where max_vel_lin_ represents the maximum linear velocity set in the parameter file, which can be dynamically adjusted, curve_factor is the curvature factor, which represents the influence of the turning arc on the linear velocity, and velocity_multiplier represents the influence of space narrowness and bubble radius on linear velocity. The weight of the environmental impact factor is represented by

$$\text{Scale} = 4 - 1.2/\text{max\_vel\_lin\_}. \qquad (6.6)$$

The greater the maximum velocity setting, the greater the weight of the environmental impact factor. The selection of this parameter is mainly determined according to the application scene of the robot. In this chapter, based on the value range [0.3, 1] of the maximum linear velocity max_vel_lin_, to constrain the value range of Scale to [0, 2.8], a linear equation (6.6) is set, which ensures that when the velocity is very low, the influence of Scale is 0. With the increase of velocity, the constraints of environmental factors gradually expand to 3 at most, so as to adapt to the application scene of robot.

In the formula

$$\text{Offset} = 0.42/\text{max\_vel\_lin\_} - 0.4, \tag{6.7}$$

Offset is set to balance the entire velocity product, limit the value range of the velocity product, and control the maximum and minimum velocity range. Since the maximum linear velocity is in the range of [0.3, 1], to achieve the maximum effect of the velocity product when the velocity is the minimum and satisfy the value range [0.02, 1], Offset in formula (6.7) is selected as a linear parameter. The formula for calculating the velocity product is as follows:

$$\text{Velocity\_multiplier} = \text{Scale} * \text{bubble\_radius} + \text{Offset}. \tag{6.8}$$

We can seen from formula (6.8) that the velocity product Velocity_multiplier is determined by Scale, bubble_radius, and Offset. When the bubble radius is the smallest, Velocity_multiplier reaches the smallest value, indicating that the surrounding environment is more complicated; on the contrary, when the bubble radius is the largest, the surrounding environment is open and safe, and at this time, the velocity product reaches the maximum; the value range of the velocity product factor is [0.4285, 1].

6.3.3 Velocity interpolation

To overcome the jumping problem of the local path target point caused by the jump of bubbles, in this chapter, we use bubble interpolation to avoid jumping. In order that the robot can more smoothly output the speed to reach the local target point and to take into account the response sensitivity of the robot to avoid obstacles, according to the motor response ability of the robot, the distance of 0.2 m in front of the robot is selected as the interpolation bubble center. The bubble interpolation is selected according to the performance of different motors, and the general range is [0.1, 0.3]. The interpolated bubble radius is calculated according to the radius of the front and rear bubbles on the band. The interpolated bubbles are iteratively updated as local path target points. The corresponding pseudo-codes are as follows:

```
While
{
  Convert the generated plan into a band;
  Take the position $0.2m$ before band[0] as the bubble center of
  the interpolated bubble;
  Take the bubble radius closest to the bubble center as its own
  bubble radius;
  Interpolate to the band;
  Set the local path target point as the center of the interpolation
  bubble;
  Release goal;
}
```

6.4 Experiment analysis

6.4.1 Experimental data set

In this chapter, the data set used is a bag recorded in the same scene, including a return corridor and several fixed obstacles, and subscribed to /cmd_vel topics and /vcl_data topic. The map size is 20 m × 20 m. In addition, this experiment ensures that other influencing factors in the recording environment can be ignored except for the return corridor and several obstacles.

6.4.2 Experimental software and hardware configuration

The optimized algorithm in this chapter and the original Eband algorithm are both experimentally verified on the same data set. The software environment is Ubuntu18.04 + ROS melodic. The hardware environment is Firefly-RK3399Pro development board.

In the path planning experiment the hardware environment adopts the RK3399Pro development board, which is a domestic high-performance AI processing chip, integrates neural network processor NPU, has a computing power of up to 3.0 Tops, and is compatible with multiple AI frameworks. This chip uses ARM dual-core Cortex-A72 and quad-core Cortex-A53 processor architecture, with the CPU clock speed of up to 1.8 GHz, and integrates quad-core ARM high-end GPU Mali-T860 MP4 graphics processor. It has powerful general-purpose computing performance and excellent overall performance.

ROS was originally developed by the Artificial Intelligence Laboratory at Stanford University in 2007. It is an open-source robot operating system software framework. After 2008, ROS was continuously developed by Willow Garage, and ROS was open sourced in 2010. Now the maintenance and development of ROS is taken over by the Open Source Robot Foundation. It provides services that are expected from the operating system, including hardware abstraction, realization of common functions, package management between processes, low-level device control, and messaging. It also provides various tools and libraries to write and run code.

The main features of ROS are as follows.

1. Distributed architecture

   ROS distributed processing architecture is a point-to-point design through nodes. Executable files can be run separately, and modules are loosely coupled. It can be operated by multi-process and multi-host, which provides convenience and possibility for multi-robot collaboration, which also disperses the calculation and operation pressure brought by the large-scale tasks, and improves the calculation speed and operation efficiency.

2. Multilingual support

   ROS is a language-neutral software framework that can be used by different language programming enthusiasts and has language interfaces for different languages. Currently, ROS supports many different languages such as C++, Python, and LISP. The compilation and operation of these languages are also very simple.

3. Rich toolkit

   The most commonly used and most important tools in ROS are visualization and debugging tools, simulation environments, etc. They are distributed in different function packages and perform different tasks in different roles. The existence of ROS tools not only facilitates the management of the software framework, but also improves efficiency.

4. Free and open source

   ROS complies with the BSD (Berkeley Software Distribution) open source agreement, so it is free and open source software that allows companies, individuals, and researchers to develop and modify. This will promote the debugging and improvement of ROS functions at all levels. Based on the above characteristics of ROS, it is very convenient to realize the robot motion path planning function in ROS. ROS can also solve the communication and coordination problems between sensor drive, display and algorithm. Ubuntu is a release version of Linux system with powerful Anti-virus function and free of charge, so the algorithms involved in this chapter are all programmed under Ubuntu 18.04 + ROS melodic robot operating system, and the code is written in C++ language. The experimental comparison chart uses the rqt-plot drawing tool that comes with ROS to compare the data curves.

6.4.3   Experimental data comparison

6.4.3.1   Smooth curvature algorithm optimization

In this section of the experiment, the problem of curvature limitation on the abrupt speed suppression effect is optimized, and the effect of speed reduction in different curvature ranges is compared, where d is the change in curvature

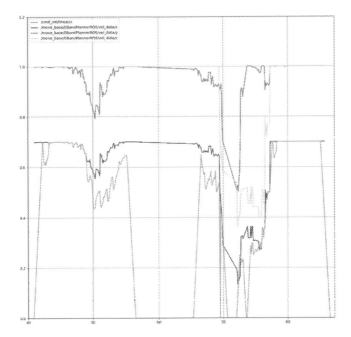

FIGURE 6.3   Curve of influence of curvature on various parameters when $d < 1.0$.

before and after and is recorded as data. The parameters are explained as follows:

1. /cmd_vel/linear/x: The linear velocity of robot movement chassis odometer;
2. /move_base/EbandPlannerROS/vel_data/x: The linear velocity of the algorithm output;
3. /move_base/EbandPlannerROS/vel_data/y: The curvature factor;
4. /move_base/EbandPlannerROS/vel_data/z: The velocity product factor.

When $d < 1.0$, we have the result shown in Fig. 6.3.

We can see that the change of the curvature factor causes a drastic change in the linear velocity. Compared with the optimized algorithm, the change of the curvature factor has a more obvious effect on the linear velocity output by the original algorithm.

When $d < 0.2$, we have the result shown in Fig. 6.4.

We can see from Fig. 6.4 that when $d < 0.2$, the cliff-like change of the curvature factor (red line – mid gray line in print version) is effectively restricted.

When $d < 0.1$, we have the result shown in Fig. 6.5, from which we can see that the fluctuation of the red line (mid gray line in print version) curvature factor is basically stable, but the effect on the linear velocity output by the original algorithm is getting smaller and smaller, and the most influential factor becomes the velocity product factor of the green line (semi-light gray in print version).

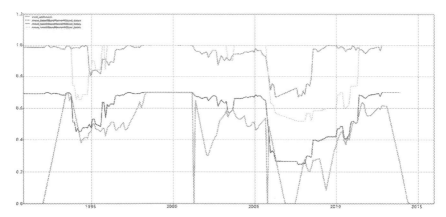

FIGURE 6.4   Curve of influence of curvature on various parameters when $d < 0.2$.

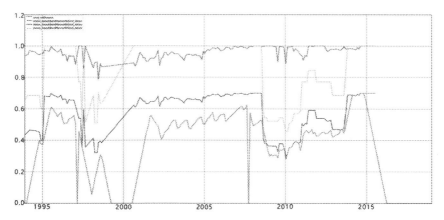

FIGURE 6.5   Curve of influence of curvature on various parameters when $d < 0.1$.

When $d < 0.15$, we have the result shown in Fig. 6.6.

Figs. 6.7(a) and 6.7(b) illustrate the comparison of the overall effect curves before and after smoothing the curvature.

We can see from Figs. 6.7(a) and 6.7(b) that the sudden appearance of obstacles causes an instantaneous change in the radius of the bubble, resulting in a sudden change in the curvature of the motion, and the curvature affects the maximum linear velocity while turning. In the algorithm of this chapter, the gradient of the curvature parameter is optimized to make the linear speed smoother when an obstacle suddenly appears or when turning around an obstacle, which effectively suppresses the sudden decrease of the robot linear speed caused by a large curvature change.

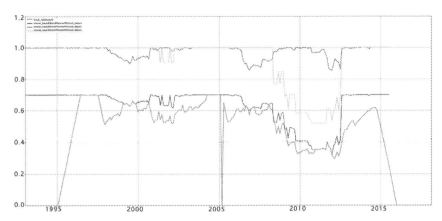

FIGURE 6.6   Curve of influence of curvature on various parameters when $d < 0.15$.

6.4.3.2   Velocity product factor denoising

When the robot moves in a narrow space, due to the narrow environment, the bubble radius generated by Eband rarely reaches the maximum. At this time the linear velocity of the robot is directly affected by the speed product factor. In this scene, if a static or dynamic obstacle suddenly appears, then the bubble radius corresponding to the band may be less than 0.1 m. In the speed setting of Eband, when the bubble radius is less than 0.1 m, the linear velocity is already quite small. In this scenario, when the bubble radius is less than 0.1 m, the linear velocity is already quite small. However, in the real scene the robot needs to continue to move forward slowly and steadily. To achieve this goal, when the bubble radius is less than 0.1 m, the influence of the velocity factor becomes a noise point for the linear velocity. This experiment performed de-noising work on this problem.

Before algorithm optimization, we have the experiment result shown in Fig. 6.8.

In Fig. 6.8(a) the experiment set the minimum linear velocity to 0.3 m/s, but due to the presence of noise, the linear velocity (blue line – dark gray in print version) output by the algorithm has many extreme points below 0.3 m/s. These extreme points are caused by the instantaneous value of the velocity product and are invalid and unusable points. They also have a direct impact on the smoothness of the linear velocity, so optimization denoising is required.

Observing Fig. 6.8(b), we can see that the linear velocity (blue line – dark gray in print version) output by the algorithm has no extreme points lower than 0.3 m/s, indicating that the noise has been removed. The velocity product mainly affects the movement of the robot when it passes through a narrow space. The influence of the velocity product is limited when the bubble is small for a short time to reduce linear velocity fluctuations and make the robot move more smoothly in a narrow space.

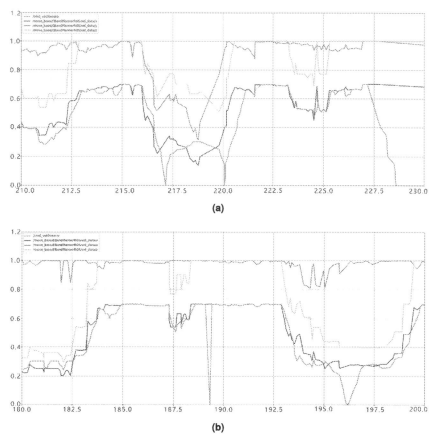

FIGURE 6.7　a) The overall effect curve before smoothing the curvature. b) The overall effect curve after smoothing the curvature.

6.4.3.3　Local target point interpolation

After the optimization of curvature and velocity product, the linear velocity output by the Eband algorithm is relatively stable. Eband adopts a segmented control strategy in local path planning and control. First, after the curve of the global path planning is obtained, the global point of the corresponding length is intercepted by the set local cost map as the local path target point. Second, the algorithm generates a band connected by bubbles, which can be transformed into a path, which is a local path. Finally, it is handed over to speed control. The speed control uses the nearest bubble center of the robot as a temporary target point for short-distance movement. This is one of the reasons why Eband can quickly respond to sudden obstacles.

However, the above methods have certain drawbacks. First of all, the local path planner is constantly planning when the robot is in constant motion. When facing dynamic obstacles, the bubbles and paths generated by the algo-

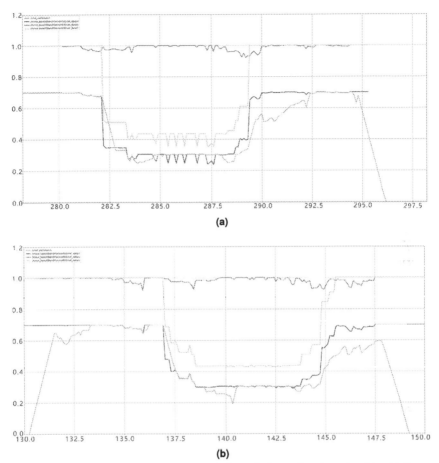

FIGURE 6.8 a) Curves of various parameters before velocity product factor denoising. b) Curves of various parameters after velocity product factor denoising.

rithm are dynamically updated, which means that local temporary target points may change. The jump of the center of the local temporary bubble directly affects the control of angular velocity and linear velocity, causing oscillation and stuttering. Observing Fig. 6.9(a), we can see that the linear velocity of the algorithm output (blue line – dark gray in print version) is tortuous and unstable. Since the bubble center of the local temporary target may jump, can we set a bubble center that does not jump sharply in the band as the temporary target point? The answer is feasible. In order not to affect the real-time performance, we perform bubble interpolation in the band closer to the robot. The center of the bubble is selected to be 0.2 m away from the robot. The bubble radius is affected by the radii of the two bubbles before and after. In this way the position of the bubble center becomes a relatively stable value, as shown in Fig. 6.9(b).

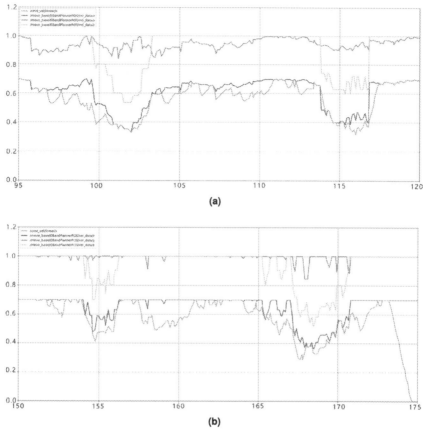

FIGURE 6.9   a) Curves of various parameters before interpolation of local target points. b) Curves of various parameters after interpolation of local target points.

Observing the linear velocity curve of the algorithm output (blue line – dark gray in print version). Compared with Fig. 6.9(a), the linear velocity smoothness of the algorithm output has been greatly improved, and the linear velocity output (purple line – light gray in print version) by the odometer also tends to be stable. The visualization effect of local target point interpolation is shown in Fig. 6.10. The bubble in red color (dark gray color in print version) represents the local target interpolation point, which can also be called the speed interpolation point. The band formed by the bubbles is the local path planned by the local path planner in real time. Comparing the output of the speed control on the left, we can see that the robot moves smoothly with a relatively smooth linear velocity, and the linear velocity calculated by the algorithm and the linear velocity measured by the odometer have a higher degree of fitting.

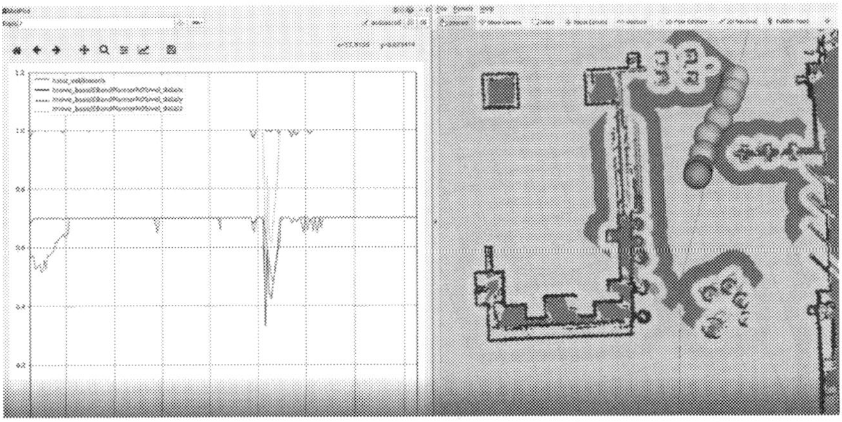

FIGURE 6.10   The visualization effect diagram of local target point interpolation.

6.5   Conclusions

The Eband algorithm has the problems of sudden bubble change, large influence of curvature on linear velocity, and robot motion freeze. This chapter mainly focused on the following three aspects to address these problems:

1. The curvature factor is smoothed to effectively alleviate the excessive suppression of the curvature factor on the linear velocity, and the calculation method of the curvature factor is replaced to make its influence on the linear velocity more reasonable.
2. The denoising of velocity product factor reduces the feeling of robot motion jamming to a certain extent and makes the motion of robot more smoothing narrow space.
3. Local target point interpolation effectively solves the problem of local target point jump and speed control change caused by bubble jump.

Of course, there are still many shortcomings in the methods of the local path planning and global path planning, which require further research and improvement. In addition, local path planning and global path planning work together so that the robot can better plan the movement path from the start point to the end point.

References

[1] F. Liu, D.Y. Yang, M.J. Lian, Z. Zhang, L.N. Wang, Development platform of intelligent industrial robot system based on ROS, Computer Systems and Applications 26 (10) (2017) 77–81.
[2] B.A. Kumar, K. Sirisha, D. Kumar, Development of robot navigation system, IOP Conference Series: Materials Science and Engineering 1057 (1) (2021) 012022.
[3] K. Balakrishnan, P. Chakravarty, S. Shrivastava, An A* Curriculum Approach to Reinforcement Learning for RGBD Indoor Robot Navigation, 2021.

[4] A. Ravankar, A. Ravankar, Y. Hoshino, et al., Transient Virtual Obstacles for Safe Robot Navigation in Indoor Environments, 2020.

[5] R. Fareh, et al., Investigating reduced path planning strategy for differential wheeled mobile robot, Robotica 38 (2) (2019) 235–255.

[6] R. Valencia Carreno, J. Andradecetto, J.P. Pleite, Path planning in belief space with Pose SLAM, Springer International Publishing, 2018.

[7] B. Chanclou, A. Luciani, Global and local path planning in natural environment by physical modeling, in: IEEE/RSJ International Conference on Intelligent Robots and Systems, IEEE, 2013.

[8] H. Kalita, S. Morad, A. Ravindran, et al., Path planning and navigation inside off-world lava tubes and caves, in: IEEE Aerospace Conference Proceedings, 2018, pp. 1311–1318.

[9] P. Marin-Plaza, et al., Global and Local Path Planning Study in a ROS-Based Research Platform for Autonomous Vehicles, Inventi Impact - Auto, 2018.

[10] D. Fox, W. Burgard, S. Thrun, The dynamic window approach to collision avoidance, IEEE Robotics & Automation Magazine 4 (1) (1997) 23–33.

[11] C. Roesmann, W. Feiten, T. Woesch, et al., Trajectory modification considering dynamic constraints of autonomous robots, in: Robotics Proceedings of ROBOTIK 2012 7th German Conference on Robotics, 2012, pp. 1–6.

[12] J. Morales, J.L. Martínez, M.A. Martínez, et al., Pure-pursuit reactive path tracking for nonholonomic mobile robots with a 2D laser scanner, EURASIP Journal on Advances in Signal Processing 2009 (2009) 1–10.

[13] J. Richalet, A. Rault, J.L. Testud, J. Papon, Model predictive heuristic control: applications to industrial processes, Automatica 14 (5) (1978) 413–428.

[14] D.Q. Mayne, J.B. Rawlings, C.V. Rao, P.O.M. Scokaert, Constrained model predictive control: stability and optimality, Automatica 36 (6) (2000) 789–814.

[15] D.Q. Mayne, J.B. Rawlings, Correction to "Constrained model predictive control: stability and optimality", Automatica 37 (3) (2001) 483–497.

[16] S. Quinlan, O. Khatib, Elastic bands: connecting path planning and robot control, in: Proc. IEEE International Conference on Robotics and Automation, Atlanta, Georgia, vol. 2, 1993, pp. 802–807.

[17] J.F. Wu, et al., An improved Timed Elastic Band (TEB) algorithm of Autonomous Ground Vehicle (AGV) in complex environment, Sensors 21 (24) (2021) 1–12.

[18] B. Magyar, et al., Timed-elastic bands for manipulation motion planning, IEEE Robotics and Automation Letters 4 (4) (2019) 3513–3520.

[19] C. Roesmann, W. Feiten, T. Woesch, et al., Efficient trajectory optimization using a sparse model, in: Proceedings of the 2013 European Conference on Mobile Robots, 2013, pp. 138–143.

[20] J.Y. Wang, Y.H. Luo, X.J. Tan, Path planning for Automatic Guided Vehicles (AGVs) fusing MH-RRT with improved TEB, Actuators 10 (12) (2021) 1–19.

[21] J. Almeida, R.T. Nakashima, F. Neves Jr, et al., A global/local path planner for multi-robot systems with uncertain robot localization, Journal of Intelligent & Robotic Systems 100 (1) (2020) 311–333.

[22] N. Wang, H. Xu, Dynamics-constrained global-local hybrid path planning of an autonomous surface vehicle, IEEE Transactions on Vehicular Technology 69 (7) (2020) 6928–6942.

Chapter 7

Modeling a modular omnidirectional AGV developmental platform with integrated suspension and power-plant

Alexander B.S. Macfarlane[a], Theo van Niekerk[a], Udo Becker[b], and Paolo Mercorelli[c]

[a]*Faculty of Engineering, Built Environment & Information Technology (EBEIT) at the Nelson Mandela University, Port Elizabeth, South Africa,* [b]*Ostfalia University of Applied Sciences, Wolfsburg, Germany,* [c]*Institute for Production Technology and Systems, Leuphana University of Lueneburg, Lueneburg, Germany*

7.1 Motivation for the use of omnidirectional AGVs

AGVs that have omnidirectional capabilities and thus increased flexibility are of great interest to the manufacturing industry, especially when compared to their non-holonomic counterparts [1]. An "omnidirectional" vehicle can move in any direction instantaneously from its current location and orientation in 2D space [2], in many regards similar to how a human being would negotiate their surroundings. Omnidirectional capabilities reduce the footprint required for vehicle manoeuvres, especially the large turning radii characteristics of Ackerman steering or differential steering [3]. This footprint reduction reduces the size of the manufacturing space overall, reducing monthly maintenance costs and capital costs associated with expansion. Omnidirectional AGVs can better function in complex working environments and older factories. These older factories were never intended to host mobile robots or AGVs and were instead designed around human kinesiology and ergonomics [4].

There exist two strategies to achieve omnidirectional motion in a vehicle [5]. The first strategy is to use a custom wheel capable of holonomic motion. Currently (circa 2022), three standard holonomic wheels exist, namely the mecanum wheel [6], the omniwheel [7], and the spherical wheel [8]. The second strategy is to use a standard wheel and rotate it about its z-axis; this is often referred to as a swerve drive system [9].

Modeling, Identification, and Control for Cyber-Physical Systems Towards Industry 4.0
https://doi.org/10.1016/B978-0-32-395207-1.00018-4

7.1.0.1 Traditional swerve drive systems

Swerve drive systems are also known as powered castors or castor drives. Swerve steering is an omnidirectional strategy based on four powered castor wheels. A powered castor wheel has powered rotation about the y-axis (rolling rotation of the wheel) and z-axis (steering axis perpendicular to the ground plane). The powered rotation about the z-axis of the castor wheels allows for omnidirectional capabilities. Two layouts are commonly used for powered castors in swerve-drive systems. The first is an in-line system where the z-axis and y-axis share a common origin point. The second is the offset system, where the z-axis and y-axis are still perpendicular to each other, but their origin points are separated by a distance called the castor offset. The in-line system is illustrated in Fig. 7.1a, whereas the offset system is illustrated in Fig. 7.1b.

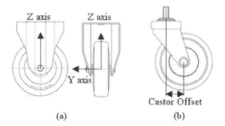

(a)   (b)

FIGURE 7.1 Power castor wheel axes.

The "castor offset" ensures that the wheels do not scuff when the wheel is rotated about its steering axis (z-axis) as this offset forces a steering arc. Without the offset, compensation must be performed by rotating the wheel about its rolling axis (y-axis) while steering occurs to form a turning arc.

For a four-wheel swerve drive system to achieve complete omnidirectional motion, the z-axis of the four wheels must be placed on the corners of a virtual square. If this is not done, then rotation about the centroid of the vehicle will be impossible without scuffing the wheels. See Fig. 7.2 for an illustration of this effect; note the orientations of the wheels, which form a tangent to a virtual circle whose center point coincides with the centroid of the AGV body. The direction of motion of each wheel in Fig. 7.2 is illustrated with a black arrow. The overall motion of the AGV body is also indicated.

The remaining omnidirectional motions of the AGV can be achieved by orienting the wheels of the AGV as shown in Fig. 7.3. All the motions swerve drive is capable of are summarized in Table 7.1.

The original patent for a swerve drive system is titled "Powered Castor Wheel Module for use on omnidirectional drive systems" [10]. It was first filed in 1998. The original system developed called for four in-line powered castor units to be placed on a pitch circle diameter (PCD) of an arbitrary radius. See Fig. 7.4.

One of the primary concerns when using a swerve drive system based on Holmberg et al.'s [10] work is the sheer number of motors that the system

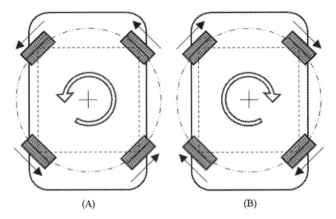

(A)   (B)

FIGURE 7.2 Swerve drive rotational operation.

TABLE 7.1 Swerve drive omnidirectional motions summary.

Figure	Figure Letter	Description
7.2	A	Anti-clockwise rotation about the centroid
7.2	B	Clockwise rotation about the centroid
7.3	C	Forward/reverse motion
7.3	D	Left/right motion
7.3	E	Forward diagonal motion
7.3	F	Reverse diagonal motion

would require. Since each unit needs steering and drive motors. The number of motors for a four-wheeled system will be eight individual motors. One approach commonly used to reduce the number of motors is to couple all the drives and steering motors together. This idea was explored by Wada et al. [11]; their design is illustrated in Fig. 7.5.

Wada et al. [11] coupled all four wheels to a single drive motor and a single steering motor using a belting system. One crucial feature that must be noted with this system is that the steering sprockets of one diagonal wheel pair must be wound opposite to the remaining diagonal pair to ensure the steering works correctly.

This design effectively reduced the number of motors from eight to two. There are, however, two significant drawbacks to this system. The first is that when the steering is coupled, as shown in Fig. 7.5, the AGV can no longer rotate about its centroid (as done in Figs. 7.2A and 7.2B). The system is thus no longer truly omnidirectional or holonomic. The second drawback is that with all of the drive motors coupled, Ackerman turning the AGV will cause the wheels to scuff without any form of a differential gear. The need for a differential, when Ackerman turning, is described in Fig. 7.6.

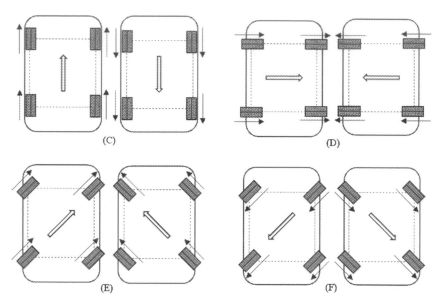

FIGURE 7.3 Swerve drive omnidirectional operation.

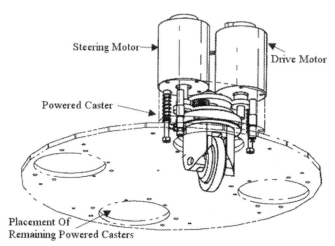

FIGURE 7.4 Holmberg & Slater's powered castor concept. Image adapted from Holmberg et al. [10].

The swerve drive system does, however, have the advantage of using traditional wheels rather than specialized holonomic wheels. The use of traditional wheels is significant as the wearing component of any AGV is the wheel, and as such cheaper wheels will ensure a cheaper running cost, even if the upfront cost is greater with this system.

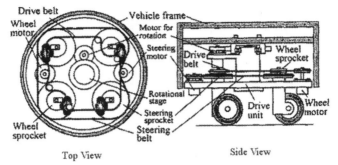

Top View   Side View

FIGURE 7.5 Wada, Takagi & Mori's synchronous powered castor concept. Image adapted from Wada et al. [11].

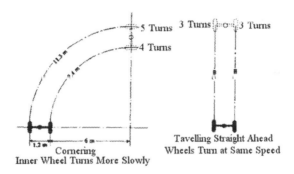

FIGURE 7.6 Differential gear justification. Image adapted from Patel [12].

7.1.1 Justifying research: two wheel swerve drive system

The research aims to produce an omnidirectional AGV for use in industry; thus this system must take industrial demands into account. The omniwheel and spherical wheel system can therefore be immediately discarded for industrial use due to their relative fragility [13,14]. Discarding these systems would leave only mecanum wheels and swerve drive systems as viable options.

Since the mecanum wheel is the wearing part of the mecanum wheel system, maintenance will be costly. Swerve drive systems do not suffer from this issue as they use conventional wheels that are cost-effective to replace. Mecanum wheels have difficulty with stability as the change in traction of a single wheel can affect the motion of the whole AGV [15]. This issue is compounded by the fact that wheel encoders cannot be easily implemented to determine slippage and compensate appropriately due to the nature of the mecanum wheel [16]. Swerve drive systems can directly compensate for slippage with an encoder and torque feedback as they use traditional wheels [17]. The final issue with mecanum wheels, when compared to the traditional wheels of the swerve drive system, is the vibration that all mecanum wheels produce as the load is transferred from one peripheral roller to the next [18].

With all of the mentioned disadvantages of mecanum wheels, it would appear that the traditional swerve drive system should be completely dominant in the manufacturing industry. Swerve drive system are, however, not dominant. For all their benefits, swerve drive systems are less common than mecanum systems in industry (when a holonomic omnidirectional vehicle is required). This lack of dominance is solely due to the initial cost of a traditional swerve drive system compared to a mecanum system. The high initial cost of swerve-drive AGVs is directly attributable to the eight motors needed to produce a functional four-wheeled system [13] when compared to the four required for an equivalent mecanum wheeled system.

This research proposes alleviating the high initial cost of the traditional swerve drive system by removing two diagonal swerve units and replacing them with unconstrained castors. This removal will reduce the number of motors from eight to four, in line with a typical mecanum wheel system.

7.1.2   Justifying research: conforming to SIL standards

Any machine that operates in an industrial environment must, by law, conform to appropriate legally binding safety standards. In South Africa, this is the SANS standard, which is based on a combination of the international standards ISO and IEC [19]. Most academic research fails to take this into account and often produces machines and prototypes that are marvels of research but need to be completely redesigned from the ground up when commercialized. This redesign is necessary as preliminary design choices made in a "safety vacuum" would not allow safety laws to be adhered to retroactively.

7.1.3   Justifying research: intergeneration of a suspension system in swerve drive

Suspension systems on factory AGVs are a rarity since most AGVs of this type are expected to work on a floor with flat ground [20]. This expectation is a common doctrine, especially in the more industrialized nations of Europe and Mainland China. This doctrine, however, cannot be extended to the South African environment. Many buildings that house factories in South Africa have floors that no longer conform to SANS 10400 [15,21]. This unevenness results from adding and removing machines over the years (leaving behind remnants of anchor points) or general wear and tear. Given the harsher conditions an AGV would likely experience in South Africa, it is advantageous to incorporate a cost-effective suspension system. This implementation has yet to be done on a swerve drive AGV.

7.2   Design of the novel two-wheel swerve drive AGV

The layout of the two-wheel swerve drive AGV had the swerve drive units located at the diagonal corners of the AGV as illustrated in Fig. 7.7.

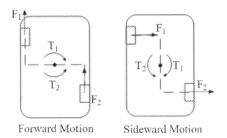

Forward Motion   Sideward Motion

FIGURE 7.7 Cancellation of net torques about the centroid.

This orientation ensured that regardless of the AGV's direction of travel, the centroid of the AGV would experience no net torque. This lack of net torque is due to the moments generated by the drive units about the centroid, canceling each other since they will always have equal magnitude but opposite directions. In practice, this is not always the case, as effects such as wheel slip will disrupt this balance, but it provides a stable starting point that allows easier correction of these issues in software. This simplifies control of the AGV significantly.

The equations that validate the effect shown in Fig. 7.7 are given in Eqs. (7.1) and (7.2).

$$\sum M = T_1 + T_2. \tag{7.1}$$

Since $T = r \times F$ in Eq. (7.1),

$$\sum M = F_1 r_1 + F_2(-r_2) = 0, \tag{7.2}$$

as $F_1 = F_2$ and $|r_1| = |r_2|$ in Eq. (7.2).

$\sum M$	= Sum of moments about the centroid	Nm
T_i	= Torque generated by tractive force i	Nm
F_i	= Tractive force generated by the ith drive unit	N
r_i	= Perpendicular distance of the ith force from the centroid	m

The two remaining wheels were implemented as unconstrained castors. The final layout of the AGV is illustrated in the block diagram of Fig. 7.8.

The design of the drive units is illustrated in Fig. 7.9. These drive units are custom-designed systems that incorporate an "inline" suspension. Traction is provided by a servo motor, whereas the steering is done via a stepper motor. Both of these motors have STO capabilities and are therefore SIL rated.

For the drive unit to function, a tractive effort must cross the boundary between the sprung and unsprung sections of the suspension system.

The torque transferal is done via a custom mechanism the author has dubbed the "vertical compromiser". Rotation can be transferred with this mechanism while the distance between the strung and unsprung mass changes. This

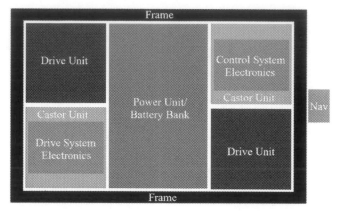

FIGURE 7.8 Block diagram layout of the AGV.

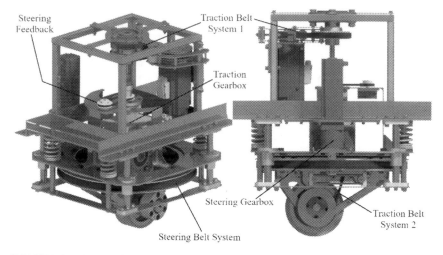

FIGURE 7.9 Finalized CAD model of the AGV drive unit.

mechanism also compensates for axial misalignment using a flexible coupling and floating bearing. The "vertical compromiser" mechanism is illustrated in Fig. 7.10.

A simplified block diagram of the traction system is illustrated in Fig. 7.11, whereas the steering mechanism is illustrated in Fig. 7.12.

7.3 Kinematics

Since the mechanical design of this vehicle drive system is so unique, very little previous work exists to describe the kinematic model, and as such, the kinematics will have to be generated from first principles. Added design features such as the castor offset and suspension further complicated this endeavor.

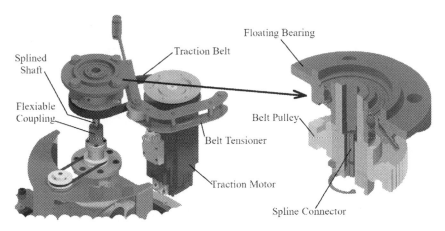

FIGURE 7.10 Final drive unit with vertical compromiser.

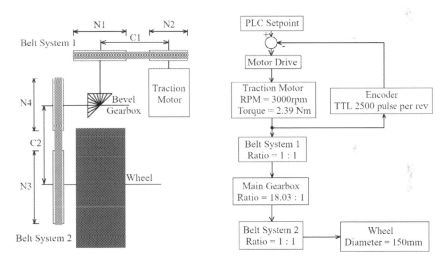

FIGURE 7.11 Tractive belt system.

7.3.1 Introduction to the drive philosophy

To produce a holonomic vehicle that is not over-contained (i.e., has more degrees of freedom than predicted by the mobility formula), using the swerve drive philosophy, at least three motors would need to be used. Using three motors would allow for the creation of a two DOF system.

7.3.2 Forward kinematics & drive unit considerations

The first task to solve in creating a kinematic model is to create the kinematics of an individual drive unit. Since the traction motors and steering motors share

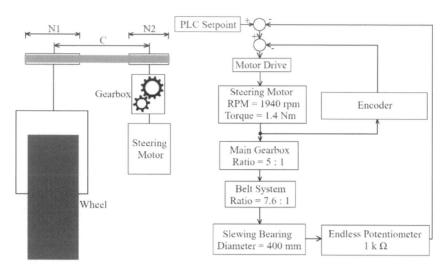

FIGURE 7.12   Steering system.

a common axis, the issue of parasitic motion rears its head (coupling of the steering and traction systems due to a shared common axis). If only the traction motor operates at ω_t and the steering motor is held still (ω_s), then the AGV will move forward. However, if the steering motor were to be actuated ($\omega_s \neq 0$), then the traction wheel will rotate as its orientation changes.

There are two strategies for solving this issue. The first is to use a feed-forward controller where an offset is applied to the traction motor angular velocity as described in the equation

$$\omega_{\text{wheel adjusted}} = \begin{cases} \omega_{\text{wheel}} + \omega_{\text{parasitic}} & \text{if } \omega_{\text{steering}} \cdot \omega_{\text{wheel}} < 0, \\ \omega_{\text{wheel}} - \omega_{\text{parasitic}} & \text{if } \omega_{\text{steering}} \cdot \omega_{\text{wheel}} > 0, \\ \omega_{\text{wheel}} & \text{if } \omega_{\text{steering}} \cdot \omega_{\text{wheel}} = 0. \end{cases} \qquad (7.3)$$

$\omega_{\text{wheel adjusted}}$ = The adjusted RPM of the wheel   rpm

ω_{wheel} = The setpoint speed of the wheel before adjustment   rpm

$\omega_{\text{parasitic}}$ = The parasitic velocity as developed by the steering mechanism   rpm

The second decoupling method is done using a mechanical compensator in the form of a decoupling gear. This idea is described by Yang et al. [22] in the thesis titled *Decoupled Powered Castor Wheel for Omnidirectional Mobile Platforms*. For this AGV, feed-forward compensation in the software will be used.

Using Fig. 7.13, it is possible to create an equation that describes the behavior of the coordinate vector of the drive unit (x, y, and the rotational velocity of the system) from the drive unit's mechanical behavior (speed of drive wheel and angular velocity of the steering mechanism), the so-called "forward kinematics". These vectors will be described as $\dot{x}_w^{\downarrow}$ and $\dot{u}_w^{\downarrow}$, respectively. The formula for this relationship is given in Eq. (7.4).

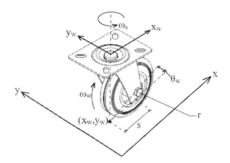

FIGURE 7.13 Coordinates of a single drive unit castor.

$$\dot{x}_w^{\downarrow} = \boldsymbol{B}_w \dot{u}_w^{\downarrow}. \tag{7.4}$$

$\dot{x}_w^{\downarrow}$	$=$ Coordinate vector of the drive unit
$\boldsymbol{B}_w$	$=$ Kinematic matrix of a drive unit
$\dot{u}_w^{\downarrow}$	$=$ Mechanical behavior vector of the drive unit

Eq. (7.4) can be expanded to

$$\begin{bmatrix} \dot{x}_w \\ \dot{y}_w \\ \dot{\theta}_w \end{bmatrix} = \begin{bmatrix} r\cos(\theta_w) & -r\sin(\theta_w) \\ r\sin(\theta_w) & s\cos(\theta_w) \\ 0 & 1 \end{bmatrix} \begin{bmatrix} \omega_w \\ \omega_s \end{bmatrix}. \tag{7.5}$$

x_w	$=$	x axis position of the steering axis	m
y_w	$=$	y axis position of the steering axis	m
r	$=$	Radius of the wheel	m
s	$=$	Castor offset of the wheel	m
θ_w	$=$	Angular orientation of the wheel	rad
$\dot{x}_w$	$=$	Velocity of a drive unit in the x direction	m/s
$\dot{y}_w$	$=$	Velocity of a drive unit in the y direction	m/s
ω_w	$=$	Angular velocity of the wheel	rad/s
ω_s	$=$	Angular velocity of the steering	rad/s

Eq. (7.5) gives the following component equations:

$$\dot{x}_w = r\cos(\theta_w)\omega_w - s\sin(\theta_w)\omega_s, \tag{7.6}$$
$$\dot{y}_w = r\sin(\theta_w)\omega_w + s\cos(\theta_w)\omega_s, \tag{7.7}$$

$$\dot{\theta}_w = \omega_w. \tag{7.8}$$

It worth noting that in Eqs. (7.6), (7.7), and (7.8) there exist common variables. Let us rearrange Eq. (7.6) in terms of

$$r = \frac{\dot{x}_w}{\omega_w \cos(\theta_w)} + \frac{s\omega_s \sin(\theta_w)}{\omega_s \cos(\theta_w)}. \tag{7.9}$$

Substituting Eq. (7.9) into Eq. (7.7) yields

$$\dot{y}_w = \left(\frac{\dot{x}_w}{\omega_w \cos(\theta_w)} + \frac{s\omega_s \sin(\theta_w)}{\omega_w \cos(\theta_w)} \right) \omega_w \sin(\theta_w) + s\omega_s \cos(\theta_w) \tag{7.10}$$

or, simplifying,

$$
\begin{aligned}
\dot{y}_w &= \frac{\dot{x}_w \sin(\theta_w)}{\cos\theta_w} + \frac{s\omega_s \sin^2(\theta_w)}{\cos(\theta_w)} + s\omega_s \cos(\theta_w), \\
\therefore \dot{y}_w \cos(\theta_w) &= \dot{x}_w \sin(\theta_w) + s\omega_s \sin^2(\theta_w) + s\omega_s \cos^2(\theta_w), \\
\therefore \dot{y}_w \cos(\theta_w) &= \dot{x}_w \sin(\theta_w) + s\omega_s.
\end{aligned}
\tag{7.11}
$$

Substituting Eq. (7.8) into Eq. (7.11) yields

$$\dot{x}_w \sin(\theta_w) - \dot{y}_w \cos(\theta_w) + s\dot{\theta}_w = 0. \tag{7.12}$$

The Pfaffian constraint matrix is generated from Eq. (7.12) as shown in Eq. (7.13), where q is the configuration $q = (x_w, y_w, \theta_w)$ of the point shown in Fig. (7.13), which can be assumed to be derived from a set of loop closure equations stating that the final position of the point q must coincide with the starting point.

$$A(q)\dot{q} = 0,$$

$$\therefore A(q)\dot{q} = A(q) \begin{bmatrix} \dot{x}_w \\ \dot{y}_w \\ \dot{\theta}_w \end{bmatrix} = 0$$

$$\implies A(q)\dot{q} = \begin{bmatrix} \sin(\theta_w) & -\cos(\theta_w) & s \end{bmatrix} \begin{bmatrix} \dot{x}_w \\ \dot{y}_w \\ \dot{\theta}_w \end{bmatrix} = 0. \tag{7.13}$$

The velocity constraint given in Eq. (7.13) ($[\sin(\theta_w) \ -\cos(\theta_w) \ s]$) cannot be integrated to give an equivalent configuration constraint matrix. Thus the system is non-holonomic in nature, with Eq. (7.12) representing the non-holonomic constraint of the system. This has the effect of reducing the space of possible velocities of the drive unit but does not reduce the space of possible configurations (i.e., positions in real space).

7.3.3 Inverse kinematics of the drive units

To control the final behavior of the drive unit, we need to calculate the steering and traction values. This calculation can be done by creating an inverse kinematic model for the drive unit from Eq. (7.4). Thus

$$\dot{u_w} = (B_w)^{-1} \dot{x_w}. \tag{7.14}$$

To determine the inverse B_w^{-1} of B_w in Eq. (7.14), according to Niku [23], we take the following steps:

1. Calculate the determinate of the matrix
2. Transpose the matrix
3. Replace each element of the transposed matrix by its own minor
4. Divide the converted matrix by the determinate

However, B_w is not a square matrix, and as such, the determinate cannot be easily found. Thus a new strategy must be used, called the pseudo-inverse (or Moore–Penrose inverse) method must be used [24]. According to Dresden [25], the pseudo-inverse matrix for $B_w \in \Bbbk^{m \times n}$ can be defined as

$$B_w^+ = (B_w^* B_w)^{-1} B_w^*. \tag{7.15}$$

If $\Bbbk = \mathbb{R}$, then the Hermitian transpose B_w^* is equivalent to the standard transpose B_w^T. Thus Eq. (7.15) becomes

$$B_w^+ = (B_w^T B_w)^{-1} B_w^T. \tag{7.16}$$

Using the pseudo-inverse matrix in Eq. (7.14) in place of the true inverse yields the following equation:

$$\dot{u_w} = \left[(B_w^T B_w)^{-1} B_w^T \right] \dot{x_w}$$

$$= \begin{bmatrix} \frac{1}{r} \cos \theta_w & \frac{1}{r} \sin \theta_w & 0 \\ -\frac{s}{s^2+1} \sin \theta_w & \frac{s}{s^2+1} \cos \theta_w & \frac{1}{s^2+1} \end{bmatrix} \begin{bmatrix} \dot{x_w} \\ \dot{y_w} \\ \dot{\theta_w} \end{bmatrix}. \tag{7.17}$$

As the system is overdetermined [24], it is very difficult to determine values for mechanical behavior vector ($\dot{u_w}$). Thus to generate a stable solution for the desired coordinate vector ($\dot{x_w}$), we can reduce the number of controlled coordinates of the coordinate vector by the number of non-holonomic constraints [24]. Since Eq. (7.12) has only one non-holonomic constraint (i.e., one formula), the coordinate vector can be reduced by one term to give a system that is not overdetermined.

As the Euclidean space is easier to work with, the rotational term in Eq. (7.5) will be released (i.e., control of the term $\dot{\theta_w}$ will be given up). This will change

Eq. (7.5) into the form

$$
\begin{bmatrix} \dot{x}_w \\ \dot{y}_w \end{bmatrix} = \begin{bmatrix} r\cos(\theta_w) & -s\sin(\theta_w) \\ r\sin(\theta_w) & s\cos(\theta_w) \end{bmatrix} \begin{bmatrix} \omega_w \\ \omega_s \end{bmatrix}.
\tag{7.18}
$$

Thus

$$
\dot{\vec{x}}_{w0} = \boldsymbol{B}_{w0}\dot{\vec{u}}_w,
$$

$$
\dot{\theta}_w = -\frac{1}{s}\dot{x}_w\sin(\theta_w) + \frac{1}{s}\dot{y}_w\cos(\theta_w),
\tag{7.19}
$$

where $\dot{\theta}_w$ is an internal variable described in terms of $\dot{x}_w$ and $\dot{y}_w$ thanks to the non-holonomic constraint discovered in Eq. (7.12). If the inverse kinematics for the new equation described in Eq. (7.19) is evaluated using the four rules defined by Niku [23] and given previously in this section, then Eq. (7.20) will result in

$$
\dot{\vec{u}}_w = \boldsymbol{B}_{w0}^{-1}\dot{\vec{x}}_{w0},
$$

$$
= \begin{bmatrix} \frac{1}{r}\cos(\theta_w) & \frac{1}{r}\sin(\theta_w) \\ -\frac{1}{s}\sin(\theta_w) & \frac{1}{s}\cos(\theta_w) \end{bmatrix} \begin{bmatrix} \dot{x}_w \\ \dot{y}_w \end{bmatrix}.
\tag{7.20}
$$

Using Eq. (7.20), the drive unit should accurately reproduce the behavior of a castor wheel; this was also proposed by Wada et al. [24]. Thus there should be no conflicts between the behavior of the driven casters in the "drive units" and the uncontrolled casters in the "castor units".

7.3.4 Kinematics of the swerve drive AGV

The kinematic model previously developed for a single wheel can be expanded to the entirety of the AGV. Since the AGV has a total of 3 DOF, two in translation (in the x and y directions) and one in rotation (about the z-axis), the AGV will need at least two drive units to control the body of the AGV relative to the universal origin. Thus the AGV will have the bare minimum of two controlled drive units.

The AGV's body and two drive units are represented in Fig. 7.14. Note that this diagram does not show the actual physical layout of the vehicle but rather a simplified layout with the drive units assumed to share a common axis, in this case, the y-axis. This simplification is done to simplify the mathematics as the distance between the wheels will not need to be broken into x- and y-components. This choice can be justified as the AGV is an omnidirectional vehicle whose front or positive x-axis is entirely arbitrary; thus a front is designated later (whatever it will be). The vehicle axis can be rotated by a determined amount about the z-axis using a transform. This simplification was proposed by Wada et al. [24].

In Fig. 7.14, the two drive units are labeled A and B; the origin of the AGV is set at the midpoint between the two drive units, which are spaced a distance

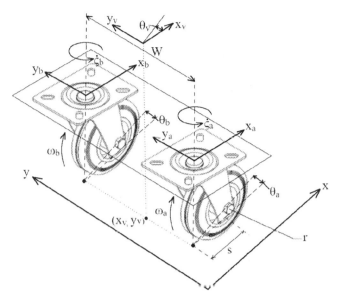

FIGURE 7.14   Coordinates of the AGV and two its drive units.

"W" apart. Since each drive unit receives $\dot{x}_w$ and $\dot{y}_w$ components of a control coordinate vector, it is necessary to differentiate between the two coordinate vectors. This adaptation is defined in the equations

$$
\begin{aligned}
\dot{x}_{w_a} &= \dot{x}_a, \\
\dot{x}_{w_b} &= \dot{x}_b, \\
\dot{y}_{w_a} &= \dot{y}_a, \\
\dot{y}_{w_b} &= \dot{y}_b, \\
\omega_{w_a} &= \xi_a, \\
\omega_{w_b} &= \xi_b.
\end{aligned}
\tag{7.21}
$$

$\dot{x}_{w_a}$	$=$	Velocity of drive unit A in the x direction	m/s
$\dot{x}_{w_b}$	$=$	Velocity of drive unit B in the x direction	m/s
$\dot{y}_{w_a}$	$=$	Velocity of drive unit A in the y direction	m/s
ω_{w_a}	$-$	Steering angular velocity drive unit A	rad/s
ω_{w_b}	$=$	Steering angular velocity drive unit B	rad/s

The forward kinematic equation to describe the AGV body, as shown in Fig. 7.14, is given in the equation

$$
\dot{\vec{x}}_v = \boldsymbol{B}_v \dot{\vec{u}}_v.
\tag{7.22}
$$

The equations of motion for the AGV's body, which are necessary to develop the matrix $\boldsymbol{B}_v$ in Eq. (7.22) are developed with the aid of the free body diagram shown in Fig. 7.15.

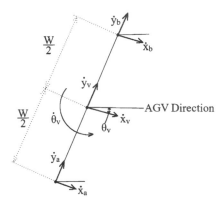

FIGURE 7.15   Free body diagram of the AGV and two its drive units.

The equation of motion for translation in the x-direction:

$$\dot{x}_v = \dot{x}_a + \dot{x}_b. \tag{7.23}$$

The equation of motion for translation in the y-direction:

$$\dot{y}_v = \dot{y}_a + \dot{y}_b. \tag{7.24}$$

The equation of motion for rotation about the z-direction:

$$\dot{\theta}_v = \dot{\theta}_{v_{x_a}} + \dot{\theta}_{v_{y_a}} - \dot{\theta}_{v_{x_b}} - \dot{\theta}_{v_{y_b}}. \tag{7.25}$$

$\dot{\theta}_{v_{x_a}}$ = AGV centroidal angular velocity due to x-component velocity of drive unit A

$\dot{\theta}_{v_{y_a}}$ = AGV centroidal angular velocity due to y-component velocity of drive unit A

$\dot{\theta}_{v_{x_b}}$ = AGV centroidal angular velocity due to x-component velocity of drive unit B

$\dot{\theta}_{v_{y_b}}$ = AGV centroidal angular velocity due to y-component velocity of drive unit B

Here

$$\dot{\theta}_{v_{x_a}} = \dot{x}_a \cos(\theta_v) \left(\frac{2}{W} \right),$$

$$\dot{\theta}_{v_{y_a}} = \dot{y}_a \sin(\theta_v) \left(\frac{2}{W} \right),$$

$$\dot{\theta}_{v_{x_b}} = -\dot{x}_b \cos(\theta_v) \left(\frac{2}{W} \right),$$

$$\dot{\theta}_{v_{y_b}} = -\dot{y}_b \sin(\theta_v) \left(\frac{2}{W} \right). \tag{7.26}$$

The equations of motion developed in Eqs. (7.23), (7.24), (7.25), and (7.26) are used to generate the matrix $\boldsymbol{B_v}$ in Eq. (7.22) to give the following equation:

$$
\begin{bmatrix} \dot{x}_v \\ \dot{y}_v \\ \dot{\theta}_v \end{bmatrix} = \begin{bmatrix} \frac{1}{2} & 0 & \frac{1}{2} & 0 \\ 0 & \frac{1}{2} & 0 & \frac{1}{2} \\ \frac{1}{W}(\cos\theta_v) & \frac{1}{W}\sin(\theta_v) & -\frac{1}{W}\cos(\theta_v) & -\frac{1}{W}\sin(\theta_v) \end{bmatrix} \begin{bmatrix} \dot{x}_a \\ \dot{y}_a \\ \dot{x}_b \\ \dot{y}_b \end{bmatrix}.
$$

$$(7.27)$$

The inverse matrix for $\boldsymbol{B_v}$ is not square, so once again, the pseudo-inverse will have to be calculated as described in Eqs. (7.15) and (7.16). Thus

$$
\vec{\dot{u}}_v = (\boldsymbol{B_v})^{-1}\vec{\dot{x}}_v
$$

$$
= \left[\left(\boldsymbol{B_v^T B_v} \right)^{-1} \boldsymbol{B_v^T} \right] \vec{\dot{x}}_v.
$$

$$(7.28)$$

Using the online mathematical tool Wolfram|Alpha [26], the pseudo-inverse matrix can efficiently be found.

$$
\vec{\dot{u}}_v = \begin{bmatrix} \dot{x}_a \\ \dot{y}_a \\ \dot{x}_b \\ \dot{y}_b \end{bmatrix} = \begin{bmatrix} 1 & 0 & \frac{W}{2}\cos(\theta_v) \\ 0 & 1 & \frac{W}{2}\sin(\theta_v) \\ 1 & 0 & -\frac{W}{2}\cos(\theta_v) \\ 0 & 1 & -\frac{W}{2}\sin(\theta_v) \end{bmatrix} \begin{bmatrix} \dot{x}_v \\ \dot{y}_v \\ \dot{\theta}_v \end{bmatrix}.
$$

$$(7.29)$$

References

[1] S. Bøgh, C. Schou, T. Rühr, Y. Kogan, A. Dömel, M. Brucker, C. Eberst, R. Tornese, C. Sprunk, G.D. Tipaldi, T. Hennessy, Integration and assessment of multiple mobile manipulators in a real-world industrial production facility, in: Proceedings for the Joint Conference of ISR 2014 - 45th International Symposium on Robotics and Robotik 2014 - 8th German Conference on Robotics, ISR/ROBOTIK 2014, no. 260026, 2014, pp. 305–312.

[2] H. Qian, T.L. Lam, W. Li, C. Xia, Y. Xu, System and design of an omni-directional vehicle, in: 2008 IEEE International Conference on Robotics and Biomimetics, ROBIO 2008, vol. 2008-Janua, 2008, pp. 389–394.

[3] J.J. Zeng, R.Q. Yang, W.J. Zhang, X.H. Weng, Q. Jun, Research on semi-automatic bomb fetching for an EOD robot, International Journal of Advanced Robotic Systems 4 (2) (2007) 247–252.

[4] J. Qian, B. Zi, D. Wang, Y. Ma, D. Zhang, The design and development of an Omni-Directional mobile robot oriented to an intelligent manufacturing system, Sensors (Switzerland) 17 (9) (2017).

[5] J.A. Batlle, A. Barjau, Holonomy in mobile robots, Robotics and Autonomous Systems 57 (4) (2009) 433–440.

[6] J. Efendi, M. Salih, M. Rizon, S. Yaacob, Designing omni-directional mobile robot with mecanum wheel, American Journal of Applied Sciences 3 (5) (2006) 1831–1835.

[7] M. Komori, K. Matsuda, T. Terakawa, F. Takeoka, H. Nishihara, H. Ohashi, Active omni wheel capable of active motion in arbitrary direction and omnidirectional vehicle, Journal of Advanced Mechanical Design, Systems and Manufacturing 10 (6) (2016) 1–20.

[8] C.W. Wu, K.S. Huang, C.K. Hwang, A novel spherical wheel driven by chains with guiding wheels, in: Proceedings of the 2009 International Conference on Machine Learning and Cybernetics, vol. 6, no. July, 2009, pp. 3242–3245.

[9] E.H. Binugroho, A. Setiawan, Y. Sadewa, P.H. Amrulloh, K. Paramasastra, R.W. Sudibyo, Position and orientation control of three wheels swerve drive mobile robot platform, in: International Electronics Symposium 2021: Wireless Technologies and Intelligent Systems for Better Human Lives, IES 2021 – Proceedings, 2021, pp. 669–674.

[10] R. Holmberg, J.C. Slater, Powered caster wheel module for use on omnidirectional drive systems, 2002.

[11] M. Wada, A. Takagi, S. Mori, Caster drive mechanisms for holonomic and omnidirectional mobile platforms with no over constraint, in: IEEE International Conference on Robotics and Automation, vol. 2, no. April, 2000, pp. 1531–1538.

[12] D. Patel, Automobile Transmission, 2017.

[13] I. Mackenzie, Omnidirectional drive system, in: 2006 FIRST Robotics Conference, 2006.

[14] C.H. Chiu, W.R. Tsai, Design and implementation of an omnidirectional spherical mobile platform, IEEE Transactions on Industrial Electronics 62 (3) (2015) 1619–1628.

[15] A.B.S. Macfarlane, Modular Electric Automatic Guided Vehicle Suspension-Drive Unit, PhD thesis, 2016.

[16] P. Xu, Mechatronics design of a mecanum wheeled mobile robot, in: K. Vedran, A. Lazinica, M. Merdan (Eds.), Cutting Edge Robotics, 1 ed., pIV pro literatur Verlag, Mammendorf, 2005, pp. 61–74, Ch. Modelling.

[17] A. Siravuru, S.V. Shah, K.M. Krishna, An optimal wheel-torque control on a compliant modular robot for wheel-slip minimization, Robotica 35 (2) (2017) 463–482.

[18] O. Diegel, A. Badve, G. Bright, J. Potgieter, S. Tlale, Improved mecanum wheel design for omni-directional robots, in: Shinji Kamiuchi and Shoichi Maeyama, "A Novel Human Interface of an Omni-directional Wheelchair", Int. Workshop on Robot and Human Interactive Communication, 2004, pp. 101–106, no. November, 2002, pp. 27–29.

[19] Siemens AG, TIA Safety, 15 ed., Siemens AG, Germany, 2019.

[20] Y. Wang, X. Lei, G. Zhang, S. Li, H. Qian, Y. Xu, Design of dual-spring shock absorption system for outdoor AGV, in: 2017 IEEE International Conference on Information and Automation, ICIA 2017, no. July, 2017, pp. 159–164.

[21] Janek, SANS10400, 2013.

[22] G. Yang, Y. Li, T.M. Lim, C.W. Lim, Decoupled Powered Caster Wheel for omnidirectional mobile platforms, in: Proceedings of the 2014 9th IEEE Conference on Industrial Electronics and Applications, ICIEA 2014, 2014, pp. 954–959.

[23] S.B. Niku, Introduction to Robotics; Analysis, Control, Applications, second ed., United States of America, John Wiley & Sons, 2011.

[24] M. Wada, S. Mori, Holonomic and omnidirectional vehicle with conventional tires, in: Robotics and Automation, 1996. Proceedings. . . . , no. April, 1996, pp. 3671–3676.

[25] A. Dresden, The fourteenth western meeting of the American Mathematical Society, Bulletin of the American Mathematical Society 26 (9) (1920) 385–396.

[26] Wolfram, WolframAlpha, 2021.

Chapter 8

Control system strategy of a modular omnidirectional AGV

Alexander B.S. Macfarlane[a], Theo van Niekerk[a], Udo Becker[b], and Paolo Mercorelli[c]

[a]*Faculty of Engineering, Built Environment & Information Technology (EBEIT) at the Nelson Mandela University, Port Elizabeth, South Africa,* [b]*Ostfalia University of Applied Sciences, Wolfsburg, Germany,* [c]*Institute for Production Technology and Systems, Leuphana University of Lueneburg, Lueneburg, Germany*

8.1   Introduction

To control the AGV using the previously developed kinematics, it is necessary first to develop a strategy to implement them.

$$
\dot{\vec{u}}_v = \begin{bmatrix} \dot{x}_a \\ \dot{y}_a \\ \dot{x}_b \\ \dot{y}_b \end{bmatrix} = \begin{bmatrix} 1 & 0 & \frac{W}{2}\cos(\theta_v) \\ 0 & 1 & \frac{W}{2}\sin(\theta_v) \\ 1 & 0 & -\frac{W}{2}\cos(\theta_v) \\ 0 & 1 & -\frac{W}{2}\sin(\theta_v) \end{bmatrix} \begin{bmatrix} \dot{x}_v \\ \dot{y}_v \\ \dot{\theta}_v \end{bmatrix}.
\tag{8.1}
$$

From Eq. (8.1) it is possible to generate the velocity vectors for drive units A and B from the desired AGV x, y velocities and angular velocity. These values can be fed into the appropriate drive unit to control accurately the desired wheel and steering angular velocities of each unit. This idea is graphically explained in Fig. 8.1.

A control system strategy can be developed from the behavior described in Fig. 8.1. This strategy is illustrated in Fig. 8.2.

8.2   Test methodology

Four operational tests were performed to validate if the AGV in this research can perform equivalently to a traditional four-wheel suspensionless AGV. These tests can be performed stably by a traditional swerve-drive AGV as validated by Holmberg et al. [1] and Chikosi [2].

These tests are:

- Straight line test

Modeling, Identification, and Control for Cyber-Physical Systems Towards Industry 4.0
https://doi.org/10.1016/B978-0-32-395207-1.00019-6

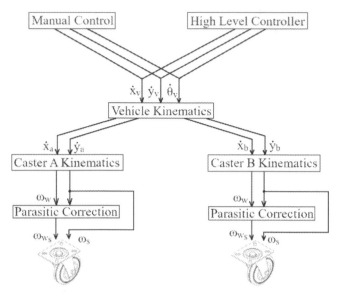

FIGURE 8.1 Control system strategy.

- Strafe test
- Ackerman steering test
- Combination test

Each test was run five times, with the average of the five tests used for analysis. Each of these five repeats contains approximately 6000 samples each. The test is created by recording a set of inputs used to perform movements in the manual mode. This recording was then streamed five times into the control system of the AGV to get a system reproducing a set of identical motions.

The variables streamed to the PLC were:

1. Pendant speed potentiometer value
2. Pendant Ackerman steering angle potentiometer value
3. Pendant strafe angle potentiometer value

The resultant values recorded during each test were:

1. Setpoint centroidal x-component velocity (m/s)
2. Setpoint centroidal y-component velocity (m/s)
3. Setpoint centroidal yaw rate (rad/s)
4. Unit A steering RPM (rpm)
5. Unit B steering RPM (rpm)
6. Unit A traction RPM (rpm)
7. Unit B traction RPM (rpm)
8. Actual centroidal x-component velocity (m/s)
9. Actual centroidal y-component velocity (m/s)
10. Actual centroidal yaw rate (rad/s)

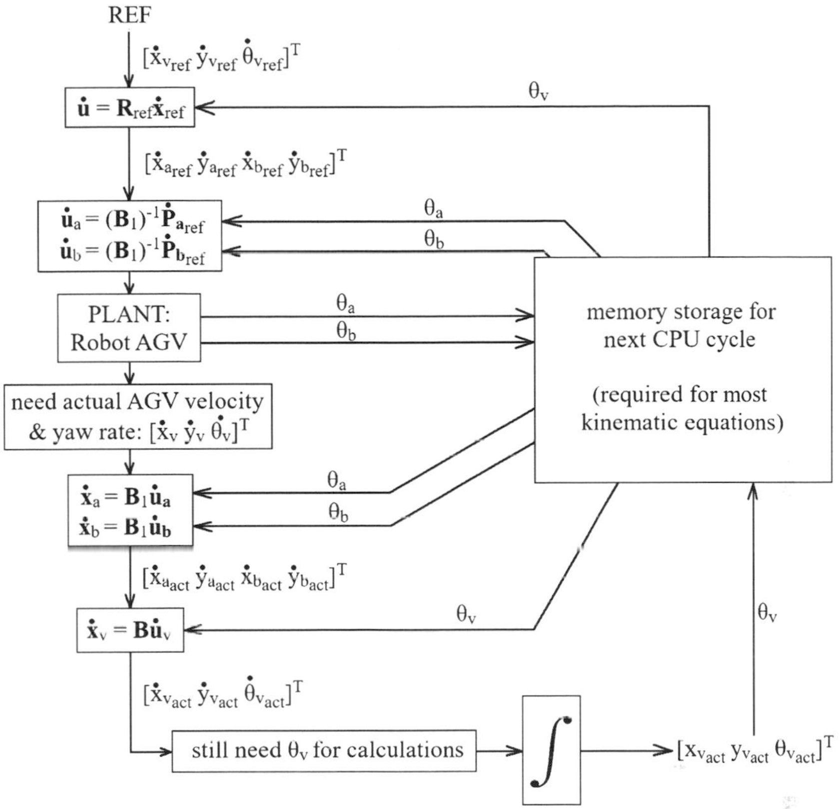

REF

$$[\dot{x}_{v_{ref}}\ \dot{y}_{v_{ref}}\ \dot{\theta}_{v_{ref}}]^T$$

$$\dot{u} = R_{ref}\dot{x}_{ref}$$ ← θ_v

$$[\dot{x}_{a_{ref}}\ \dot{y}_{a_{ref}}\ \dot{x}_{b_{ref}}\ \dot{y}_{b_{ref}}]^T$$

$$\dot{u}_a = (B_1)^{-1}\dot{P}_{a_{ref}}$$ ← θ_a
$$\dot{u}_b = (B_1)^{-1}\dot{P}_{b_{ref}}$$ ← θ_b

PLANT:
Robot AGV → θ_a → θ_b →

memory storage for
next CPU cycle

(required for most
kinematic equations)

need actual AGV velocity
& yaw rate: $[\dot{x}_v\ \dot{y}_v\ \dot{\theta}_v]^T$

$$\dot{x}_a = B_1\dot{u}_a$$ ← θ_a
$$\dot{x}_b = B_1\dot{u}_b$$ ← θ_b

$$[\dot{x}_{a_{act}}\ \dot{y}_{a_{act}}\ \dot{x}_{b_{act}}\ \dot{y}_{b_{act}}]^T$$

$$\dot{x}_v = B\dot{u}_v$$ ← θ_v θ_v

$$[\dot{x}_{v_{act}}\ \dot{y}_{v_{act}}\ \dot{\theta}_{v_{act}}]^T$$

still need θ_v for calculations → $\int$ → $[x_{v_{act}}\ y_{v_{act}}\ \theta_{v_{act}}]^T$

FIGURE 8.2 Kinematic control strategy.

11. Actual unit A steering angle (deg)
12. Actual unit B steering angle (deg)
13. Actual centroidal angle (deg)

8.3 Results

The results of the tests described in Section 8.2 are recorded in this section.

8.3.1 Straight-line test

The straight-line test locks the steering (Ackerman and strafe) to zero degrees, i.e., the AGV is traveling in a straight line. Only the speed potentiometer on the pendant was used; the other two pots were left in the "neutral" position. This resulted in the centroid setpoint velocities shown in Fig. 8.3.

Note that in Fig. 8.3 the only value that changes is the AGV setpoint centroidal x-component velocity. The setpoint centroidal y-component velocity and

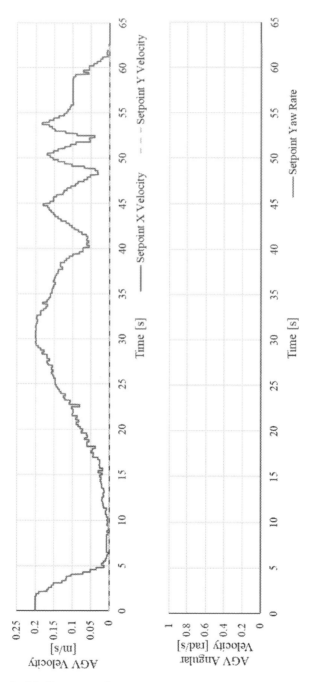

FIGURE 8.3   Straight-line test setpoints.

setpoint centroidal yaw rate remain zero. This behavior is expected as the AGV is only moving in the forward (x-axis) direction.

This test was started at 0.2 m/s as illustrated in Fig. 8.3. This starting point was chosen to saturate the AGV acceleration, as it would have to jump from 0 m/s to 0.2 m/s immediately (obviously, acceleration limits were imposed on using the servo drives configuration). Next, the system was ramped down to near 0 m/s, before being very gradually being ramped back up to 0.2 m/s. Following this, the speed of the AGV was ramped up and down three more times, with each successive ramp accelerating and de-accelerating faster than the last. These ramps are visible as the last three spikes between times 40 s and 55 s in Fig. 8.3. The AGV was run at a fixed speed between 55 s and 60 s. The test was stopped at approximately the 65 s mark.

From the setpoint values developed in Fig. 8.3 the kinematic model calculated the required traction and steering speeds to ensure that the centroid of the AGV moved per these desired setpoints.

Since there is no centroidal component velocity in the y-direction or a yaw rate caused by Ackerman steering, the steering motors remain at zero RPM throughout the entire test. Hence why these graphs were not included in the results. The two traction motors, however, had non-zero RPMs as illustrated in Fig. 8.4.

The tractive RPM values shown in Fig. 8.4 represent the actual wheel RPM. This value was determined by reading the motor encoder's RPM and dividing this value by the gearbox ratio.

As illustrated in Fig. 8.4, the immediate jump from 0 m/s to 0.2 m/s at the < 1 s mark results in the highest acceleration of the wheels. Both drive wheels go from 0 rpm to 25 rpm in under 1 s (as illustrated in Fig. 8.4). After 1 s, the acceleration of the traction motors will not be saturated, so for the most part, the curve of the rpm of the wheels can be matched to the centroidal setpoint x-velocity curve in Fig. 8.3. This behavior is expected, as when the steering is not used, the speed potentiometer on the pendant essentially becomes the setpoint for the wheel speed.

From the actual RPM of the steering and traction motors it was possible to use the kinematic model to calculate the actual centroidal x-component velocity, actual centroidal y-component velocity, and actual centroidal yaw rate.

These results were plotted in the graphs contained in Fig. 8.5. In Fig. 8.5, we can see that the actual y-component velocity is zero throughout the test, which is entirely expected as the AGV traveled in a straight line during this test.

The centroidal x-component velocity is almost an exact match for the setpoint x-component velocity (shown in Fig. 8.3), which is to be expected as the actual value should closely follow the setpoint. However, there is a slight discrepancy between the two graphs at the <1 s mark. This discrepancy was where the traction motor acceleration was saturated, and as a result, there was a delay between when the setpoint dictated that the AGV should be moving at 0.2 m/s and when the AGV was moving at 0.2 m/s.

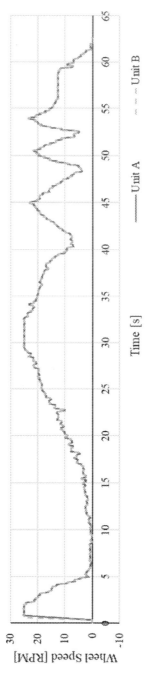

FIGURE 8.4 Straight-line test tractive wheel RPMs.

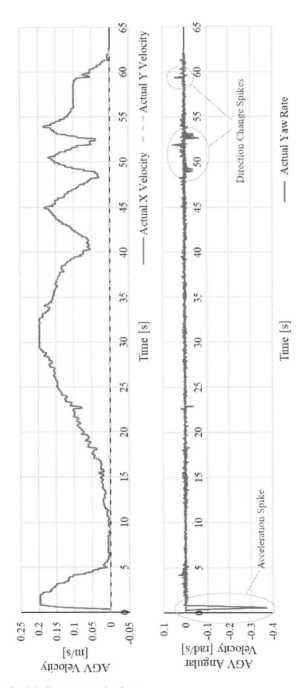

FIGURE 8.5 Straight-line test actual values.

Also note the yaw rate graph shown in Fig. 8.5. We would expect this graph to be constant at zero, like the setpoint graph in Fig. 8.3, but this is not the case. Since the traction and steering mechanism share a common axis, rapid changes in the behavior of the traction can cause the steering mechanism to shift from its setpoint position (zero in this case). The closed-loop control of the steering mechanism then will quickly bring the steering back to its correct setpoint position. This behavior is visible as spikes on the actual yaw graph of the AGV that correspond time-wise to inflexions of the actual centroidal x-component velocity. These spikes are highlighted in red (mid gray in print version) on the actual yaw rate graph in Fig. 8.5. The worst offender is the spike that occurs during acceleration saturation, named the "acceleration spike", and the remaining spikes, called "direction change spikes", have minimal effect and are not noticeable in the real world.

8.3.2 Strafe tests

During the strafe test, the setpoint speed of the AGV was kept as constant as possible. The speed pot was ramped up from zero to the point where the tractive wheels were rotating at 16 rpm (see Fig. 8.7). The strafe angle of the AGV (holonomic motion with AGV reference frame remaining fixed relative to the universal frame) was then adjusted so that the machine was strafing either left or right. Strafing was done using the strafe potentiometer on the pendant. The sequence of movements is as follows:

- turn anticlockwise 90° quickly, then back to 0° quickly
- turn clockwise 90° quickly, then back to 0° quickly
- turn anticlockwise 90° slowly, then back to 0° slowly
- turn clockwise 90° slowly, then back to 0° slowly
- turn anticlockwise 180° slowly, then back to 0° slowly
- turn clockwise 180° slowly, then back to 0° slowly

The motions described in the previous itemized list will change the ratio between the x- and y-centroidal component velocities. The angle describes the ratio change between these two developed as a result of the strafe potentiometer setting, which is illustrated in the equations

$$
\begin{aligned}
x_v &= v_{\text{AGV}} \cos(\xi_{\text{strafe}}), \\
y_v &= v_{\text{AGV}} \sin(\xi_{\text{strafe}}).
\end{aligned}
\tag{8.2}
$$

x_v	=	x-centroidal component velocity	m/s
y_v	=	y-centroidal component velocity	m/s
v_{AGV}	=	Setpoint velocity of AGV (from speed pot)	m/s
ξ_{strafe}	=	Setpoint strafe angle (from strafe pot)	rad

The yaw rate of the AGV will remain zero for this test as no Ackerman steering takes place, and as such, the reference frame of the AGV will maintain its orientation with reference to the universal frame.

The setpoint x- and y-centroidal component velocities are illustrated in Eq. (8.2), along with the zero yaw rate throughout the test. The results of these measurements are illustrated in Fig. 8.6. Note that this test took 2 minutes (120 s) to run.

The setpoint values shown in Fig. 8.6, when fed into the kinematic model, caused the wheels and steering to move as shown in Fig. 8.7 to follow these setpoints.

As we see in Fig. 8.7, the tractive (wheel) RPM remains relatively constant after the AGV is ramped up from zero RPM, with only minor disturbances. These disturbances coincide with fast steering motions and are caused by the traction and steering mechanisms coupling due to their shared axis. Coupling only occurs at high steering speeds as at these speeds the compensation algorithm cycle time is too slow to compensate correctly. These spikes are temporary, and the compensation algorithm will return the speed to steady state conditions given time. These disturbances are highlighted in red circles (mid gray circles in print version) on the wheel speed graph in Fig. 8.7.

On the steering speed graph in Fig. 8.7, the steering motions of the AGV have been highlighted. When enclosed in a red (mid gray in print version) dotted block, the steering is attempting to move the strafe direction of the AGV in a counter-clockwise direction and then back to zero. The direction is evident as the AGV steering motion (for both drive units) increases in the positive direction (CCW id defined as positive), peaks and then decreases to zero RPM. This Zero RPM point represents the steering angle zenith and the inflexion point where the steering angle will return to zero (the negative RPM after this point). The sections of the graph enclosed in a dotted blue (dark in print version) line represent a clockwise motion for the AGV, as the RPM accelerates in a negative direction, decelerates to the inflexion point (maximum steering angle point), and then accelerates in the positive direction to the zero-degree angle point.

The motions in the previous paragraph can be more easily understood if viewed as an angle value rather than a velocity value. This analysis can be done with the aid of Fig. 8.8, which represents the steering angles of the drive units with reference to time. Fig. 8.8 gives the angle in degrees as measured directly from the relative encoder on the stepper motors (multiplied by a gear ratio).

From the actual values produced by the encoders (both on the tractive system and steering system) the real-world values were captured and represented in Fig. 8.7. When these values are fed into the kinematic model in reverse, the actual centroidal x-component and y-component velocities can be determined along with the actual centroidal yaw rate as illustrated in Fig. 8.9. It is then possible to compare the actual values to the setpoint values given in Fig. 8.6 to determine the accuracy of the control system.

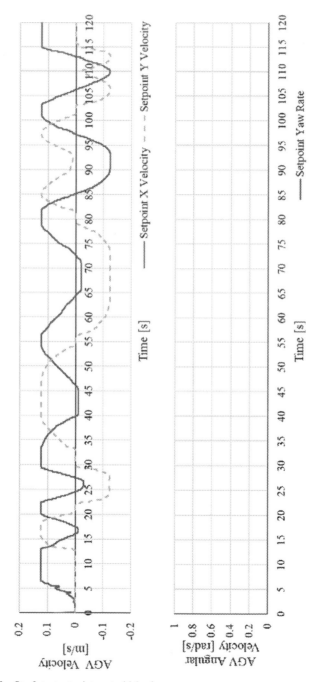

FIGURE 8.6 Strafe test setpoint centroidal values.

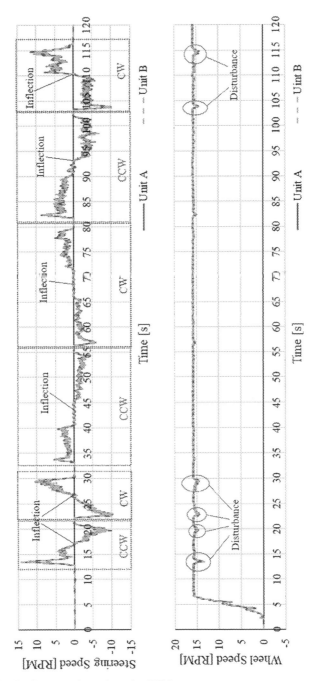

FIGURE 8.7 Strafe test tractive and steering RPMs.

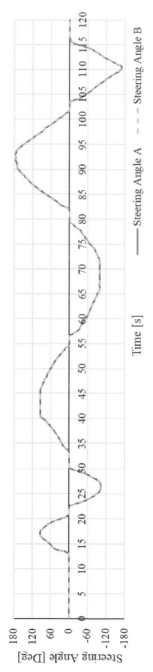

FIGURE 8.8 Strafe test drive unit steering angles.

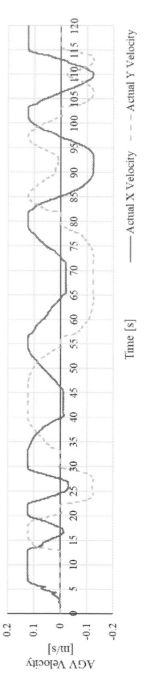

FIGURE 8.9 Strafe test actual centroidal values.

When comparing the actual (Fig. 8.9) and setpoint (Fig. 8.6) centroidal motion values, we can see that the actual value very closely matched the setpoint value. This result was expected since the drives were not saturated and were able to keep their speed in sync with the setpoint speed.

The last set of values that need to be examined is the behavior of the yaw of the AGV. These values include the actual yaw rate and the actual yaw angle of the AGV. In Fig. 8.6 the setpoint yaw rate is shown to be zero throughout the test; however, the actual yaw rate does not remain at zero. This effect is illustrated in Fig. 8.10.

Although there is plenty of jitter about the actual centroidal yaw rate (as illustrated in Fig. 8.10), this jitter is centralized and balanced around the 0 rad/s point. With the peaks extending to only ~ 0.01 rad/s (0.095 rpm). Thus this jittery is negligible and not noticeable under normal observation for the most part.

Of concern, however, is the drift experienced by the yaw angle. This angle should have remained zero throughout the entirety of the test; however, as illustrated in Fig. 8.10, the yaw angle drifted consistently by approximately 1° every 110 seconds in the clockwise direction. The yaw angle is produced by integrating the actual yaw rate. Since the actual yaw rate is relatively consistent about the zero point (see yaw rate graph in Fig. 8.10), likely, the time interval used for the numerical analysis-based integration (trapezoidal method) is not as stable as it should be. That is to say, there exists jittery about the sample point times. Stabilizing the sample times is not easy to fix as the cyclic interrupt system (used for the sampling) on Siemens PLCs is closed-source, and the programmer can only set an interval time.

The easiest solution to this issue would be to run the integrated value through an error correction control loop that used the navigation sensor data (from the NAV 350) to correct this drift or abandon the integration altogether and only use feedback from the NAV 350 sensor for the yaw angle.

8.3.3 Ackerman steering test

Ackerman steering involves changing the heading of the AGV (i.e., the AGV frame will rotate relatively to the universal frame) like a car. The sequence of motion for this 85-second test is listed in the itemized text:

- hard steer anti-clockwise (left), then return to zero
- hard steer clockwise (right), then return to zero
- gradual steer clockwise (right), then return to zero
- gradual steer anti-clockwise (left)
- continue to hard steer anti-clockwise (left) and return to zero
- extremely hard steer clockwise (right) and return to zero

During the steering test, the AGV linear velocity was kept constant in the x-direction (relative to the AGV reference frame) at 0.9 m/s. This behavior is illustrated in the setpoint centroidal velocity graph contained in Fig. 8.11. Since Ackerman steering is performed during this test, the setpoint yaw rate, for the

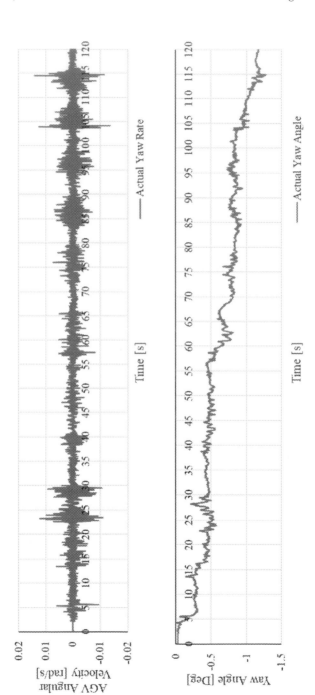

FIGURE 8.10 Strafe test yaw rate and yaw angle.

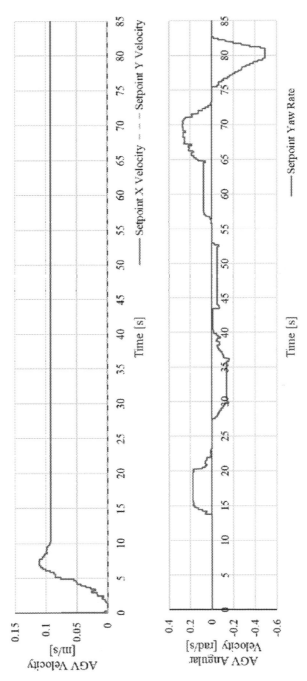

FIGURE 8.11 Ackerman steering setpoint centroidal speeds.

first time, is non-zero. The yaw rate value (Fig. 8.11 is determined by the Ackerman steering pot found on the pendant).

The setpoint values described in Fig. 8.11 were fed into the kinematic model by the AGV control system to produce a set of tractive and steering RPMs (shown in Fig. 8.12). To perform Ackerman steering correctly, both the tractive velocity of the wheels themselves and the steering angle they moved to had to be varied. If only the steering angle were to change and the tractive rpm of the wheels were to remain constant, then scuffing would occur. Thus the RPMs of the wheels were varied to produce a virtual differential effect, as we can see in the wheels speed graph in Fig. 8.12.

To simplify understanding, the motions described in the previous itemized list (for this test) are highlighted by enclosing them in dotted boxes. Red (mid gray in print version) dotted boxes represent clockwise steering relative to the universal frame (i.e., the AGV is turning right), whereas blue (dark gray in print version) dotted boxes represent counter-clockwise steering relative to the universal frame (i.e., the AGV is turning left). When interpreting the steering speed, it is important to note that the speed will "ramp up" for a deterministic amount of time to move the steering to the desired steering angle. Once this angle has been reached, the steering speed will drop to zero RPM. The AGV will turn about the yaw until the steering is "ramped down" to the zero angle. The "ramp up" and "amp down" are labeled in Fig. 8.12 for clarity.

The direction of yaw rotation in Fig. 8.12 can be determined by establishing which unit is rotating in a positive direction (for Ackerman steering, the other unit will always rotate counter to this, i.e., in a negative direction). If the AGV is turning anti-clockwise relative to the universal frame (i.e., turning left), then the "ramp up" would have been positive for unit A and negative for unit B. The opposite will be true if the AGV turns clockwise (i.e., right) relative to the universal frame.

This steering direction effect is also visible in the wheels speed graph in Fig. 8.12, where one of the wheel speeds is boosted relative to the other (to create a virtual differential). The unboosted unit remains at the nominal speed of 12 RPM. For clarity:

Counter-clockwise (left) steering	Unit A speed boost
Clockwise (right) steering	Unit B speed boost

To prevent scuffing, the RPM of the wheels themselves will have to change to replicate a virtual differential. This behavior, however, breaks down when the boost speed reaches saturation (i.e., maximum allowable wheel RPM). When this happens, the non-boosted unit will also experience a speed boost, as can be seen in the last CCW and CW motions of the wheel speed graph in Fig. 8.12. Note that the last CW (counter-clockwise motion) attempts to turn the AGV left at such a sharp turning radius for the given speed that the speed of both units becomes saturated, and the actual and setpoint values will deviate.

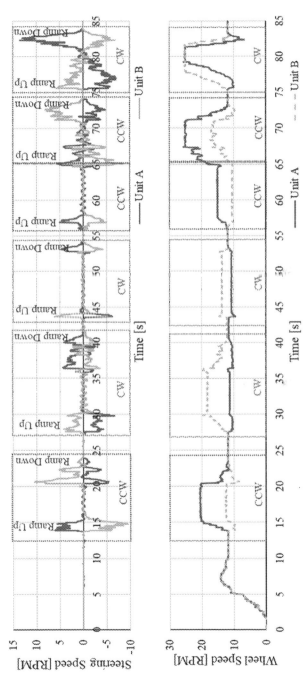

FIGURE 8.12 Ackerman steering and tractive system RPMs.

The actual wheel and steering RPMs (described in Fig. 8.12) were fed into the reverse kinematic model to determine the actual centroidal x-component velocity, y-component velocity, and yaw rate. These results are represented in Fig. 8.13.

For the most part, the actual speed values mirror the setpoint values given in Fig. 8.11. As previously mentioned, this relationship only breaks down when the traction speed is saturated. Between 65 and 75 seconds, a mark on the AGV velocity graph in Fig. 8.13 indicates a minor instability occurring on the x-component velocity. At this point the speed of Unit A was saturated, and as such, the speed of Unit B had to be increased to maintain the desired steering arc. However, even with the minor instability, the actual yaw rate could be maintained to match the desired yaw rate shown in Fig. 8.11. However, the system falls apart when the speeds of both units are saturated (between 77 and 83 seconds). At this point the actual yaw rate (AGV angular velocity graph in Fig. 8.13) deviates from the setpoint yaw rate in Fig. 8.11. This deviation is marked with a red (mid gray in print version) deviation bar in Fig. 8.13.

The saturation of the steering speed between 77 and 83 seconds also caused the AGV to inadvertently strafe when it was not supposed to as it could not maintain the desired steering arc. This behavior can be seen in Fig. 8.13 as the y-centroidal component deviating away from zero RPM and peaks at the 80-second mark.

The steering angles of the drive units during this test are illustrated in Fig. 8.14. It is important to note that the steering angle does not have the same magnitude but opposite direction as would intuitively be expected. As we see in Fig. 8.14, the angles have different magnitudes. These magnitudes diverge according to a fixed constant resulting from the speed of the AGV and the desired yaw rate. The higher the desired yaw rate (and thus the tighter the turning arc), the larger this constant.

When the steering angle plateaus at a non-zero value in Fig. 8.14, the AGV turns at a constant rate relative to the universal frame. That is to say, the steering arc of the AGV remains constant, and if the steering angle does not change, the AGV turns in a circle whose radius is determined by the steering arc/yaw rate. This is illustrated in Fig. 8.15.

Nothing prevents the AGV from performing a full 360° rotation relative to the universal frame. For Fig. 8.15, a full 360° is mapped between −180° and 180°. Thus, whenever the AGV yaw angle crosses over the 180° mark, it is mapped onto the opposite sign spectrum. This effect is visible in Fig. 8.15 as a vertical straight line that jumps from −180° to 180° or vice versa.

Note that although the integration drift is not easily visible in Fig. 8.15, it still exists as described in Fig. 8.10. This drift is 1° every 110 seconds in the clockwise (negative) direction.

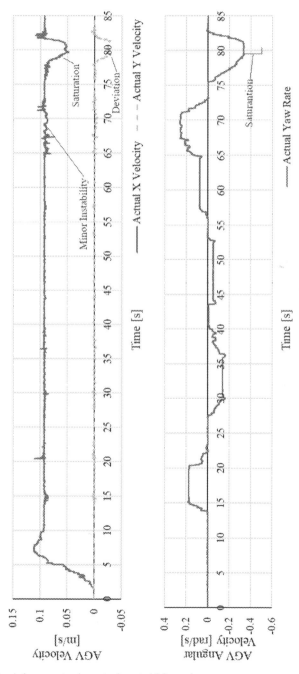

FIGURE 8.13 Ackerman steering actual centroidal speeds.

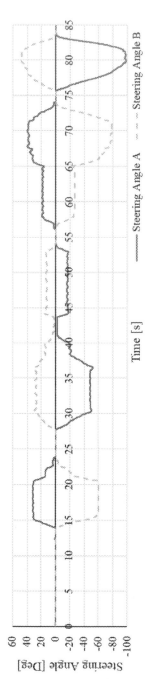

FIGURE 8.14 Ackerman test drive unit steering angles.

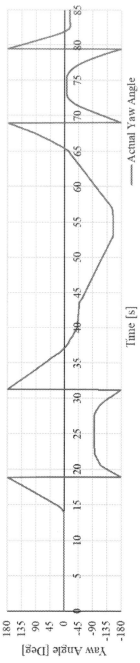

FIGURE 8.15 Ackerman test AGV yaw angle.

8.4 Combination test

The combination test combines the three previous tests to check if there is any unexpected behavior when running these different manoeuvrers together. The sequence of the test is as follows:

- Ramp up to constant velocity (50° max velocity)
- Strafe in the anticlockwise (left) direction, then return to zero
- Strafe in the clockwise (right) direction, then return to zero
- Ramp speed up to max speed, then back down to 50°
- Ackerman steer anticlockwise (left), then return to zero
- Ackerman steer clockwise (right), then return to zero

Therefore this test involves all three control potentiometers on the pendant: the speed, strafe, and steering pots. The values of these pots with respect to time were used to develop the centroidal setpoint velocities and centroidal yaw rate of the AGV. These centroidal velocities and the yaw rate are given with respect to time in the graphs in Fig. 8.16. This test took 130 s to complete.

In Fig. 8.16 the changes in linear speed of the AGV are enclosed in green blocks (light gray blocks in print version), the strafe manoeuvrers are enclosed in blue blocks (dark gray blocks in print version), and Ackerman steering is enclosed in red blocks (mid gray blocks in print version).

These three setpoint values (in Fig. 8.16), once fed into the kinematic models, would produce a set of steering and traction RPMs that the tractive and steering motors would run at. The resultant wheel and steering speeds, as recorded by encoders, are given in Fig. 8.17.

In Fig. 8.17 the different manoeuvrers are once again highlighted in different colored blocks (green [light gray in print version] = speed change, blue [dark gray in print version] = strafe, and red [light gray in print version] = Ackerman steering). From this it is possible to deduce that a speed change will only affect the wheel RPMs, with both units having the same RPM magnitude and sign of wheels (see Fig. 8.17), and the steering RPMs remain constant. When the AGV strafs left or right, the wheel speed remains constant, and only the steering RPMs are affected. Steering RPMs of Both units have the same magnitude and sign during this operation. Finally, both the steering and wheel RPMs are affected when a manoeuvrer is performed using Ackerman steering. Ackerman steering causes the steering RPMs to have opposite signs and whose magnitude is a constant scaling dependent on the setpoint yaw rate. The wheel RPMs during Ackerman steering have the same sign, but one of the unit speeds will be boosted relative to the other to create a "virtual differential". The unit whose speed is boosted depends on whether the Ackerman steering turns the AGV left or right (left = unit A speed boost, right = unit B speed boost).

The reverse kinematic model, when used on the actual RPM values recorded in Fig. 8.17, will produce the actual centroidal x-component velocity, y-component velocity, and centroidal yaw rate. These actual centroidal values are given in Fig. 8.18.

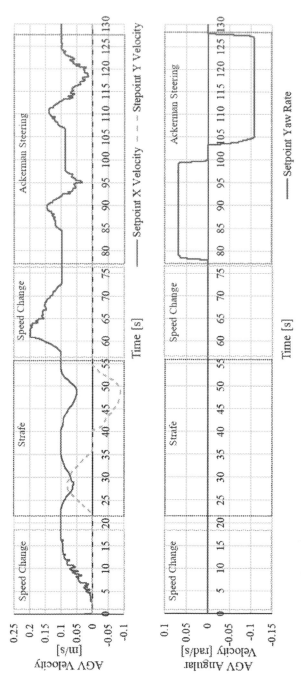

FIGURE 8.16 Combination test centroidal setpoint values.

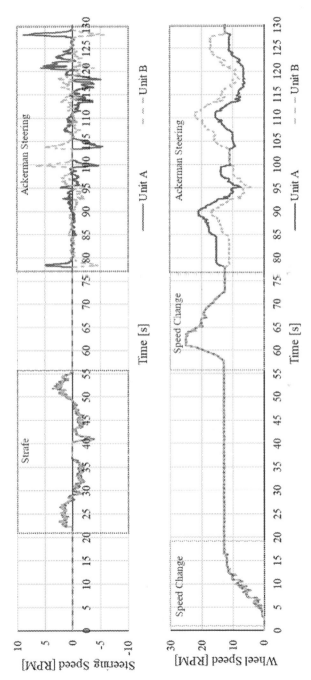

FIGURE 8.17   Combination test wheel and steering RPMs.

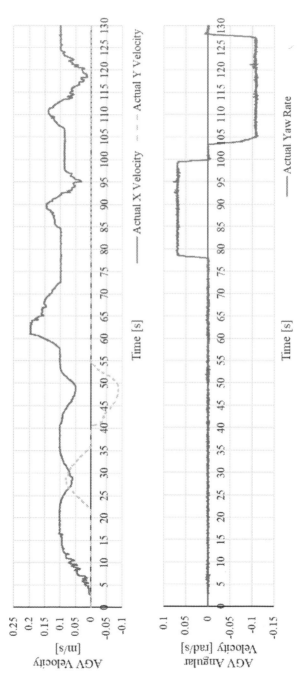

FIGURE 8.18 Combination test centroidal actual values.

When in Fig. 8.16 the setpoint values are compared to the actual values in Fig. 8.18, they are almost identical except for some minor noise. This effect was expected as during this test, there was no acceleration or velocity of any motors driven to saturation.

The steering angles of the drive units, captured using the stepper motor encoders and a gear ratio calculation, are given in Fig. 8.19. The angle of the wheels during a strafe manoeuvrer is enclosed in a blue box (dark gray box in print version) in Fig. 8.19, whereas during an Ackerman steering manoeuvrer, it is enclosed in a red box (mid gray box in print version).

Note in Fig. 8.19 that when the AGV is strafing, the angles have the same magnitude and sign. When the AGV is performing Ackerman steering, the angles of the wheels have opposite signs and will be scaled relative to each other by a constant determined by the desired yaw rate, mirroring the behavior seen in Fig. 8.17 for the steering RPMs.

A change in the setpoint speed of the AGV has negligible effects on the steering angle.

The yaw angle of the AGV is determined using integration of the actual yaw rate (from Fig. 8.18) and is given in Fig. 8.20. The yaw angle measures the angle between the AGV frame and the universal frame; as such, it will only change from zero when Ackerman steering is used. This behavior can be seen between the 75 and 130 seconds on Fig. 8.20, which corresponds to a set of Ackerman manoeuvrers during this test. It is important to note that the yaw angle drift of 1° every 110 seconds in the clockwise (negative) direction (discovered in Section 8.3.2) is still present but not noticeable due to the scale used in Fig. 8.20.

8.5   Conclusions

This section discusses important observations made from the testing section. These tests were used to confirm the validity of the design, kinematics, and control system. Confirmation of these tests validates the functionality of the two-wheel swerve-drive system with an integrated suspension system.

In terms of confirming validity, the AGV behaved as desired for the four tests: straight-line test, strafe test, Ackerman test, and combination test. The setpoint values of the AGV were calculated and updated every 50 ms with the speed of the AGV kept below the specified maximum of 1.3 m/s. The minimum number of control inputs that the AGV required to achieve omnidirectional operation was three. These control inputs were:

- Centroidal x-velocity
- Centroidal y-velocity
- Centroidal yaw rate

The system was controlled in the velocity domain, as seen from the listed control inputs. An attempt was made to control the steering in the position domain; however, this control strategy proved very jerky and unsuitable for stable operation.

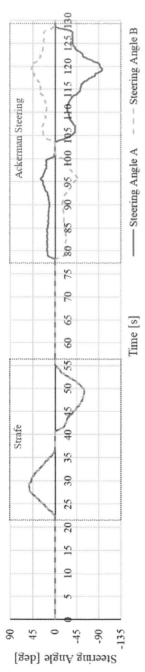

FIGURE 8.19 Combination test drive unit steering angles.

FIGURE 8.20 Combination test AGV yaw angle.

The primary control principles for the system are as follows:

- Velocity in a straight line. If the AGV is to be driven straight forward, then the magnitude of the centroidal x-velocity is set to the desired linear velocity. The centroidal y-velocity and centroidal yaw rate are kept at zero.
- Strafe at a given angle and speed. The strafe behavior refers to the AGV moving linearly at an angle not aligned with the front of the AGV (the AGV frame of reference will not change its orientation relative to the universal frame). This behavior is achieved by setting the centroidal x- and y-velocities to a calculated component magnitude of the desired linear velocity. The magnitude of these components is calculated using the desired strafe angle. During a purely strafe operation, the centroidal yaw rate of the AGV is kept at zero.
- Ackerman steering at a given speed. For a purely Ackerman turn where the strafe angle is zero, the centroidal x-velocity is set to the desired linear velocity, whereas the centroidal y-velocity is zero. The turning arc radius is determined by the yaw rate of the centroid coupled with the linear velocity of the AGV.
- Combination manoeuvres. A combination of the previously mentioned op erations can be achieved by superimposing these manoeuvres to achieve the centroidal x-velocity, centroidal y-velocity, and centroidal yaw rate set.

During the tests, it was found that when the acceleration of the AGV was purposefully saturated or rapid changes between deceleration to acceleration were performed (near acceleration saturation), the steering angle of the AGV would be pulled off of its desired angle. The closed-loop control corrected this effect; however, it illustrated that the shared axis between the tractive and steering systems could cause issues even with mathematical compensation. However, this issue is unlikely to affect normal operation as saturating the acceleration or rapidly changing between accelerating and deceleration in the manner needed to cause this effect will likely damage the payload long before the coupling effect is noticed. For more details on this issue, we refer to Section 8.3.1.

The tractive system affects the steering system, but the steering system was observed to affect the tractive system; this was counter to expectations. This interaction was shown in the second test (strafe), where the tractive velocity was kept constant while the steering was actuated. In Section 8.3.2, it was observed that during a fast steering angle change, a minor disturbance occurred on the tractive speed. However, this disturbance was comparably minor compared to the effect that rapid acceleration of the tractive system can have on the steering system (as previously discussed). This behavior is because the steering velocity is significantly smaller than the tractive velocity. The only way the steering system could have a notable effect on the tractive system was when the Ackerman steering velocity was saturated, as observed in Section 8.3.3. In this case the AGV cannot turn at the desired turning arc relative to the AGV linear speed. Thus to compensate for this and maintain the turning circle at the desired arc radius, the linear velocity decreases.

The last effect of prominence observed during the testing was the drift of the yaw angle. The yaw angle is determined using the numerical trapezoidal method to approximate the integral of the yaw rate. If the sample time is not precisely constant, then drift can occur. This yaw drift is invisible for most tests because it is rather small compared to the actual yaw angle change. However, during the swerve test (Section 8.3.2), when the yaw angle remains zero for the entirety of the duration of the test, the drift can be observed. It is approximated to be 1° every 110 seconds. If the drift proves to be constant, then it could be compensated by using a compensation algorithm; however, an extended duration test would have to be run to determine if this drift is as linear as it appears to be. This issue could also be eliminated by removing the reliance of this value on integration and instead determining it by using feedback from the navigation sensor; this concept is discussed further in Section 8.3.2.

8.5.1 Research conclusion

To recap, the primary research goal of this project was to produce a novel two-wheel swerve-drive AGV with an integrated suspension system. Therefore to validate the research into this system, it must be proved to function in a manner equivalent to a traditional four-wheel swerve-drive system. Therefore the AGV had to pass the four tests denoted in the testing section (Section 8.3).

As discussed in Section 8.5, the AGV passed all four validation tests and therefore can act as a replacement for traditional four-wheel swerve-drive AGVs such as Chikosi's [2] or Holmberg et al.'s [1] AGV. The AGV was still able to perform omnidirectional manoeuvres even with an included suspension system; this was due to the unique mechanism called the "vertical compromiser".

Since the AGV researched in this project has only four motors as opposed to the eight found in a traditional swerve-drive system, the cost of implementation of a swerve-drive AGV could be reduced by as much as 50%. However, this is closer to 40% as the cost of implementing "castor units" in place of the two removed "drive units" is not negligible.

The AGV is entirely SIL safety compliant, which means that this AGV can be directly implemented into the manufacturing industry; Wada et al.'s [3] AGV (similar two-wheel swerve-drive idea) was not compliant with and therefore could not be directly implemented into industry.

This type of AGV is costs less than a traditional swerve-drive system while still performing the same tasks and having an integrated suspension system. These facts make the AGV researched in this paper an ideal candidate for replacing traditional swerve-drive AGVs in the manufacturing industry, where omnidirectional capabilities are needed with poor floor conditions. This AGV has the benefit of being economically viable compared to mecanum wheel AGVs as this system has the same number of motors, four in total. Due to the added complexity of the mecanum wheel, any economic advantages that the system might have at initial capital outlay (due to a more straightforward gearing topography) are negated.

There are a couple of concerns about this system related to behavior near motor speed saturation points. At these points the control system performs erroneously, as discussed in Section 8.5. This erroneous behavior only occurs when the AGV is operated outside normal operating conditions and therefore cannot invalidate this AGV as a replacement for the traditional four-wheel swerve-drive system.

References

[1] R. Holmberg, J.C. Slater, Powered caster wheel module for use on omnidirectional drive systems, 2002.
[2] G. Chikosi, Autonomous Guided Vehicle for Agricultural Applications, Tech. Rep., 2014.
[3] M. Wada, S. Mori, Holonomic and omnidirectional vehicle with conventional tires, Robotics and Automation, 1996. Proceedings. ..., no. April (1996) 3671–3676.

Chapter 9

Mecanum wheel slip detection model implemented on velocity-controlled drives

C.L. Gilfillan[a], Theo van Niekerk[a], Paolo Mercorelli[b], and Oleg Sergiyenko[c]

[a]Faculty of Engineering, Built Environment & Information Technology (EBEIT) at the Nelson Mandela University, Port Elizabeth, South Africa, [b]Institute for Production Technology and Systems, Leuphana University of Lueneburg, Lueneburg, Germany, [c]Applied Physics Department of Engineering Institute of Baja California Autonomous University, Mexicali, BC, Mexico

9.1 Introduction

Automated/automatic guided vehicles or AGVs, as they are commonly known, are an indispensable tool within industry [1]. Modern AGV applications lie in intra-logistics, which involve internal handling, organization, and optimization of material within an industrial, production, or commercial environment. AGVs offer unique benefits over conventional automated systems, like conveyor belts and operator centric systems [2]. AGVs are utilized for part handling in a flexible manufacturing system. A flexible manufacturing system is a manufacturing process that is not "hard wired" to accommodate one production process. It possesses the potential to handle multiple production processes [3]. In the context of an AGV, it is a goods-handling system whereby any change in the production process of a component can be accounted for, without a complete redesign of the handling system [4]. AGVs may be designed in various ways to achieve a specific purpose. Typically, the design directions are focused on manoeuvrability, efficiency, and serviceability. In an ideal situation, an AGV would have all the design directions without any trade-offs. This means that the AGV can travel in any direction; the AGV would be modular and can run efficiently on its battery supply.

The pinnacle of manoeuvrability is achieved with an omnidirectional drive train. Although it is a non-omnidirectional drive train counterpart, it is not constrained by an instantaneous center for curvature (ICC). The ICC is a point around which a non-omnidirectional drive train must rotate to achieve a given position. This imposes constraints on the motion of the non-omnidirectional drive train that typically manifests as the necessity to change the drive trains

Modeling, Identification, and Control for Cyber-Physical Systems Towards Industry 4.0
https://doi.org/10.1016/B978-0-32-395207-1.00020-2
199

orientation to achieve the position. Examples of such drive trains are the differential drive train and the Ackermann steering based drive train [5,6]. In contrast, an omnidirectional drive train, such as that of the mecanum wheel-based AGV, which is the subject of this paper, is not bound by an ICC and thus is unconstrained along its plane of motion.

The way a mecanum wheel-based AGV achieves omnidirectional capabilities is through the superpositioning of the force vectors produced by the wheels (F_{yi}, F_{xi}, where i is the corresponding wheel number, i.e., 1,2,3,4), as seen in Fig. 9.1. The superpositioning of the forces (F_{net}) take place at the geometric center of the AGV body assuming an equal distribution of weight across the wheels. The F_{net} applied at the geometric center of the AGV would determine the trajectory of the AGV [7,8]. If there is an unbalanced force applied to the mecanum wheel drive train (MWDT) through any of the wheels, then the AGV would deviate from the planed trajectory.

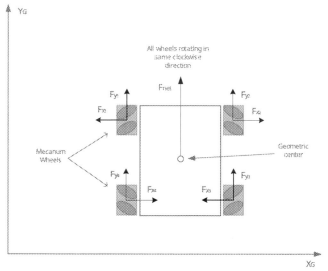

FIGURE 9.1 Figure illustrating the forces produced by the mecanum wheel and the net force produced as the center of the MWDT.

To achieve simplistic movements such as rotations about the geometric center and linear translations along the ground plains the MWDT, at least two mecanum wheels are required to work with each other again. To move in the direction of F_{net} in Fig. 9.1, all the forces along the x-axis are required to be neutralized. To neutralize the forces in the x direction, wheel 1 opposes wheel 2, and wheel 3 opposes wheel 4.

In its most basic form, this relationship can be defined by the forward kinematics of the MWDT AGV [7,9,10].

- ω_i [rad/s] is the angular velocity of the ith wheel.

- $v_x v_y$ describes the robot's linear velocity with respect to the global co-ordinate frame X_G and Y_G
- r [m] is radius of the roller

The longitudinal velocity in the x-direction:

$$v_x(t) = (\omega_1 + \omega_2 + \omega_3 + \omega_4) \cdot \frac{r}{4}. \tag{9.1}$$

The transverse velocity in the y-direction:

$$v_y(t) = (-\omega_1 + \omega_2 + \omega_3 - \omega_4) \cdot \frac{r}{4}. \tag{9.2}$$

The angular velocity of the AGV around its z-axis:

$$\omega_z(t) = (-\omega_1 + \omega_2 - \omega_3 + \omega_4) \cdot \frac{r}{4(l_x + l_y)} \tag{9.3}$$

Once this relationship is understood, it can be leveraged to create a cross coupled controller that can be used to mitigate the effect a single wheel slipping has on the MWDT AGV. This controller is known as the slip mitigation controller form here onwards.

9.2 Hardware considerations

The slip mitigation controller was designed to interface with the existing control hardware. This hardware included a Siemens s7-1516f PLC, coupled to four Festo CMMS-ST-C8-7-G2 drives, which drove the four stepper motors connected to the mecanum wheels of the MWDT AGV.

The data needed from the drives was the velocity of each motor (ω_i, $i = 1, 2, 3, 4$) and the current required by each motor (τ_i, $i = 1, 2, 3, 4$). However, due to the limitations of the hardware, namely the inability for the drives to output more than one drive parameter over PROFIBUS, an Arduino was used to acquire the missing parameter via an analogue output pin on the drive.

To control the drives effectively during normal operations, the drives were commissioned in "velocity control mode", which ensures that the drives reach the setpoint velocity and remains there by controlling the applied torque to each wheel. The torque value is relayed as current to the Arduino, which in turn relays this information to the PLC.

9.3 Slip mitigation controller

The slip mitigation controller in made up of two distinct parts that work in tandem to achieve slip mitigation. The first part is the slip detection algorithm, and the second part is the mitigation algorithm. It goes without saying that slip must be detected before it is mitigated.

9.3.1   Slip detection

When slip occurs once the drives have reached the setpoint speed, it creates a notable drop in the current the motor requires to maintain the setpoint speed. This fact was proven during a slip test conducted on the wheel 1, the results of which can be seen in Fig. 9.2. Slip was induced on wheel 1 at the start of the brown box in Fig. 9.2 and removed after 3 s elapsed, marking the end of the brown box.

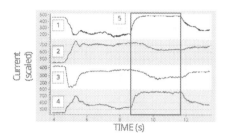

FIGURE 9.2   Current values when wheel 1 slipped.

The drop in current of one wheel resulted in a cascade effect across the other wheels, which unsurprisingly, resulted in more current being required by the non-slipping wheels to maintain the setpoint speed. The connection between slip and current was used to formulate a slip detection algorithm. Since the drives varied the torque to maintain a given wheel velocity, any sudden change in torque without an apparent change in the actual wheel velocity indicated the wheel was slipping. It does this by measuring the rate of change, or the derivative, of the current for each wheel. If the derivative was above some value designated as "SlipDerivativeValue", then the wheel was slipping. The SlipDerivativeValue was determined by experimental means; see Fig. 9.3.

The slip detector checked whether the derivative (CurrentChange) was larger than the threshold for slip detection (SlipDerivativeValue); if so, then the wheel was slipping, setting "Slipping" to true. If not, then the wheel was not slipping, setting "Slipping" to false. The variable "Slipping" was used later to activate and deactivate the slip mitigator.

9.3.2   Slip mitigator

The four wheels were separated into wheel pairs as illustrated in Fig. 9.4.

As a proof of concept, it was only necessary to develop the slip mitigation controller for the front wheels in pair 1 for a linear movement in the x direction. In pair 1, both wheels were connected to a slip detector and connected to a controller that applied the mitigation technique.

The goal of the controller was to ensure that the torque between the wheels in pair 1 were the same if one of the wheels slipped. This would ensure that the net force acting on the AGV was still pointing in the direction of travel. When

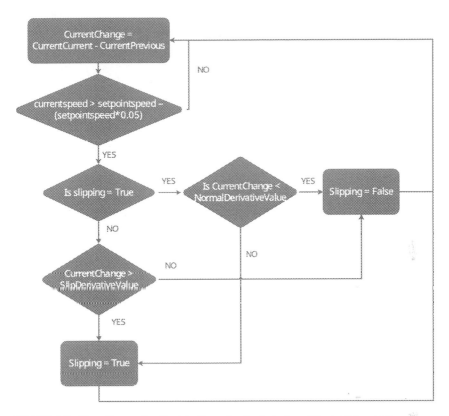

FIGURE 9.3 Flow diagram for slip detection using the derivative of the motors current.

slip was detected, the controller would modulate the set point speed of the non-slipping wheel to reduce the torque applied to it, to match that of the slipping wheel. Thus the wheels in pair 1 still maintained a force equilibrium.

In Fig. 9.5, A_P is the current draw from a slipping wheel in a wheel pair, A_i is the current draw of a non-slipping wheel in a wheel pair, e_{Api} is the error between the current values, and $e_{c,I}$ is the error between the set point speed of the wheel ω_t and the current speed of the wheel ω_i. The "*conventional motor control*" is the controller within the ith drive that ensures the wheel reaches its speed set point. The *Cross-Coupled Motor Control* is the controller that alters the non-slipping wheel speed to reduce the current draw of the non-slipping wheel to match the current draw of the slipping wheel. The final slip mitigation controller and its implementation on the physical AGV is shown as a flow diagram in Fig. 9.6.

The AGV begins its movement forward by generating a wheel velocity set point for the two wheels $(\omega_{s1}, \omega_{s2})$ and delivering the set point to the drives through the "Decision" blocks. The AGV accelerates the slip, and traction de-

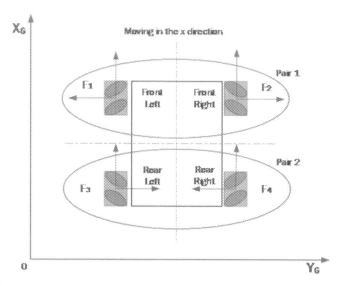

FIGURE 9.4 Wheel pairing when the AGV is moving in the x direction.

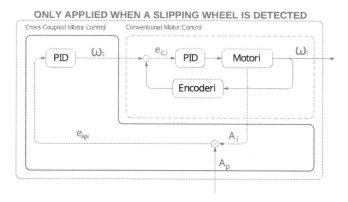

FIGURE 9.5 Slip mitigation control loop deployed on the MWDT AGV.

tectors are deactivated. Once the actual wheel speeds (ω_{a1}, ω_{a2}) are within the set point speed threshold, the slip and traction detectors come online. While this is occurring, the actual currents from the drives (τ_{a1}, τ_{a2}) are fed into the closed-loop controller. The closed-loop controller continuously generates a controlled set point wheel velocity for each wheel (ω_{c1}, ω_{c2}). The controller always generates a controlled wheel velocity because if a wheel slips, then a corrective velocity would have already been generated from the live torque values. This worked well because the detectors often detect the slip after it had occurred. With the controller continuously generating a potential correction, a corrected wheel velocity was readily available when needed.

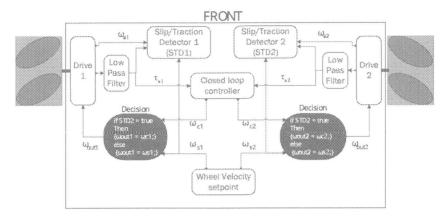

FIGURE 9.6 Flow diagram of the slip mitigation controller deployed on the physical AGV.

When slip was detected at one of wheels, it would prompt the "decision" blocks to take the controlled set point velocity and pass it to the drive that was not slipping, thus mitigating the effect that one of the wheels may have had on the AGV.

9.4 Test methodology

Slip was introduced on the front left wheel (Slip wheel). Slip was introduced at a set period and triggered once per test for a total of three tests per slip period. Each test was conducted by moving the physical AGV in the forward direction at 0.400 m/s. The period of the slip disturbances were 100 milliseconds, 500 milliseconds, and one second. The results from these tests behave as a point for comparison between the mitigated and unmitigated slips.

Three main test points are investigated. These points are illustrated in Fig. 9.7; the start of the disturbance is denoted as n, the end of the disturbance effect is denoted N, and the altered trajectory is shown for $t > N$. These points are characterized by the AGV rate of change away from Y_G axis, denoted as Δy, and its change in orientation or rotation around R_Z, denoted as $\Delta\varphi$, with Δy and $\Delta\varphi$ formalized as

$$\Delta y = y(t) - y(t-1)$$

and   (9.4)

$$\Delta\varphi = \varphi(t) - \varphi(t-1).$$

When the AGV was moving before the slip disturbance was introduced to the front right wheel at $t < n$, the Δy and $\Delta\varphi \approx 0$, signaling no change in the trajectory of the AGV. At $t = n$ the slip disturbance was introduced for a period. Once the disturbance has been introduced, there was an observable change in Δy and $\Delta\varphi$; thus Δy and $\Delta\varphi \neq 0$.

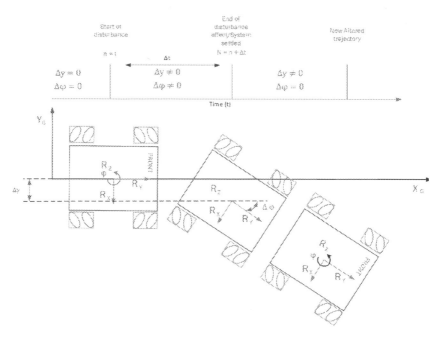

FIGURE 9.7 Points of interest during testing.

The end of the disturbance period was not relevant as the effects of the disturbance still acted on the AGV after the disturbance period had elapsed. What was interesting was the full impact of the disturbance on the relative position of AGV; thus the second point of interest was the end of the disturbance effect at N.

The end of the disturbance effect was characterized by $\Delta\varphi \approx 0$ but $\Delta y \neq 0$. This means that the AGV was no longer rotating about R_z, signaling the end of the effect the disturbance had on the AGV. However, since the disturbance had changed the trajectory of the AGV, the AGV moved away from the Y_G.

The results outlined in Tables 9.1, 9.2, and 9.3 set out the total change of AGV in a position between the start of the disturbance n and the end of the disturbance effect N. Thus y_t and φ_t were generated and were formalized as

$$y_t = y(N) - y(n)$$

and (9.5)

$$\varphi_t = \varphi(N) - y(n).$$

9.5 Discussion

The two metrics that were observed was the change of angle of the AGV and the change of the AGV Y_G position before and after the disturbance was introduced. The tests were conducted with and without slip mitigation.

Figs. 9.8 and 9.9 provide the movement of the AGV with a disturbance period of 1 second. The graphs provide a visual indication of effectiveness of the slip mitigation. Concerning Fig. 9.8, in the graph of angle vs time with a disturbance of 1 second, we can see that at 3.61 s the disturbance was applied to the AGV. Without slip mitigation, the AGV behaved unfavorably, resulting in an angle change after the AGV settled at 4.9 s of −6.07 degrees (Table 9.1) according to the trend line. With slip mitigation, the AGV followed its original trajectory better, resulting in an angle change of only 0.27 degrees. Considering Fig. 9.9, after the disturbance was introduced to the unmitigated AGV, the Y_G position was significantly altered by as much as 70.54 mm. With slip mitigation, this was reduced to 12.64 mm. Implementing slip mitigation improved the AGVS ability to reject angle disturbance by 103% and movement away from its original trajectory by 88,19% (Table 9.1).

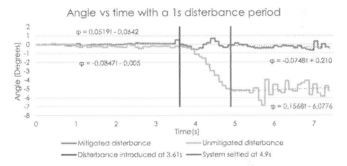

FIGURE 9.8   Graph of angle vs time with a disturbance of 1 second.

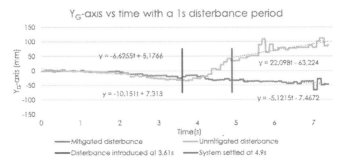

FIGURE 9.9   Graph of y-axis vs time with a disturbance of 1 second.

Figs. 9.10 and 9.11 describe the movement of the AGV with a disturbance period of 1 second. Fig. 9.10: Graph of angle vs time with a disturbance of 500 milliseconds provides a similar behavior to Fig. 9.8. At 3.69 s a disturbance is applied to the AGV, and the AGV settled at 5.16 s. The unmitigated AGV has a angle change of 3.19 degrees (Table 9.2). Unsurprisingly, this is roughly half

TABLE 9.1 Results of unmitigated and mitigated tests with a disturbance period of 1 second.

Degree vs time with a 1 s disturbance period

	Before disturbance	After disturbance	rate of change delta	change %	Offset delta	Change %
Unmitigated Results	$\varphi = -0,0847(t) - 0,0052$	$\varphi = 0,1568(t) - 6,0776$	0,2415	285,12	−6,07	−116777
Mitigated Results	$\varphi = -0,0519(t) - 0,0642$	$\varphi = -0,0748(t) + 0,2105$	−0,0229	−44,12	0,27	427,1

Result comparison

	Slope difference	offset difference	slope change%	Offset change%
Unmitigated vs mitigated results	−0,23	−6,29	309,63	103,46

Y-axis vs time with a 1 s disturbance period

	Before disturbance	After disturbance	rate of change delta	change %	Offset delta	Change %
Unmitigated Results	$y = -10,151t + 7,3135$	$y = 22,098t - 63,224$	0,2415	317,69	−70,54	964,48
Mitigated Results	$y = -6,6255t + 5,1766$	$y = -5,1215t - 7,4672$	−0,0229	22,69	−12,64	244,25

Result comparison

	Slope difference	offset difference	slope change%	Offset change%
Unmitigated vs mitigated results	−27,22	55,76	531,48	88,19

that of the 1 second disturbance. With slip mitigation, the angle change was reduced to 1.12 degrees. Fig. 9.11 indicates an unmitigated change of the AGV Y_G position by 83.43 mm, and with slip mitigation, this is reduced to 6.07 mm. Implementing slip mitigation improved the AGVS ability to reject angle disturbance by 65.34% and movement away from its original Y_G position by 97% (Table 9.2).

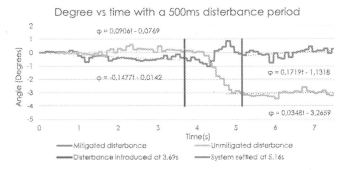

FIGURE 9.10 Graph of angle vs time with a disturbance of 500 milliseconds.

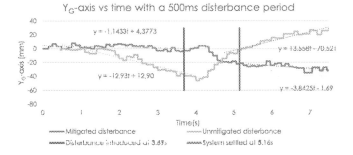

FIGURE 9.11 Graph of y-axis vs time with a disturbance of 500 milliseconds.

The final test was conducted using a disturbance period of 100 ms injected at 2.78 s, and as seen in Figs. 9.12 and 9.13, the AGV settled at 3.23 s. The unmitigated AGV has an angle change of 1.18 degrees, whereas the slip-mitigated AGV experienced a change of 0.12 degrees. Fig. 9.13 indicated that the unmitigated AGV experienced a deviation from its original Y_G position by 79.4 mm, whereas the mitigated AGV experienced a change of only 0.52 mm. According to Table 9.3, the angle disturbance rejection improved by 98.25%, and Y_G position disturbance rejection improved by 106.24%.

9.6 Conclusions

A slip mitigation controller was developed to mitigate the effects that the loss of traction experienced by a single mecanum wheel used on a mecanum

TABLE 9.2 Results of unmitigated and mitigated tests with a disturbance period of 500 milliseconds.

Degree vs time with a 500 ms disturbance period

	Before disturbance	After disturbance	Slope difference	Change %	Offset difference	Change %
Unmitigated results	$\varphi = 0,0906t - 0,0769$	$\varphi = 0,0348t - 3,2659$	−0,0558	61,59	−3,19	−4146,94
Mitigated results	$\varphi = -0,1477t - 0,0142$	$\varphi = 0,1719t - 1,1318$	0,3196	216,38	−1,12	−7870,42

Result comparison

	Slope difference	Offset difference	Slope change %	Offset change %
Unmitigated vs mitigated results	0,1371	−2,1341	79,76	65,34

Y-axis vs time with a 500 ms disturbance period

	Before disturbance	After Disturbance	Slope difference	Change %	Offset difference	Change %
Unmitigated results	$Y = -12,93t + 12,907$	$Y = 13,558t - 70,521$	26,49	204,86	−83,43	646,38
Mitigated results	$Y = -1,1433t + 4,3773$	$Y = -3,8425t - 1,69$	−2,7	−236,09	−6,07	138,61

Result comparison

	Slope difference	Offset difference	Slope change %	Offset change %
Unmitigated vs mitigated results	−17,4005	68,831	452,84	97,6

TABLE 9.3 Results of unmitigated and mitigated tests with a disturbance period of 100 milliseconds.

Degree vs time with a 100 ms disturbance period

	Before disturbance	After disturbance	Slope difference	Change %	Offset difference	Change %
Unmitigated results	$\varphi = -0,017t + 0,034$	$\varphi = -0,3064t - 1,1477$	$-0,2894$	$-1702,35$	$-1,1817$	$3475,59$
Mitigated results	$\varphi = 0,1333t - 0,1393$	$\varphi = 0,0175t - 0,0201$	$-0,1155$	$86,84$	$0,1192$	$85,57$

Result comparison

	Slope difference	Offset difference	Slope change %	Offset change %
Unmitigated vs mitigated results	$0,3239$	$-1,1276$	$1850,86$	$98,25$

X-axis vs time with a 100 ms disturbance period

	Before disturbance	After disturbance	Rate of change delta	Change %	Offset difference	Change %
Unmitigated results	$Y = 0,4364t - 2,6884$	$Y = 31,525t - 82,102$	$-31,0836$	$-7123,88$	$-79,4116$	$-2953,86$
Mitigated results	$Y = -9,8123t + 5,6492$	$Y = -9,5974t + 5,1262$	$-0,2149$	$2,19$	$-0,523$	$9,26$

Result comparison

	Slope difference	Offset difference	Slope change %	Offset change %
Unmitigated vs mitigated results	$-41,12$	$87,23$	$428,47$	$106,24$

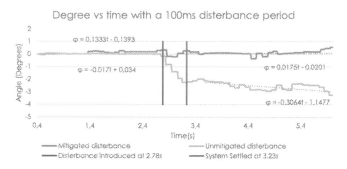

FIGURE 9.12   Graph of angle vs time with a disturbance of 100 milliseconds.

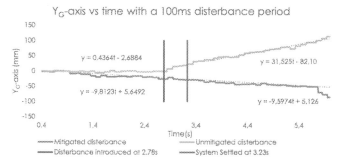

FIGURE 9.13   Graph of y-axis vs time with a disturbance of 100 milliseconds.

wheel-based AGV. The controller was designed to work with the existing AGV hardware, Siemens s7-1516f PLC, coupled to four Festo CMMS-ST-C8-7-G2 drives.

The controller used a combination of current and velocity feedback from the drives to detect when a mecanum wheel has slipped. Upon slip detection, the slip mitigator reduced the negative impact the slipping wheel had on the AGV trajectory. The slip mitigation controller achieved this by cross coupling the torque between two wheels in a wheel pair. The cross coupling in the event of wheel slippage ensured that the net force produced on the AGV remained in the intended trajectory, thereby reducing the negative effects of wheel slippage.

The slip detection and compensation techniques effectively reduced the impact that a slipping wheel had on the trajectory of the MWDT AGV. It improved the ability of the AGV to reject angle changing disturbance by an average of 88.9% and trajectory changing disturbance by an average of 97.1% considering 1 second, 500 millisecond, and 100 millisecond disturbance durations.

References

[1] S. Kaliappan, L. Lokesh, P. Mahaneesh, M. Siva, Mechanical design and analysis of AGV for cost reduction of material handling in automobile industries, International Research Journal of Automotive Technology (2018).

[2] C. Benevides, The Advantages and Disadvantages of Automated Guided Vehicles (AGVs), 2019.

[3] V. Serebrenny, D. Lapin, A. Mokaeva, The concept of flexible manufacturing system for a newly forming robotic enterprises, in: Proceedings of the World Congress on Engineering 2019, in: Lecture Notes in Engineering and Computer Science, 2019.

[4] V. Chawla, A. Chanda, S. Angra, Automatic guided vehicles fleet size optimization for flexible manufacturing system by grey wolf optimization algorithm, Management Science Letters 8 (2) (2018) 79–90.

[5] A. Calvó, Automated Guided Vehicles for commercial aircraft manufacturing industry, in: Aerospace Science & Technology, Universitat Politècnica de Catalunya, 2015.

[6] H. Soonshin, S.C. Byoung, M.L. Jang, A precise curved motion planning for a differential driving mobile robot, Mechatronics 18 (9) (2008) 486–494.

[7] N. Tlale, M. de Villiers, Kinematics and dynamics modelling of a mecanum wheeled mobile platform, in: 2008 15th International Conference on Mechatronics and Machine Vision in Practice, IEEE, 2008.

[8] D. Swan, R. Tang, A look at holonomic locomotion, in: Servo Magazine, T & L Publications, 2013, pp. 40–48.

[9] P.F. Muir, C.P. Neuman, Kinematic modeling for feedback control of an omnidirectional wheeled mobile robot, in: Robotics and Automation. Proceedings. 1987 IEEE International Conference on, IEEE, 1987.

[10] H. Taheri, B. Qiao, N. Ghaeminezhad, Kinematic model of a four mecanum wheeled mobile robot, Journal of Computer Applications 113 (3) (2015) 6–9.

Chapter 10

Safety automotive sensors and actuators with end-to-end protection (E2E) in the context of AUTOSAR embedded applications

Horia V. Căpriță[a,b] and Dan Selișteanu[c]

[a]*Continental Automotive Systems Sibiu, Sibiu, Romania,* [b]*University of Craiova, Craiova, Romania,* [c]*Department of Automatic Control and Electronics, University of Craiova, Craiova, Romania*

10.1 Introduction

The automotive industry represents the totality of the design, development, production, market, and sales processes of current vehicles. Currently, the design and production processes are more complex than before. These processes are based on the newest technologies from design tools, mechanics, hardware, or software perspectives. Controlling the functionality of a car demands more and more hardware resources and the corresponding software able to use them in an effective manner.

An automotive hardware/software assembly is called an electronic control unit (ECU). The architecture of a current car specifies dozens to hundreds of ECUs, each of them having the role of controlling a certain functionality of the car. All these ECUs form the car control system; the interaction between ECUs is done through communication networks (LIN, CAN, FlexRay, Ethernet, etc.). These networks are used by ECUs to transmit information received from the sensors they interact with, to send status or control commands, and respectively to receive such information from other ECUs [1]. In this way the effective control of various functionalities in the car is achieved, such as vehicle dynamics, environment interpretation (including pedestrian or traffic signs detection), intra- and inter-car communication (which requires strong cyber security algorithms), power control, stability, and so on.

The control logic of the functions in the vehicle is mostly implemented in software. Software applications are increasingly complex, and that is why a standard architecture has been defined for automobiles, Automotive Open System Architecture (AUTOSAR). By using AUTOSAR the aim is to increase productivity in the development of applications, to standardize the structure of software applications, and to ensure the level of functional safety required for machine control. The functional safety of the road vehicles is defined and standardized in ISO 26262, an international standard for functional safety of electrical and/or electronic systems installed in road vehicles [2].

To ensure the safety of the passengers, it is necessary that both hardware and software components comply with the ISO 26262 standard. Also, the communication mechanism between the ECUs must ensure the level of safety claimed by the controlled functionality.

This chapter extends paper [3], which presents a method of hardware migration of some software mechanisms to protect communication in the car. The proposed hardware model must ensure the same robustness of functionality as in the case of the software variant, with an optimization of the use of available resources. To do this, it is proposed to migrate the end-to-end (E2E) software library from the AUTOSAR architecture in the sensors and actuators used in automotive [4]. This method allows us to connect the basic sensors/actuators directly to the communication busses in the car, which will make the data provided/consumed by them directly available to all ECUs. The validation of the proposed model was made by using a Xilinx Spartan 7 FPGA electronic device [5].

10.2 Architecture of a car

10.2.1 Hardware overview

Currently, each functionality within a car is controlled by means of one or more electronic control units that are based on a microcontroller-type processing unit. These ECUs perform basic functions such as braking, steering, or engine power control systems. In addition to these basic systems, indispensable for the operation of a vehicle, more and more is being invested in the direction of the realization of self-driving vehicles and electric vehicles. This leads to more complex software applications and to an increase in the number of ECUs that interact in the car (Fig. 10.1).

The current distributed architecture of cars lays into the model presented in Fig. 10.2 [6]. This architecture evolved from the first generation of electronically controlled vehicles with independent ECUs control isolated functions inside the car. The second generation exposes an architecture in which several ECUs were organized in a cluster to control a specific functionality. Only one ECU in a cluster was responsible for data communication with another ECU placed in a different cluster (one-to-one communication). In the third generation the communication between clusters of ECUs is coordinated either by an ECU or by a central gateway.

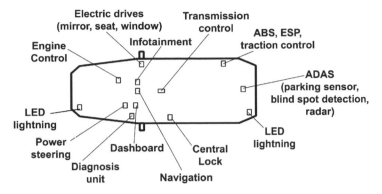

FIGURE 10.1 Functionalities inside a car.

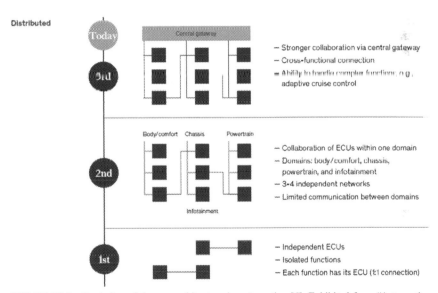

FIGURE 10.2 Evolution of the car architecture in automotive [6]. Exhibited from "Automotive software and electronics 2030 Mapping the sector's future landscape", July 2019, McKinsey & Company, www.mckinsey.com. Copyright (c) 2022 McKinsey & Company. All rights reserved. Reprinted by permission.

There is a trend in that, until 2030, the car architecture will evolve from a decentralized architecture towards more centralized systems. The fourth generation (Fig. 10.3), on which automotive companies are working already, will be domain centralized. In this architecture a domain controller (DCU) will communicate with specialized ECUs and will be able to handle more complex functions.

The aim of the fourth generation is to prepare the transition to the fifth generation (Fig. 10.4), in which the control unit runs functions or services of different

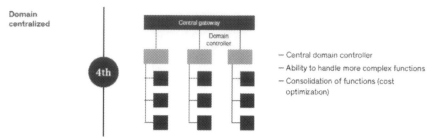

FIGURE 10.3   The fourth generation of the car architecture based on domain controllers [6]. Exhibited from "Automotive software and electronics 2030 Mapping the sector's future landscape", July 2019, McKinsey & Company, www.mckinsey.com. Copyright (c) 2022 McKinsey & Company. All rights reserved. Reprinted by permission.

functionalities. Such a control unit will have an advanced operating system (like Linux) able to manage a more complex hardware like HPC (High Performance Computer). We are in the digital age where mobile phones and mobile applications have become irreplaceable for the current generations. This has also influenced the functionalities in current cars, and we expect that in the future, most of these applications will be found directly in the car. This will increasingly influence the hardware and software structure of the control subsystems in the car.

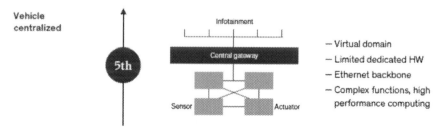

FIGURE 10.4   The fifth generation of the car architecture based on virtual domains [6]. Exhibited from "Automotive software and electronics 2030 Mapping the sector's future landscape", July 2019, McKinsey & Company, www.mckinsey.com. Copyright (c) 2022 McKinsey & Company. All rights reserved. Reprinted by permission.

Despite of the car architecture, the electronic or domain control units are connected to each other directly or through gateway type units. Gateway units have the role of allowing the transfer of information from one communication network to another communication network regardless of the communication protocol used by these networks. Currently, the communication protocols most often used in automotive are LIN (Local interconnection network), CAN (Controller Area Network), and/or CAN-FD (CAN with flexible data rate), FlexRay and Ethernet.

LIN is a serial communication protocol (ISO/AWI 17987-8) between car components. It is based on a single electric wire through which data can be transmitted at a maximum speed of 19.2 kbps on a bus with a maximum length of 40 meters. The protocol works on the master–slave principle. Only the master node can initiate communication at a given time, the slave component responding to its request. Using the LIN protocol, an ECU (master) can be interconnected with a maximum of 15 slave nodes (for example, sensors that have a LIN interface). The detailed structure of the LIN frame is shown in Figs. 10.5 and 10.6. The frame header contains 4 fields (Fig. 10.5): synchronization break field (at least 13 bits), a delimiter (1 bit), synchronization field (value 0x55), and protected identifier field (PID) having 10 bits, one start bit, eight ID bits, and one stop bit. This frame header is sent on the LIN bus by the master node to initiate the communication. All the slaves receive the frame and based on the PID field, will decide if they must answer or not. The frame response is filled by one slave only, with maximum 8 bytes of data (Fig. 10.6).

The CAN/CAN-FD protocols (ISO 11898/ISO 11898-1) allow the interconnection of several master ECUs that can initiate communication on the bus (message-driven multi-master protocol). The messages transmitted on the bus are characterized by a maximum length of 8 bytes (CAN), respectively 64 bytes (CAN-FD). Within the message, an identifier is defined that allows prioritizing access to the bus.

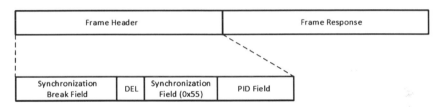

FIGURE 10.5 Structure of a LIN frame: header details.

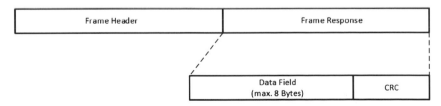

FIGURE 10.6 Structure of a LIN frame: response details.

In the case of concurrent transmissions, the bus will be won by the ECU that will transmit the message with the lowest identifier (as value). The CAN protocol allows data transfer rates of maximum 1 Mbps, whereas CAN-FD supports, in the data rate phase, a data transfer rate of up to 5 Mbps. The CAN/CAN-FD frame is presented in Fig. 10.7. The fields inside the frame have the following meaning:

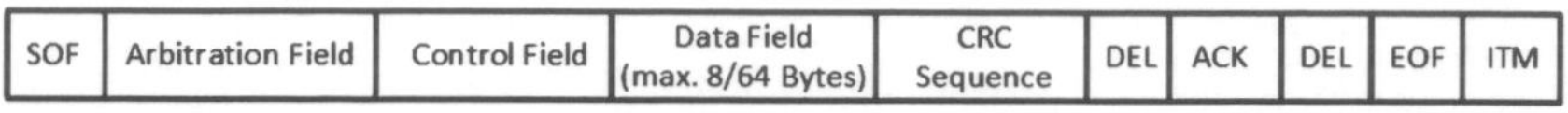

SOF	Arbitration Field	Control Field	Data Field (max. 8/64 Bytes)	CRC Sequence	DEL	ACK	DEL	EOF	ITM

FIGURE 10.7 Structure of a CAN/CAN-FD frame.

- Start of Frame bit (SOF)
- Arbitration field contains an identifier (11 bits) or an extended identifier (29 bits) of the message
- Control field basically contains the data length code (DLC), which specifies the length of the data (in bytes). This value is coded in case of a CAN-FD frame for data lengths greater than 8 bytes
- Data field represents the data to be sent on the bus. CAN frames can transport up to 8 bytes of data, whereas CAN-FD frames up to 64 bytes of data
- CRC field is computed by the hardware driver itself over the previous fields in the frame (excepting SOF)
- Acknowledge bit (ACK) is used by all receiving nodes on the network to signalize that the frame on the bus can be received with no errors
- End of frame bit (EOF)
- DEL bits represent delimiters between ACK and EOF bits
- Intermission (three bits) represent the delimiter between two consecutives frames on the bus. This is the minimum time needed between current frame and SOF bit of the next frame (usually, the communication bus will go idle between two consecutive frames)

The FlexRay protocol (ISO 17458-1 to ISO 17458-5) is a high-speed (up to 10 Mbps) fault-tolerant protocol. FlexRay is a time-driven protocol organized in repetitive cycles of five milliseconds each. The FlexRay communication cycle consists in 64 cycles (Fig. 10.8). From timing perspective a cycle is split into four parts: static segment, dynamic segment, symbol window, and network idle time (NIT). Periodic messages are transmitted in the static segment, whereas those generated by events are transmitted in the dynamic segment. The symbol window is used to transmit collision avoidance symbol to indicate the start of the first communication cycle initiated by a node. The NIT part is used to separate two 5-ms consecutive cycles.

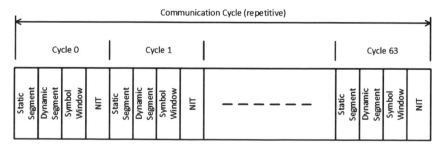

FIGURE 10.8 Communication cycle on FlexRay protocol.

The FlexRay frame (Fig. 10.9) consists of a header, a payload, a trailer, and a channel idle delimiter (CID). The header field is used in identifying the type of the frame, the slot of time allocated inside of a 5-ms cycle, or the transported data length.

Header	Payload (max. 254 Bytes)	Trailer	CID

FIGURE 10.9   Structure of a FlexRay frame.

The payload field contains the data to be transported (up to 254 bytes). The trailer contains a CRC calculated by the hardware driver over the entire frame. An ECU connected to the FlexRay bus has an internal clock that is permanently resynchronized with the other ECUs. In this way, all ECUs on the FlexRay bus are synchronized with each other, and the operating system can use this timing to activate its own software tasks. For this reason, the FlexRay protocol is used in applications that claim a high degree of safety in the automotive industry.

The Ethernet protocol was introduced in automotive because of the need to allow higher transfer rates and large amounts of data, especially in situations where video cameras are integrated into the system. In the future, this protocol will be used, for example, to support wireless communication between interconnected cars in traffic in the case of autonomous driving.

The communication in Ethernet networks is based on packets. A packet consists of 7 bytes of preamble (filled with 0x55 values), a start frame delimiter (SFD) on 1 byte (0xD5), and an Ethernet basic or tagged frame (Figs. 10.10 and 10.11). The Ethernet basic frame (Fig. 10.10) contains MAC addresses for sender and receiver, the type of the frame (basic or tagged), the data to be transported (46–1500 bytes), a padding field (PAD) to ensure the minimum length of the data (or the frame), and the CRC field calculated by the hardware driver over the entire frame. In addition, the Ethernet tagged frame (Fig. 10.11) contains the VLAN tag field to specify a virtual network. Although it was adopted in automotive to replace the FlexRay protocol, Ethernet is used in non-safety critical applications.

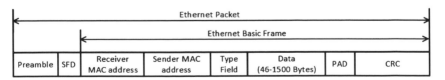

FIGURE 10.10   Structure of an Ethernet basic frame.

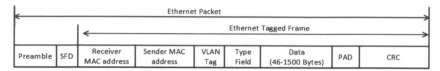

FIGURE 10.11   Structure of an Ethernet tagged frame.

10.2.2   Safety overview

Automotive applications inside of a car are designed to handle non-safety- or safety-related functionalities. Safety in automotive is related to the injuries that can occur in case of a malfunctioning of a system inside of the car [2]. ISO 26262 specifies several safety levels (ASIL, Automotive Safety Integrity Level) to be fulfilled by a hardware/software/system defined for every controller in automotive. Each controller is ranked according to the determined ASIL level on the analysis phase; this level is strictly related to the following factors (Table 10.1): severity (damages in case of system failure), exposure (probability of failure occurrence), and controllability (the control degree of the system made by the driver or external actions).

TABLE 10.1 Levels of safety factors used to determine the ASIL level in automotive applications.

	Severity		Exposure		Controllability
S0	No injuries	E0	Incredibly unlikely	C0	Controllable in general
S1	Light to moderate injuries	E1	Very low probability (injury could happen only in rare operating conditions)	C1	Simply controllable
S2	Severe to life-threatening (survival probable) injuries	E2	Low probability	C2	Normally controllable (most drivers could act to prevent injury)
S3	Life-threatening (survival uncertain) to fatal injuries	E3	Medium probability	C3	Difficult to control or uncontrollable
-	-	E4	High probability (injury could happen under most operating conditions)	-	-

Four ASIL levels denoted from A to D are defined in ISO 26262. ASIL A is the minimum level of risk, and ASIL D is the maximum. The standard also defines an additional level for not safety relevant applications, QM (Quality management). The ASIL level of an automotive application (ECU) is determined by correlating all the factors defined in Table 10.1 and using the matrix depicted in Fig. 10.12.

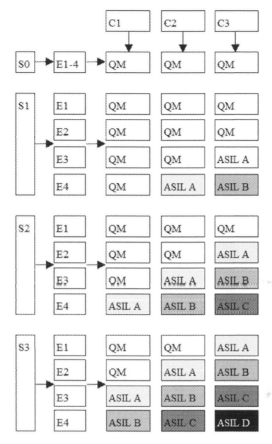

FIGURE 10.12   Determining the ASIL level based on severity, exposure, and controllability [2].

Another cause that can affect the integrity of automotive safety is given by external cyber-attacks, which are becoming more frequent since cars will be increasingly connected externally via the Internet. These tendencies to penetrate the safety mechanisms in the car must be prevented and stopped by specific cybersecurity mechanisms. The AUTOSAR standard already provides prevention and protection mechanisms against external attacks. The hardware also comes with support for the implementation of these cyber-security mechanisms. Soon all cars will have these protection mechanisms implemented as a standard.

10.2.3   Software overview

The complexity of the automotive software requires hardware resources able to perform many tasks in parallel having limited resources of computing power and memory. The current ECUs are based on multicore processors integrated inside microcontrollers. The management of an ECU is performed by a multicore real-time operating system (RTOS). The software that controls an ECU must be robust to fulfill the ASIL level required by its functionality. The robustness can be achieved if the software is standardized, which in turn will improve the performance and ensure the required level of safety. This is the reason why the automotive manufacturers defined a standardized software architecture AUTOSAR to be used in automotive applications [7].

AUTOSAR is a layered software architecture with benefits in decreasing the dependence between hardware and software, decoupled software development, and software reusability. Currently, two platforms of AUTOSAR are defined, Classic Platform (CP) and Adaptive Platform (AP).

The AUTOSAR Classic Platform (CP) is used for embedded systems with hard real-time and safety constraints, whereas AUTOSAR Adaptive Platform (AP) is used for high-performance computing (HPC) to build fail-operational systems (e.g., autonomous driving) [4].

The layered software architecture defined in AUTOSAR Classic Platform is depicted in Fig. 10.13 [8]. There are three main software layers: Basic Software layer (BSW), Runtime Environment (RTE), and Application Software layer (ASW). BSW layer is used for controlling the hardware of the ECU and to provide services for: communication, diagnostics, safety assurance, or access to different types of memories (Fig. 10.14). Microcontroller abstraction layer contains software drivers with direct access to the microcontroller (MCU) and its peripherals. By using this layer the higher software layers will be independent of MCU. ECU abstraction layer provides a set of APIs for access the peripherals/devices regardless of their location (MCU internal/external) and may contain drivers for external devices. The service layer provides basic services for BSW and ASW like OS functionality, network management and communication, memory management, diagnostics, state management, etc.

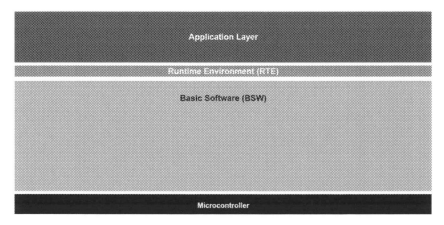

FIGURE 10.13 AUTOSAR classic layered software architecture [8].

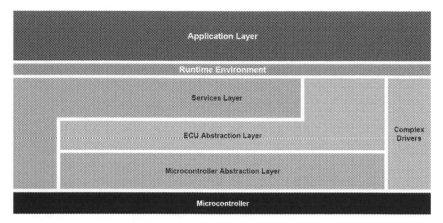

FIGURE 10.14 A view inside BSW layer in AUTOSAR classic [8].

AUTOSAR runtime environment (RTE) is a "glue" layer that allows communication between BSW and ASW modules called software components (SWCs), states transitions, tasks mapping, and more. Using this layer creates independence between basic and application software development. The AUTOSAR metamodel is a necessary input for both parties involved in development of the mentioned software layers (BSW and ASW). This metamodel contains the interfaces, ports, runnable, and other structures used in BSW/ASW communication and is also used as input for RTE software generation process. For this generation, there are several applications on the market [9] developed by automotive stakeholders like Vector, Elektrobit, Continental, etc.

The AUTOSAR Adaptive Platform (AP) provides mechanisms for high performance computing [10]. The need to define a new automotive architecture arose due to the significant increase in computing power required to implement complex functionalities (including algorithms for artificial intelligence or software update using over-the-air protocols). In the current ECUs the increase in computing power was made by increasing the number of cores, resulting in a homogeneous processing of applications. The AUTOSAR Classic Platform already offers multicore synchronization mechanisms, but the future generations of machines will use heterogeneous multicore systems composed of co-processors, GPUs, FPGAs, and accelerators to process applications. Heterogeneous processing is exploited by current HPC systems to increase the processing performance; the automotive industry must turn to these modern systems to achieve the proposed goal of creating intelligent cars.

The AP consists of two main parts (Fig. 10.15): AUTOSAR Runtime for Adaptive Applications (ARA) and Adaptive Applications (AA), which consist of user applications and customized services. ARA consists of functional clusters like operating system, communication management, cryptography, etc. and provide C++ interfaces so far. The operating system in this case is required to provide multi-process POSIX OS capability (e.g., Linux).

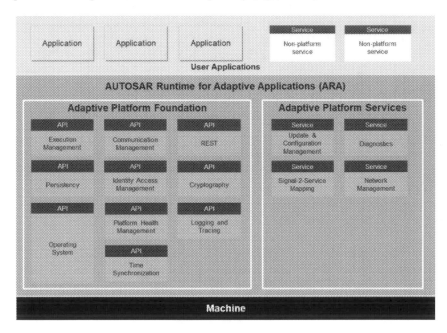

FIGURE 10.15 AUTOSAR adaptive software architecture [10].

10.3 Communication stack in AUTOSAR

Regardless of the integrated control units (ECUs or DCUs) and the AUTOSAR platform used in development, there is a need to ensure communication with the other hardware entities in cars. AUTOSAR platforms specify standards of communication modules that must meet the ASIL safety level demanded by the controlled functionality.

Communication stack (COM) is used to implement the logical communication between control units in a car. It consists of three different modules placed in BSW layers: communication drivers, communication hardware abstraction (interfaces), and communication services (Fig. 10.16). Communication drivers represent software modules that have protocol-specific implementations and that control the protocol-specific hardware (e.g., LIN, CAN, FlexRay). Communication hardware abstraction is a group of modules (protocol-specific) that provide an equal mechanism to access a bus channel regardless of its location (on-chip / on-board) [8]. Communication services consist of a set of software functions used to send/receive data, pack/unpack signals in/from a message, etc. These services are independent on the bus channel or the used protocol in communication.

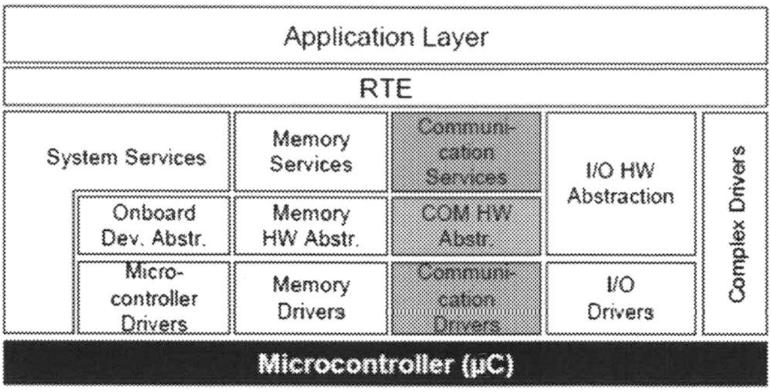

FIGURE 10.16   BSW COM stack in AUTOSAR classic [8].

From the safety point of view the communication between ECUs must be protected on the two levels, hardware and software. Communication is protected on the hardware level by using a specific bus driver (ASIC). As an example, in case of a transmission ordered by the communication layer, this integrated circuit will pack the message (Data Field) into a frame to be sent on the bus as defined by the standard protocol. On the receiving side the error detection is made in hardware, based on the CRC (Cyclic Redundancy Code) Sequence field.

The hardware verification mechanism is not enough from the safety perspective. As an example, a received message (Data Field in frame) can be validated on the destination. As a result, it will be provided to the software modules in the communication stack, which in turn will provide the received data to ASW. In such case the data received and validated by the hardware can be altered along the transfer between software modules due to several factors: wrong configured identifiers, wrong configuration of the message inside AUTOSAR modules, wrong configured data paths, etc. To prevent such faults and to ensure the required safety level, AUTOSAR introduced a mechanism responsible with communication protection on the software level, end-to-end communication protection (E2E).

10.4   End-to-end protection (E2E) in AUTOSAR embedded applications

10.4.1   End-to-end (E2E) communication protection in AUTOSAR

End-to-end communication concept was first presented in 1973 by Branstadt [11]. The concept has been implemented on the large scale into banking application systems for high-level auditing procedures as a matter of policy and legal requirement [12]. The E2E concept evolved over the years, and the number of areas of applications has been increased. Therefore E2E has been adopted by the industry as a standardized functions library. This library contains algorithms for data protection; the responsibility for the correct use of the library lies with the calling software module.

Using AUTOSAR E2E library in software development guarantees a higher level of safety communication and allows the runtime detection of either hardware or software-related errors (Fig. 10.17).

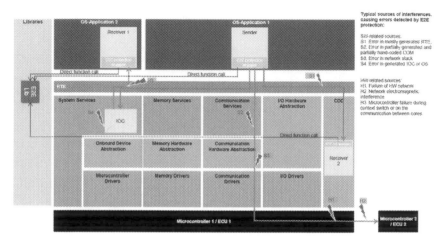

FIGURE 10.17 Examples of faults mitigated by E2E protection in AUTOSAR applications [14].

The protection realized by E2E library consists of computing the CRC over the data to be sent/received in BSW [13,14]. E2E data protection is based on the CRC calculation method defined in CRC-8-SAE J1850 standard [15,16], which specifies computing procedures for different lengths of the CRC (8-bit, 16-bits, 24-bits, etc.). According to the E2E standard, the data must contain a counter incremented on each transmission. This counter will be used on the destination for timeout detection (incomplete received sequence of data) [17,18]. Moreover, to calculate the CRC, a data identifier is needed (fixed value defined for the application). This data ID can be implicitly or explicitly transmitted on the bus.

E2E defines several profiles to be used in AUTOSAR applications:

- E2E Profile 1 (1A, 1B, and 1C variants) with 8-bits CRC, 16 bits of data ID used for CRC computation (implicit or explicit transmitted on the bus), a 4-bits Counter (for timeout detection), using the polynomial value 0x1D (Figs. 10.18 and 10.19).
- E2E Profile 2–8-bits CRC, a list of 8-bits data ID addressed by the Counter, 4-bits Counter, the polynomial value 0x2F.
- E2E Profile 4–32-bits CRC, 32-bits data ID explicitly transmitted on the bus, 16-bits Counter, the polynomial value 0x1F4ACFB13.
- E2E Profile 5–16-bits CRC, 16-bits data ID implicitly transmitted on the bus, 8-bits Counter, the polynomial value 0x1021.

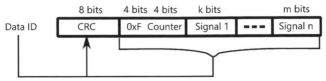

8 bits
4 bits
4 bits
k bits
m bits
Data ID
CRC
0xF
Counter
Signal 1
Signal n
CRC := CRC8 over (1) Data ID, (2) all serialized signals (including empty areas 0xF, excluding CRC byte itself)

FIGURE 10.18 Message content according to E2E Profile 1A (implicit transmission of Data ID) [13].

8 bits
4 bits
4 bits
k bits
m bits
Low byte of Data ID
CRC
*)
Counter
Signal 1
Signal n
CRC := CRC8 over (1) Data ID, (2) all serialized signals (including empty areas 0xF, excluding CRC byte itself)
*) Low nibble of High byte of Data ID

FIGURE 10.19 Message content according to E2E Profile 1C (explicit transmission of Data ID) [13].

- E2E Profile 6–16-bits CRC, 16-bits data ID implicitly transmitted on the bus, 8-bits Counter, the polynomial value 0x1021, 16-bits-length field for variable length data.

The principle of using E2E mechanism in AUTOSAR is depicted in Fig. 10.20.

On the sender side the data to be transmitted is protected by calling E2E_Protect function that will increment the counter signal (Counter) and will calculate the CRC over the entire data, including Counter.

The message sent on the bus will contain, along the data produced by the application (BSW or ASW), also the calculated CRC and Counter. On the receiver side a similar E2E mechanism is applied. The consumer will call E2E_Check function that will check the Counter to be in the right sequence (for timeout detection) and will calculate the CRC over the received data (excluding received CRC).

Finally, it compares the calculated CRC against received CRC; if these values are equal, then the data is safe to be used, and otherwise it is not. The result is communicated to the consumer application that will use the data or not according to the safety requirements.

10.4.2 Migrating E2E communication protection library in hardware

10.4.2.1 Hardware E2E module for basic sensors

At this time the basic sensors in automotive are connected directly on an ECU through a standard interface like SPI, I2C, PSI5, ADC, etc. (Fig. 10.21) [19]. In this case the physical data is read by the ECU1, which transforms and protects

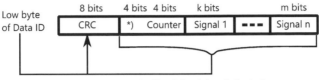

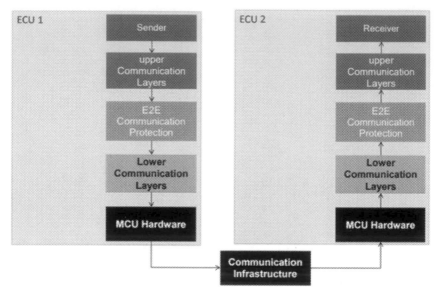

FIGURE 10.20   E2E communication protection principle in AUTOSAR [13].

the data (using E2E mechanism) to use it and to be sent on the configured bus channel. The other ECUs will read the data from the bus, will check the E2E status, and if the latter is valid, then will use it inside the BSW or ASW according to the specification.

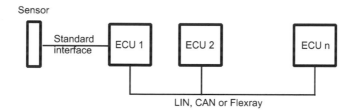

FIGURE 10.21   Connection of a basic sensor inside the car.

The aim of this chapter is to design and evaluate an E2E hardware module for basic sensors [20,21] able to send E2E protected data on an automotive communication bus. Fig. 10.22 depicts the integration of the proposed sensor in an automotive communication network.

A customized message provided by the sensor has been defined. It includes CRC, Counter fields (according to the E2E Profile 1A), and Signal A, which represents the raw value of the physical value measured by the sensor. The format of the message on the bus is depicted in Fig. 10.23. For example, a mapping of this customized message on a CAN/CAN-FD frame is presented in Fig. 10.24.

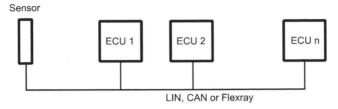

FIGURE 10.22 Basic sensor with E2E protection communicating on an automotive network.

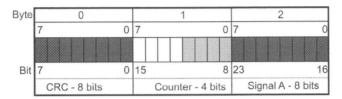

FIGURE 10.23 Sensor message format according E2E Profile 1A.

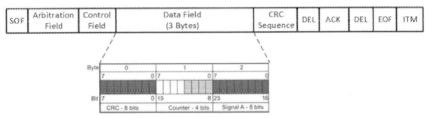

FIGURE 10.24 Example of a CAN/CAN-FD frame containing the sensor message (E2E Profile 1A).

To realize the integration of the new sensor into an automotive network, several changes are required:

- The sensor must have a converter able to transform physical value into raw value (logic value) for generating Signal A.
- Add logic able to handle the Counter value (rolling value).
- Add logic able to provide the Data ID (fixed value).
- To implement a fast CRC calculation algorithm, the sensor must have an internal ROM memory to store a look-up table with 256 elements [22]. The look-up table values will be used in XOR operations for generating the CRC.
- Adding a hardware bus driver/transceiver (LIN, CAN or FlexRay) used in communication on the bus (this is not in the scope of this chapter).

The CRC computing algorithm is performed by the sensor according to the E2E Profile 1A specifications. The logic diagram of the algorithm is presented in Fig. 10.25.

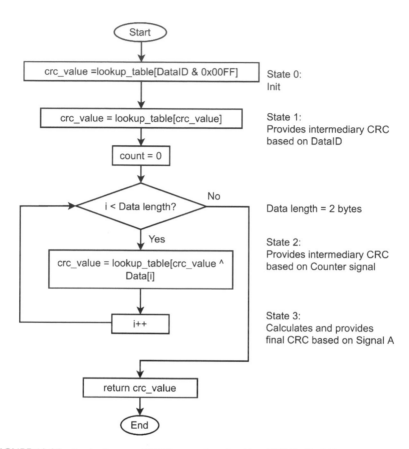

FIGURE 10.25 Logic diagram of CRC computing algorithm (E2E Profile 1A).

10.4.2.2 Detailed design of hardware E2E module

A field programmable gate array (FPGA) is an integrated circuit able to be programmed using a hardware description language (HDL) to be used for custom-specific applications. An FPGA contains an array of hierarchical configurable logic blocks (CLBs) and reconfigurable interconnects, which can be configured according to the desired application [23].

The Xilinx Vivado 2021.2 tool [24] has been used to design, synthesize, validate, and integrate the E2E module in an FPGA device. Vivado allows the user to design the hardware using basic block elements like basic registers, logic gates, ALU units, etc. Moreover, Vivado allows the user to develop new logical blocks (user customized) by supporting VHDL and Verilog hardware description languages. After the design phase, the user have the possibility to simulate the design, to analyze the results, to synthesize it onto FPGA, and finally to run it directly on an FPGA device [25,26]. The results presented in this chapter were obtained using a Xilinx Spartan 7 FPGA device, which is the highest density device of the Spartan-7 family [27].

The block diagram of the E2E module for basic sensors is presented in Fig. 10.26 and consists in input buffers (Data ID and Signal A), clock generator, a 4-state machine, a CRC computing logic (including look-up table), and output buffers for generated signals (CRC, Counter, and Signal A).

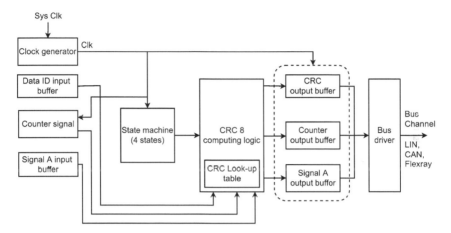

FIGURE 10.26   Block diagram of the E2E module for basic sensors (Profile 1A).

The 4-state machine controls the CRC computing logic. The meanings of the states mentioned in Fig. 10.25 are as follows:

- State 0: the initialization value is read from the LUT based on the Data ID and start value 0xFF.
- State 1: the first intermediary CRC is calculated.
- State 2: the second intermediary CRC is calculated using Counter value.
- State 3: the final CRC value is calculated using Signal A and stored onto the output buffer.

Each state contains four-clock cycles along the value of the CRC is calculated (Fig. 10.27):

- Clock 0: the address of a location in look-up table (LUT) is calculated (XOR operation)
- Clock 2: the data addressed in the LUT is read and provided in an internal buffer for the next step (state)
- Clock 3 on state 3: the final values of the CRC, Counter, and Signal A are stored into output buffers. In this step the values of the signals are ready to be processed by the bus driver to be sent on the communication bus.

There are five steps necessary to be performed in Vivado tool to implement the E2E hardware module in FPGA (Figs. 10.28, 10.29, 10.30):

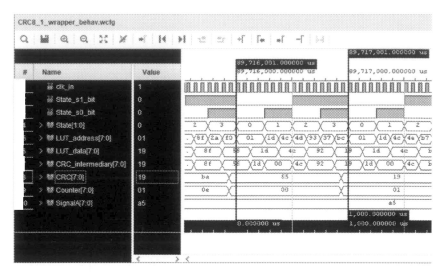

FIGURE 10.27   Waveforms of the cycles of states and CRC values (E2E Profile 1A).

1. Create block design based on the block diagram (Fig. 10.26). The design contains predefined or customized logical blocks. Verilog hardware description language has been used in designing the E2E module to describe the run-time libraries (RTL) of inner customized blocks. The detailed design of the CRC 8 computing logic block (Fig. 10.26) is shown in Fig. 10.28.

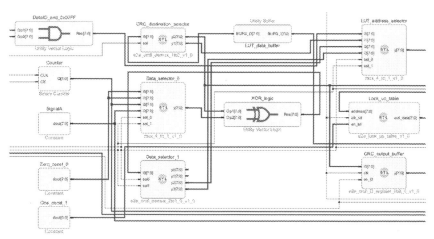

FIGURE 10.28   Block design of the E2E module for basic sensor. A view inside CRC8 computing logic block.

2. Create elaborated design. In this phase the block design is mapped on standard logical basic blocks (Fig. 10.29).

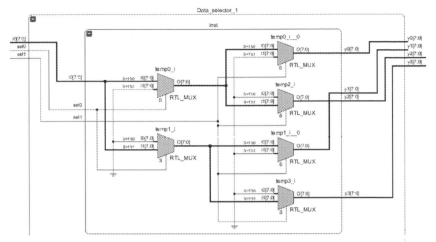

FIGURE 10.29 Elaborated design of the E2E module for basic sensor. A view inside Data_selector_1 block.

3. Synthesize the design. The next step consists in mapping the synthesized design onto the FPGA structure. Logical basic blocks are mapped onto the configurable logic blocks (CLBs) inside the configured FPGA in the project. The schematic is complex and hard to depict in one figure.
4. Implement the design. The synthesized design is placed into FPGA onto specific CLBs. In this phase, it is possible to make changes related to placement of the modules inside FPGA area. Fig. 10.30 depicts the implemented design inside Xilinx Spartan 7 FPGA, where the CLBs are marked with black rectangles.
5. Generate the bitstream and program the FPGA. It is necessary to have the development kit connected to the computer and to "write" the generated bitstream inside the FPGA device.

E2E block design is based on profile 1A, in which DataId is contained by the CRC. Other E2E profiles require DataId to be explicitly sent to the bus, which leads to change of the design (due to the change of the sensor message).

10.4.3 Experimental results and discussions

The simulation of the E2E module (Vivado tool) is based on a clock signal with a period of 62.5 microseconds. Using this timing, the sensor will send a message on the bus on every millisecond, which is a reasonable data rate for a CAN bus for example. The results reflect the functionality of the E2E module. Generation of Signal A and the implementation of the bus driver are outside the scope of this chapter. The range of the Counter signal is 0x00...0x0E (0x0F represents an invalid value). Signal A had a constant value 0xA5 during the experiments and

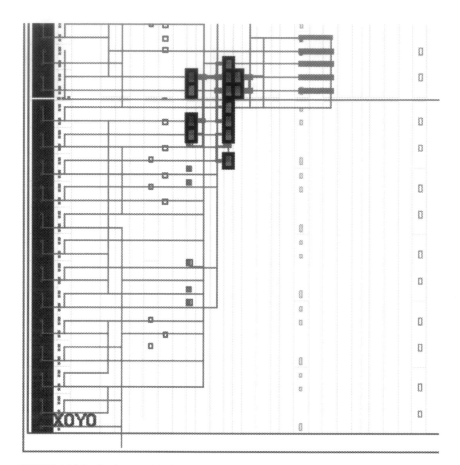

FIGURE 10.30 Implemented design of the E2E module in Xilinx Spartan 7 FPGA.

is represented on 8 bits (see Fig. 10.23), whereas Data ID was set to value 1. The output of the hardware module E2E are presented in Table 10.2 and Fig. 10.31.

These results were validated also using a different environment based on Vector CANalyzer tool [28]. In this environment the computer is the sender (via a Vector VN 7610 hardware), whereas the receiver (and checker) is a real ECU. The communication is made using a real CAN network. The values provided by the software E2E library, implemented on computer using the CAPL scripting language [29], were captured from the real CAN bus. Fig. 10.32 represents the waveforms of the signals captured on the CAN bus (the values of the signals are displayed in decimal), whereas Fig. 10.33 shows the data values of the signals (hexadecimal representation).

A comparison of the data shown in Table 10.2 and Fig. 10.33 shows that the values are similar, which validates the proposed hardware model designed in Vivado tool. Also, on the receiver side the ECU reported no errors on checking

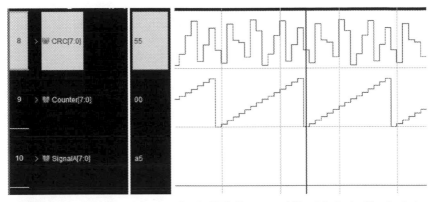

FIGURE 10.31 Waveforms of output signals CRC, Counter, and Signal A obtained by simulating the E2E hardware model.

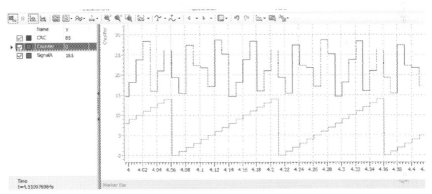

FIGURE 10.32 Waveforms of the signals CRC, Counter, and Signal A captured on a real CAN bus.

the E2E status of each received message, which means that the values generated by the E2E module are correct.

The utilization rate of the CLBs of FPGA is presented in Fig. 10.34. It is observed that for E2E hardware module, there is a need of 46 look-up-table (LUT) elements for logic blocks, 30 Flip-Flop registers, 1 block RAM, and 54 IO pins.

Having in view that one CLB inside Xilinx Spartan 7 FPGA contains 8 LUTs, 16 Flip-Flop registers, and 256 bits of RAM [27], it results in that there is a need of 13 CLBs (Fig. 10.30) to integrate E2E module in FPGA. It is worth mentioning that the implementation of an E2E module able to handle more than 2 bytes of data will require supplementary hardware resources.

The results presented in this chapter were obtained using a Xilinx Spartan 7 FPGA device placed on Spartan-7 SP701 Evaluation Board (Fig. 10.35). The SP701 evaluation board is based on the XC7S100FGGA676 device, a member

TABLE 10.2 Output values of CRC, Counter, and Signal A.

Cycle	CRC	Counter	Signal A
0	0x55	0x00	0xA5
1	0x19	0x01	
2	0xCD	0x02	
3	0x81	0x03	
4	0x78	0x04	
5	0x34	0x05	
6	0xE0	0x06	
7	0xAC	0x07	
8	0x0F	0x08	
9	0x43	0x09	
10	0x97	0x0A	
11	0xDB	0x0B	
12	0x22	0x0C	
13	0x6E	0x0D	
14	0xBA	0x0E	

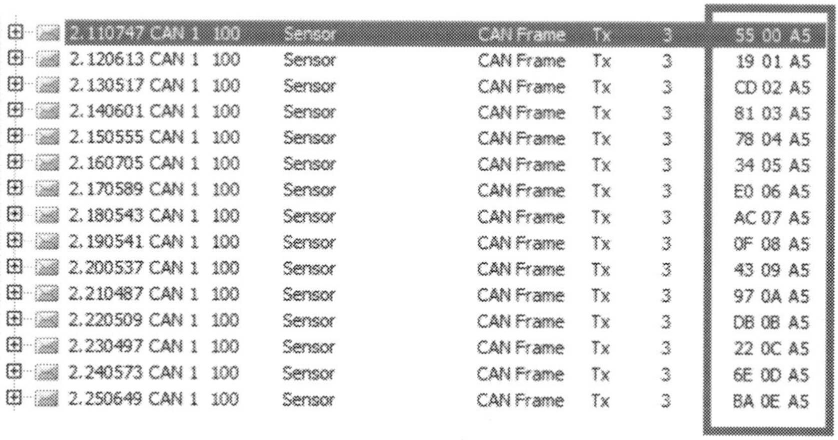

⊞ 🖿 2.110747 CAN 1 100	Sensor	CAN Frame	Tx	3	55 00 A5
⊞ 🖿 2.120613 CAN 1 100	Sensor	CAN Frame	Tx	3	19 01 A5
⊞ 🖿 2.130517 CAN 1 100	Sensor	CAN Frame	Tx	3	CD 02 A5
⊞ 🖿 2.140601 CAN 1 100	Sensor	CAN Frame	Tx	3	81 03 A5
⊞ 🖿 2.150555 CAN 1 100	Sensor	CAN Frame	Tx	3	78 04 A5
⊞ 🖿 2.160705 CAN 1 100	Sensor	CAN Frame	Tx	3	34 05 A5
⊞ 🖿 2.170589 CAN 1 100	Sensor	CAN Frame	Tx	3	E0 06 A5
⊞ 🖿 2.180543 CAN 1 100	Sensor	CAN Frame	Tx	3	AC 07 A5
⊞ 🖿 2.190541 CAN 1 100	Sensor	CAN Frame	Tx	3	0F 08 A5
⊞ 🖿 2.200537 CAN 1 100	Sensor	CAN Frame	Tx	3	43 09 A5
⊞ 🖿 2.210487 CAN 1 100	Sensor	CAN Frame	Tx	3	97 0A A5
⊞ 🖿 2.220509 CAN 1 100	Sensor	CAN Frame	Tx	3	DB 0B A5
⊞ 🖿 2.230497 CAN 1 100	Sensor	CAN Frame	Tx	3	22 0C A5
⊞ 🖿 2.240573 CAN 1 100	Sensor	CAN Frame	Tx	3	6E 0D A5
⊞ 🖿 2.250649 CAN 1 100	Sensor	CAN Frame	Tx	3	BA 0E A5

FIGURE 10.33 Data values generated by the software E2E library on the real CAN bus.

of the Xilinx 7 series FPGA family. It is optimized for low cost, low power, and high I/O performance.

The board comes with advanced high-performance FPGA logic based on real 6-input look-up table (LUT), 36-Kb dual-port block RAM, support for DDR3L interface up to 1866 Mb/s, XADC with 12-bit 1 MSPA ADC with on-chip thermal and supply sensors, and powerful clock management tiles (CMTs). The

Resource	Utilization	Available	Utilization %
LUT	46	64000	0.07
FF	30	128000	0.02
BRAM	1	120	0.83
IO	54	400	13.50

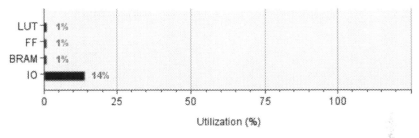

FIGURE 10.34 Utilization rate of the Xilinx Spartan 7 FPGA resources.

board is designed for high performance and lower power with a 28 nm, 1 V core voltage process [30].

The standard AUTOSAR E2E library is used in automotive applications to guarantee the safety of the communication in automotive networks. Right now, this mechanism is implemented in software on sender and receiver sides. This chapter presents a hardware model of the software AUTOSAR E2E library for using it inside of the basic sensors in automotive. The new hardware module will allow the sensor to send the protected data according to the safety requirements in communication with the ECUs via the automotive network (Fig. 10.22). This new approach has the advantage that all the ECUs will have direct access to the protected data provided by the sensor. In this case, there is no need for a specific ECU to compute the values of the protection signals (CRC and Counter) in software and to forward it towards other ECUs (nodes) into the network. This mitigates the risk of error occurrence in software or incorrect message sending on the bus.

Regarding the timing, the output data rate of the model depends on the data rate on the communication bus (LIN, CAN, or FlexRay), which is few milliseconds or tens of milliseconds. Either SW or HW implementation must fulfill these timings; in both cases the internal timing of producing data is not relevant for the performance itself as long as this timing is not greater than the data communication rate. Another advantage is that the new sensors can be placed anywhere inside the car because they can provide data straightforward onto the communication bus. In case of accelerations sensors for example, they are placed in some cases directly on the ECU, and this ECU should be placed in the car on a certain position to be sure that they make the right measurements. By using the new

FIGURE 10.35 Spartan-7 SP701 Evaluation Board [30].

sensor the placement of the ECU is not a constraint. It is important in this case that only the sensor (small size) is placed on the right position.

10.5 Conclusions

After all of work, the conclusion is that the AUTOSAR E2E protection communication mechanism is feasible to be implemented in hardware since, in case of a message of 3-bytes length (3 signals) in FPGA, it requires 46 look-up-table (LUT) elements for logic blocks, 30 Flip-Flop registers, 1 block RAM, and 54 IO pins. The utilization rate of the internal modules of FPGA is 0.1%, whereas for CRC look-up table, it is 0.83%, and for I/O pins, it is 13.5%. By increasing the number of bytes/message only the utilization rate of internal modules will be affected (increased number for Flip-Flop registers) since the CRC look-up table and I/O pins will remain the same. The proposed hardware is able to provide data on each millisecond (using a clock of 62.5 microseconds), which is an acceptable data rate for CAN communication. This data rate can be adjusted (according to the requirements) by increasing the clock timing inside of the FPGA.

Future work will consist in designing an E2E module able to process many signals of a message for calculating the CRC, which will have impact on the hardware complexity (e.g., a memory module to store the signals is needed). Computing the final CRC value will require many clocks per state in the 4-state machine (1 clock/byte). Another direction in this research to improve the presented concept is to migrate all the E2E defined profiles into the hardware on the sending site [31]. Moreover, there is room to exploit this experience to also integrate the checking part on the receiver side, e.g., by the integration of an E2E checking module into actuators. The new models of actuators can benefit by such a hardware module to check the safety data received from the communication bus.

Another direction in future research is investigation of the possibility to integrate the E2E protection mechanism along with the communication driver inside an FPGA [32] and make qualitative and quantitative evaluations of the new hardware. If the results will fulfill the expectations, then the proposed model could be directly used as a baseline of future sensors/actuators in automotive industry.

References

[1] C. Falconi, S. Mandal, Interface electronics: state-of-the-art, opportunities and needs, Sensors and Actuators A: Physical 296 (2019) 24–30, https://doi.org/10.1016/j.sna.2019.07.002.

[2] Road Vehicles - Functional Safety, International Standardization Organization ISO 26262-1:2018.

[3] H.V. Căpriță, D. Selișteanu, Improvement of automotive sensors by migrating AUTOSAR end-to-end communication protection library into hardware, Elektronika Ir Elektrotechnika 28 (5) (2022) 34–44, https://doi.org/10.5755/j02.eie.31154.

[4] AUTOSAR Standards. [Online]. Available: https://www.autosar.org/standards/.

[5] Spartan-7 FPGAs: Meeting the Cost-Sensitive Market Requirements, Xilinx Inc. [Online]. Available: www.xilinx.com.

[6] O. Burkacky, J. Deichmann, J.P. Stein, Automotive Software and Electronics 2030. Mapping the Sector's Future Landscape, McKinsey & Company, July 2019.

[7] M. Staron, Automotive Software Architectures. An Introduction, 2nd ed., Springer International Publishing, Cham, Switzerland, 2021, https://doi.org/10.1007/978-3-319-58610-6.

[8] AUTOSAR Layered Software Architecture, AUTOSAR organization. [Online]. Available: https://www.autosar.org.

[9] S. Piao, H. Jo, S. Jin, W. Jung, Design and implementation of RTE generator for automotive embedded software, in: Proc. of 2009 Seventh ACIS Int. Conf. on Software Engineering Research, Management and Applications, Haikou, China, 2009, pp. 159–165, https://doi.org/10.1109/SERA.2009.35.

[10] AUTOSAR Adaptive, AUTOSAR organization. [Online]. Available: https://www.autosar.org.

[11] D.K. Branstad, Security aspects of computer networks, Proc. AIAA Computer Network Systems Conference, in: Proc. AIAA Computer Network Systems Conference, Huntsville, Alabama, 1973, AIAA Paper no. 73-427.

[12] J.H. Saltzer, D.P. Reed, D.D. Clark, End-to-end arguments in system design, ACM Transactions on Computer Systems 2 (1984) 277–288, https://doi.org/10.1145/357401.357402.

[13] Specification of SW-C End-to-End Communication Protection Library, AUTOSAR standards, https://www.autosar.org, Document ID 428: AUTOSAR_SWS_E2ELibrary.

[14] E2E Protocol Specification, AUTOSAR FO R20-11, AUTOSAR standards, https://www. autosar.org, Document ID 849: AUTOSAR_PRS_E2EProtocol, 2020.

[15] Standard SAE-J1850 8-bit CRC, SAE, 2006.

[16] P. Koopman, T. Chakravarty, Cyclic Redundancy Code (CRC) polynomial selection for embedded networks, in: Proc. of Int. Conf. on Dependable Systems and Networks, Florence, 2004, pp. 145–154, https://doi.org/10.1109/DSN.2004.1311885.

[17] T. Forest, M. Jochim, On the fault detection capabilities of AUTOSAR's End-to-End communication protection CRC's, in: SAE World Congress & Exhibition, Detroit, USA, 2011, https://doi.org/10.4271/2011-01-0999.

[18] T. Arts, S. Tonetta, F. Koornneef, C. van Gulijk, Safely using the AUTOSAR End-to-End protection library, in: 2014 Int. Conf. on Computer Safety, Reliability, and Security, Delft, in: LNCS, vol. 9337, Springer, 2015, pp. 74–89, https://doi.org/10.1007/978-3-319-24255-2_7.

[19] W.J. Fleming, Overview of automotive sensors, IEEE Sensors Journal 1 (2001) 296–308, https://doi.org/10.1109/7361.983469.

[20] M.A. Pillai, S. Veerasingam, S.D. Yaswanth, Implementation of sensor network for indoor air quality monitoring using CAN interface, in: Proc. of Int. Conf. on Advances in Computer Eng., Bangalore, 2010, pp. 366–370, https://doi.org/10.1109/ACE.2010.85.

[21] Y. Xie, Y. Guo, S. Yang, J. Zhou, X. Chen, Security-related hardware cost optimization for CAN FD-based automotive cyber-physical systems, Sensors 21 (2021) 6807, https://doi.org/10.3390/s21206807.

[22] X. Dong, Y. He, CRC algorithm for embedded system based on table lookup method, Microprocessors and Microsystems (ISSN 0141-9331) 74 (2020) 103049, https://doi.org/10.1016/j.micpro.2020.103049.

[23] S.M.S. Trimberger, Three ages of FPGAs: a retrospective on the first thirty years of FPGA technology: this paper reflects on how Moore's law has driven the design of FPGAs through three epochs: the age of invention, the age of expansion, and the age of accumulation, IEEE Solid-State Circuits Magazine 10 (2018) 16–29, https://doi.org/10.1109/MSSC.2018.2822862.

[24] Vivado Design Suite UserGuide, Xilinx Inc. [Online]. Available: www.xilinx.com.

[25] J. Guajardo, S.S. Kumar, G.J. Schrijen, P. Tuyls, FPGA intrinsic PUFs and their use for IP protection, in: Int. Workshop on Cryptographic Hardware and Embedded Systems CHES 2007, Vienna, in: LNCS, vol. 4727, Springer, 2007, pp. 63–80, https://doi.org/10.1007/978-3-540-74735-2_5.

[26] I.Z. Mihu, H.V. Căpriţă, Architectural improvements and FPGA implementation of a multi-model neuroprocessor, in: Proc. of 9th Int. Conf. Neural Inform. Proc., Singapore, vol. 4, 2002, pp. 1749–1753, https://doi.org/10.1109/ICONIP.2002.1198975.

[27] 7 Series FPGAs Configurable Logic Block, Xilinx Inc. [Online]. Available: www.xilinx.com.

[28] CANalyzer the Tool for Comprehensive ECU and Network Analysis, Vector GmbH. [Online]. Available: www.vector.com, 2016.

[29] CAPL for CANalyzer and CANoe, Vector GmbH. [Online]. Available: www.vector.com, 2015.

[30] SP701 Evaluation Board User Guide, Xilinx Inc. [Online]. Available: www.xilinx.com.

[31] J. Mitra, T. Nayak, Reconfigurable very high throughput low latency VLSI (FPGA) design architecture of CRC 32, Integration VSLI Journal 56 (2017) 1–14, https://doi.org/10.1016/j.vlsi.2016.09.005.

[32] Y. Son, J. Park, T. Kang, Design and implementation of CAN IP using FPGA, Journal of Institute of Control, Robotics and Systems 22 (8) (2016) 671–677, https://doi.org/10.5302/j.icros.2016.16.0089, Institute of Control, Robotics and Systems.

Part III

Motion and control of autonomous unmanned aerial systems as a challenge in Industry 4.0 process

Paolo Mercorelli[a], Hamidreza Nemati[b], and Quanmin Zhu[b]

[a]*Institute for Production Technology and Systems, Leuphana University of Lueneburg, Lueneburg, Germany,* [b]*Department of Engineering Design and Mathematics, University of the West of England, Bristol, United Kingdom*

> *Drones overall will be more impactful than I think people recognize, in positive ways to help society.*
>
> —Bill Gates

Autonomous unmanned aerial systems are technologically advanced aircrafts that operate without a pilot aboard. They can be used by the military and law enforcement to monitor vantage points, transmit real-time video surveillance, and even conduct reconnaissance missions at a fraction of the cost of manned aircraft. Many people associate drones with unmanned aerial vehicles used by the U.S. military. These aircraft are often called drones because of the drone aircraft that make up their bases [1]. However, there are other types of drones as well. For example, sensor platforms are also often called drones since they are unmanned aerial systems that carry sensors for collecting data. There is also a distinction between unmanned aerial systems (UAS) and unmanned ground systems (UGS). These terms refer to how many vehicles you are talking about when discussing drones.

The fourth industrial revolution refers to a period of rapid technological advancement and social change that we are entering. This revolution is driven by new technologies and changes in the manufacturing sector. It will also bring new job opportunities, improved manufacturing processes, and heightened military capabilities. This 4.0 is the name given to the current phase of the industrial revolution [2]. This phase brings about mass digitization and connectivity in manufacturing, transportation, information, and communication industries. It is characterized by autonomous mobility, machine learning, cyber-based, and III generative systems. By creating a better world, technology will lead us to a cleaner environment and improved living standards. The fourth industrial revolution focuses on intelligent, digital, and connected industries. It emphasizes rapidly creating high-quality products using advanced technologies such as artificial intelligence and machine learning. Furthermore, these systems can be interconnected to create smart factories that are also digitally controlled. This allows manufacturers to produce goods more efficiently while reducing costs through advanced technology.

Motion and control planning of autonomous unmanned aerial systems is the process of designing and executing a plan that will enable an unmanned aerial system to move from one location to another while avoiding obstacles and maximizing efficiency. This process can be broken down into several steps, including path planning, path execution, and obstacle avoidance. Path planning involves determining the best route from the current location to the desired destination, taking into account factors such as the terrain, weather, and availability of resources [1]. Path execution involves following the plan and making necessary adjustments along the way. Finally, obstacle avoidance is a critical part of this process, as it ensures that the unmanned aerial system does not collide with any obstacles along the way.

There are various challenges that are faced while trying to motion and control autonomous unmanned aerial systems in an Industry 4.0 process. One such challenge is the management and handling of big data. The data sets generated by autonomous systems are often large and complex, making it difficult to manage and control them effectively. Another challenge is the need for real-time decision making. In 4.0 processes, decisions need to be made quickly and accurately to keep up with the fast-paced nature of the process. Autonomous systems can be challenging, as they often take longer to make decisions than humans. Finally, another challenge is the coordination of different autonomous systems. There are often many different autonomous systems working together. This cannot be easy to coordinate, as each system may have different objectives and goals.

References

[1] P. Fust, J. Loos, Development perspectives for the application of autonomous, unmanned aerial systems (UASs) in wildlife conservation, Biological Conservation 241 (2020) 108380.
[2] M. Xu, J.M. David, S.H. Kim, The fourth industrial revolution: opportunities and challenges, International Journal of Financial Research 9 (2) (2018) 90–95.

Chapter 11

Multibody simulations of distributed flight arrays for Industry 4.0 applications

Lewis Yip[a], Hamidreza Nemati[a], Paolo Mercorelli[b], and Quanmin Zhu[a]
[a]*Department of Engineering Design and Mathematics, University of the West of England, Bristol, United Kingdom,* [b]*Institute for Production Technology and Systems, Leuphana University of Lueneburg, Lueneburg, Germany*

11.1 Introduction

Distributed Flight Arrays (DFAs) are a development upon modern micro-air vehicle developments and trends in drone swarm technology, which utilize the cooperative capabilities of drones to achieve a variety of tasks efficiently. DFAs have the unique ability to autonomously, and rigidly, connect to "n" number of other modules, creating drone formations of different proportions as required for a specific task. Previous DFA designs share many characteristics such as the hexagonally shaped chassis, which due to its symmetry, allows for tessellation while maximizing the number of possible neighboring connections. A previously proposed design also included co-axial propellers rotating in opposite directions to counteract the torque generated by a single drone module without increasing the footprint and a swashplate-less rotor design [1] to improve responsiveness and allow for solo flight, without greatly increasing the mechanical complexity, as found in traditional swashplates.

One of the many applications of DFAs is in logistics, specifically material handling, due to their potential as lifting platforms. While DFAs are not suited to long-distance flight, their modularity and flexible arrangements would allow them to perform specialized material handling tasks within a warehouse setting. Often in warehouses the task of material handling/order picking is carried out manually or with the assistance of vehicles, such as forklifts, and more technologically advanced warehouses implement automated storage and retrieval systems. Recent years have seen automated vehicles becoming more common in warehouse operations as a way to improve the efficiency and flexibility of order picking operations, which are considered the most important task within warehouse logistics. Material handling by aerial vehicles has previously taken two main forms in regard to the payload attachment system, fixed payload, and

Modeling, Identification, and Control for Cyber-Physical Systems Towards Industry 4.0
https://doi.org/10.1016/B978-0-32-395207-1.00023-8

slung payload. Both options have their advantages and disadvantages such as footprint, control, and flexibility.

The modularity of DFAs allows them to be easily adapted to either payload system; however, the mechanisms that allow the modules to form multibody structures must be considered as the payload system and its state can cause great variation in the distribution of forces across the DFA platform, which if not accounted for could cause catastrophic structural failure. Previous DFA designs have largely relied on a magnet-based docking system to bond the DFA structure together; however, the strength and structural capabilities of such mechanisms have not been tested while under the increased weight of a payload, in static or dynamic conditions.

Physical experimentation with drone prototypes in a controlled environment is a very time-consuming and resource-heavy method of testing, especially when considering the numerous ways that the DFA can be arranged. Alternatively, physics simulations can be used to complete more test iterations while sacrificing some accuracy. A physics simulation can be used to prototype and test the effectiveness of different payload systems and the effects of external excitations of the system such as wind or collisions much faster than real-world experiments.

Multibody physics simulations may also allow for a model-based approach for the research and design of complex control systems that would be found within a DFA platform, allowing further extensive simulations investigating the functionality and viability of various drone tasks, such as obstacle avoidance, docking, perching strategies, and similar energy conserving manoeuvres, and payload transportation throughout different array configurations.

11.2   Aims, objectives, and scopes

This research focuses on two main objectives, the first being identifying the effects of various slung payload systems on the structure of different array setups under static and dynamic load cases using computer-based physics simulations. By using sensors to measure the shear forces between module chassis it is possible to assess the overall viability of different array structures for material handling in a warehouse setting and to identify any unforeseen behavior caused by the payload system.

This would be achieved by simulating various DFA setups using identical hexagonal bodies with properties approximating their real-world counterparts (dimensions, mass, inertia, thrust force) and attaching a standardized mass via a rope/tether simulating the slung payload. This simulation could then undergo external excitation in the form of force vectors acting on the payload. The research findings can then be used to supplement further design and research into the specifics of DFA docking mechanisms and payload systems by highlighting the systems behavior and the magnitude of forces they will potentially experience during operation.

The second objective of this research is to investigate the multibody simulation software's flight and control system capabilities to assess the suitability for further simulation of DFA tasks. By constructing simple control systems and simulating controlled flight tasks it is possible to evaluate the software's viability and potential for larger complex simulations. This would be achieved by simulating a DFA control system using traditional linear control methods (such as PID) to direct a single drone module to complete simple tasks such as a controlled hover, executing a defined flight path, and docking. This research and documentation of the process would then serve to supplement further design and simulation research of DFAs using similar software techniques.

11.3 Background research

Distributed flight arrays are described as modular vertical take-off and landing vehicles, capable of self-assembling into an unlimited number of arbitrary configurations [2]. This level of functionality is achieved through the integration of several key subsystems, which appear throughout many previously proposed designs [2–5].

11.3.1 Chassis

The purpose of the module chassis is twofold: firstly, it is to act as a container that can securely house all the other subsystems of the module protecting them from impacts, and secondly, to provide rigidity and structure to the module array. These design criteria encourage the use of durable, lightweight materials, with previous designs featuring low-density expanded polypropylene (EPP) foam and Polyamide-12 engineering plastic. Some of these materials also have the ability to be 3D printed, which has many benefits such as rapid prototyping, faster production and repairs, and a higher level of accessibility. Similar drones of this caliber, such as modern quadcopters, have their form factor dictated mainly by minimizing their size and material usage to effectively house their components without any unnecessary weight; however, the design of a DFA must also satisfy the structural requirements to support docking with adjacent modules.

This factor has led the majority of designs to take the form of hexagonal prisms. This particular shape has numerous benefits; hexagons are very space efficient and can form a perfect tessellation pattern with itself – this allows the module array to have no gaps and require no alternate chassis shapes to tessellate, which maximizes the area of the docking surfaces, improving their potential adhesion to one another. The hexagon also has the highest number of faces out of all polygons that can form regular tessellations (triangles, squares, hexagons). This means that each module is able to dock with up to six adjacent modules. While the bulk of the module chassis is usually designed to be made from plastics, some designs feature a carbon fiber tube spanning the width of the module as both a mounting point for the motor/propeller assembly, but also for added rigidity. However, this use of composites has not yet been applied to the whole module chassis.

11.3.2 Thrust mechanism

The thrust mechanism is one of the most important systems of the DFA functionality as it is the primary mode of transport, and its performance will determine a large amount of the overall effectiveness of a module. All previous designs have featured propeller powered flight as it is the most appropriate form of flight for this application offering precision, control, and vertical take-off and landing capability, all without compromising the hexagonal prism profile of the chassis, unlike a fixed wing or other, more exotic, methods.

DFAs have previously featured propeller thrust as a single fixed-pitch propeller at the center of each module. This method is lightweight and both mechanically and computationally simple; however, the use of a single propeller causes the modules to generate a torque. This problem as if not addressed can cause uncontrollable rotation of the DFA structure. Previously, this has been handled by having half of the array's rotors traveling in opposing directions to counter this effect. While this solution is quite simple, it has the potential to cause issues with structures using an odd number of modules; however, previous personal research of the DFA design involved the proposed use of coaxial propellers, two rotors stacked one on top of the other, rotating in opposite directions. This change nullifies the need to have modules with opposing rotors; as the torque generated is neutralized individually increasing the flexibility of configurations, the extra propeller also increases the maximum thrust output of the modules.

An additional change was proposed alongside the coaxial design to utilize a "swashplateless" rotor. In a swashplateless rotor, each of the propeller blades are connected to a hinged rotor hub via an asymmetrically oriented angled linkage to create a lag pitch coupling. This allows the pitch of the blades to be controlled by modulating the speed of the rotor. The change in speed causes the propellers to lag behind or over rotate, which causes the pitch to change. This mechanism combined with co-axial rotors has been shown to be capable of emulating "fully actuated" flight with only two actuators [1]. This greatly improves the solo-flight capabilities of the modules, as single-propeller modules (one actuator) are severely underactuated for flight without being docked in an array with multiple other modules.

11.3.3 Driving mechanism

The DFA design is intended to be capable of fully automated operation; this includes the processing of docking with one another, which requires the modules to accurately position and orient themselves. Since individual DFA modules are traditionally incapable of controlled solo-flight, they require the use of some other form of movement to allow them to carry out the docking process. Previous DFA designs utilize wheels to allow the modules to move while grounded, specifically omni wheels and kiwi drive in particular, kiwi drive being a three-wheel system, with each wheel equally spaced radially around a point, the center in the case of DFA modules, oriented tangentially. Omni wheels have perpendicularly oriented rollers instead of traditional treads, allowing the wheel to be driven regularly but also to slide laterally. This setup requires less wheels than a traditional four-wheel system, reducing complexity and the amount of actuation required, while also allowing the modules to have zero turning radius, which improves the manoeuvrability, simplifying the docking process physically and computationally.

11.3.4 Docking mechanism

The core mechanism of the DFA is its ability to physically connect with other agents to form structures of arbitrary formation and size. This physical interface can also be used to act as a connection to transfer data through, allowing the connected modules to communicate and share sensor data, which is superior to alternatives such as wireless/optical systems, which can be more unreliable. Whereas there is a large variety of ways to mechanically join two objects, DFAs have a few specific requirements: The modules must form a rigid connection ideally with a low tolerance, must be able to be engaged and disengaged on command, and must fit within the chassis (six times in the case of the hexagon chassis). Additionally, the design must also consider the weight of the mechanism, its complexity, and any extra requirements it may have such as power for actuation. These requirements have previously led towards a common solution of using pairs of oppositely aligned permanent magnets on the docking faces, and additionally, in some cases, the docking faces have had interlocking protrusions and grooves [2]. This solution is extremely simple, lightweight, and is a passive system requiring no additional power/control to function. The system is activated by the proximity of two docking modules and is disengaged by the opposing force generated by the module drive system. While the tolerance is not very precise, the magnets assist with self-alignment along with the addition of grooves. The docking faces can also be made effectively genderless by containing both north and south poles across the face and maintaining chirality.

However, for the modules to be capable of lifting payloads several times, the weight of the array and the strength, bending, and shear stiffness of the joint needs to be considered. While permanent magnets are a very practical docking solution, as mentioned above, the force required to separate them in the shear direction is largely dependent on the frictional force between the two faces, and in certain cases estimated by some to be only 15–20% of the magnet's theoretical adhesion force (in the normal direction) [6]. Alternatively, the usage of an actuated, physically interference-based, docking mechanisms could drastically increase the structural strength of the array, improving the stability and load bearing capabilities. These present two major design challenges: the physical form of the locking mechanism and the method of actuation.

When designing a locking mechanism for the distributed flight array, there are several constraints that must obeyed such as having a small mechanism profile that does not extend far outside of the chassis while disengaged and able to be contained within the limited space of the module chassis. The mechanism would also ideally be "genderless" to allow for the docking to function with any combination of the six docking faces. Operating all six docking mechanisms individually is equally as challenging, as the weight and cost of the Distributed flight array are important factors, which disincentivize the usage of additional actuators, and therefore reducing the number of required actuators below six would be ideal. This can be achieved a number of ways such as reducing the input required to operate the mechanisms by designing them to be self-locking, or monostable, or through the usage of a passive restoring forces such as spring-loading, which enables single actuators to operate multiple discrete docking mechanisms. This concept is fundamentally similar to that of multiplexing, a method used in telecommunications to transmit multiple signals over a single shared communication channel.

11.3.5 Sensors

The intention of the Distributed flight array is for it to be capable of operating autonomously. Therefore it requires ways to gather information about its position and surroundings. Drones typically do this through the use of on-board sensors. Previous Distributed flight array designs found success using a combination of three-axis rate gyro sensors to measure angular rates and either pressure sensors or downward facing infrared sensors to detect altitude [2,5]. The infrared transverse mounted on the sides of the modules have also been used to establish communication between one another.

Alternatively, such information can also be collected through external means and the data provided through a wireless connection. This can be achieved through external sensors such as indoor real-time locating systems (RTLS) and image tracking, which are able to calculate the position and orientation of the modules. The advantages of a system like this is that the positional calculations can be extremely accurate, with some ultra-wideband-based RTLS achieving accuracy within 10 cm, and others under certain conditions can get as accurate as sub-millimeter [7], allowing for better and more consistent path planning and obstacle detection, and the data for these calculations can be more easily accessed by a dedicated external processing unit, rather than being handled onboard or transmitted back and forth to be processed. This also relieves the module system from the weight of the sensor components.

11.3.6 Previous usages of DFAs

The initial motivation that inspired the Distributed flight array design was to create a drone research platform able to explore distributed estimation problems, which involve the logistical issues that occur within sensor networks due to network architecture, communications, and data fusion algorithms and in "Distributed Control Systems", which are computerized control systems made up of multiple control system processors, over which the computational load is distributed, rather than being centralized [8].

The Distributed flight array's design achieves this by allowing them to be tasked with a number of different scenarios, such as assembly, flight and navigation, with each module acting as a potential node in the network for data and communication. Manoeuvring and interacting with the modules effectively by utilizing each of their sensory and positional data outputs is a problem of scalable complexity. Subsequent research projects involving the distributed flight array have produced their own sets of module design iterations as research platforms either to study the design process or for the purpose of testing different distributed control algorithms and control strategies [5]. This project research into the payload carrying capabilities of DFAs is relevant in the context of previous research as it aims to explore the potential logistical uses, which will naturally expand upon the capabilities of the platform and broaden the types of distributed estimation problems for research and development purposes.

11.3.7 Warehouse applications and material handling

In the industry of warehouse logistics, there is an ongoing shift towards automation. A warehouse's throughput can affect the efficiency of the entire supply chain, and their improvements and optimizations are driven by the growing demand for faster and more efficient management of products. This has caused great amounts of time, money, and research to be invested into streamlining all aspects of the warehouse system, such as management techniques, the buildings architecture, layout of facilities, storage systems, and order processing operations, which is often regarded as being the most important.

Order picking refers to the order processing operation of identifying and retrieving products as requested by an order before shipment and is one of the most fundamental and sometimes referred to as the most critical warehouse operation [9] as it tends to be the most expensive and labor-intensive activity that has wide-reaching consequences if underperformed. Many modern-day warehouses still employ human workers as their main order picking operators [9], who are tasked with manoeuvring in the most effective paths, either on foot or by vehicles such as forklifts, around the warehouse facility either to store or collect product according to a pick-order and deliver to a set location; however, this is an extremely complex mathematical problem, which falls under the field of combinatorial optimization. The issue of routing efficiency in warehouses is comparable to the "Traveling salesman problem", which involves determining the shortest route between a set of nodes while visiting each node only once, considered an "NP-complete" problem under computational complexity theory [10]. This implies that perfect solutions to the routing problem are difficult to compute and exact solutions require brute force algorithms, which are computationally slow and inefficient as they check all permutations, which leads to many practices being subject to heuristic algorithms and techniques to simplify the solutions and the computation required, but losing out on potential efficiency [9].

As research explores more optimal solutions to the routing efficiency issue, and technology improves, the usage of more complex warehouse management systems and computer-controlled order picking operators, such as automated storage and retrieval systems (AS/RS) and automated vehicles (AV), become more appealing. Although computer algorithms are still subject to restrictions such as computational load and processing time, they are able to achieve much more optimal solutions and utilize and develop more complex order-picking strategies, [9] such as complex storage, batching, zoning, wave picking, and routing practices, than what can be accomplished with a purely human work-force. The most common examples of modern Automated Storage and Retrieval Systems (AS/RS) come in the form of rows of large palette storage racks, often extending up to the height of the warehouse, being managed by automated aisle bound lift modules able to travel the length and height of the storage racks, retrieving and delivering products as requested. These systems, although being very expensive to install and maintain, are very time and space efficient. However, systems like the one described have hard limitations, such as being aisle bound, which restricts the variety of tasks they can be used for. This lack of flexibility must be handled by other operators, such as human workers, or in some cases a fleet of autonomous vehicles. Autonomous vehicles combine the high accuracy, speed, and precision found in computer-driven systems like AS/RS, with the unbound flexibility of human operators. This makes them uniquely equipped to carry out a variety of non-standard tasks such as working around developing areas, and dynamically responding to the changes in operations while still being able to contribute towards regular order-picking operations at a high level, and their innate flexibility allows them to be more easily integrated into existing warehouse systems than AS/RS. A combination of traditional AS/RS and AV systems theoretically allows warehouses to handle a high variety of situations while maintaining high levels of efficiency.

Appropriately designed DFAs could be used as a special class of AV systems that offer a unique dimension to travel through their flight capabilities. The modularity of DFAs is also suited towards warehouse logistics as it allows the fleet to be divisible into smaller or larger work groups as required, allowing better multitasking with the system. Individual modules also make up a much smaller percentage of the overall fleet, meaning that repairs and maintenance of modules have a smaller impact on the overall performance.

11.3.8 Payload systems

There are two main payload attachment systems to consider for the design of a Distributed flight array, a fixed payload system, or slung payload system. A fixed payload system involves rigidly attaching the payload to the module array, which can be achieved by either using an on-board storage compartment or some kind of interfacing mechanism between the module exteriors and payload. This type of system has much more predictable flight physics than a slung payload; however, the exact method and location the payload is attached to the array can negatively affect the efficiency of the array. Assuming that the payload cannot be divisible into parts of equal mass, the sole attachment point must be approximately aligned with the center of lift. In the case of DFAs, attaching to the location of the center of lift, either directly above or below, will partially block the array propellers. Solving this issue requires the module array to form around the attachment point/payload compartment, increasing the array footprint. This complicates the process of computing the optimal array structure, navigation of the environment, and the loading/unloading process due to clearance with obstacles such as warehouse equipment/structures, and potentially requiring specialized compartments to handle different sized payloads, limiting the DFAs flexibility.

A slung payload system would consist of "N" number of tethers (cables/ropes) attached to the array converging to designated attachment points on the payload exterior/pallet system. This type of system has much more complex flight physics as the payload is subjected to excitations from the flight manoeuvres that must be considered and controlled. The tethers suspending the payload below the array allow for much better clearance for the propellers, allowing for smaller footprints than a fixed attachment system. This also allows this system to be more flexible than a fixed system as it does not require bulky equipment like storage compartments, and the reduced footprint creates better clearance for the loading and unloading process. Using multiple tethers simultaneously across different attachment points on the array could potentially better distribute the load, reducing the stress on the module joints; however, this causes more complex oscillation behavior, which must be understood. Similar to the fixed payload system requiring select storage compartments for different payload sizes and weights, the harmonic mechanics of the slung system mean the properties of the cables, such as elasticity, dampening, and length, must be considered for each payload to avoid resonance.

11.3.9 Navigation

Airborne navigation of the drone modules is a foundational element of a Distributed flight array functionality. While existing literature thoroughly covers the control and flight dynamics of more traditional drone platforms, such as quadrotors, the modularity and variability of the Distributed flight array create unique challenges for many aspects of controlled flight that are highly niche, such as unique array sizes, imbalanced thrust distribution. However, this concept has previously been explored [11] and details a proposed "Parameterized Control Methodology" targeted at controlling distributed flight arrays. Topics such as the flight dynamics, expected disturbances, and implementation are discussed at length with the developed control strategy aimed at achieving a controlled hover with any flight feasible Distributed flight array configuration. A robust flight controller will allow the manoeuvrability of the drone array to be maximized, increasing the flexibility of flightpaths and other manoeuvres. Accuracy and flexibility of flight paths are important factors when efficiency is a priority, such as in a warehouse environment. Responsive flight planning, for on-the-fly path alterations and adjustments for obstacle avoidance, and minimized flight times for energy conservation are both highly desirable.

11.3.10   Perching

The length of flight times is an important metric to control for maximizing operation up-time by conserving energy, which is primarily achieved through streamlining flight paths to reducing unnecessary time spent in the air. However, there may occur situations in a given flight operation where it is necessary for the drone to stay in a location of temporary stand-by. In such scenarios, the ideal action would be to allow the drone array to land to maximize energy conservation, but when this is not possible due to space restrictions, the drone must remain airborne. If the drone enters an assisted hover, by contacting environmental features such as walls or ledges to partially support the weight of the drone array, otherwise known as "Perching", then it may act as a middle ground between maintaining a hover and fully landing, allowing the drone to conserve energy.

Previous research on the energy efficiency of perching tested quadrotors featuring a variety of specialized landing gear to improve contact with, or grip, environmental features such as ledges or branches. The study by Hang et al. [12] covered the design and testing of an actuated landing gear consisting of metal legs with some additional contact modules attached to allow the drone to perch in various ways. Some methods allowed the landing gear to be hooked around the (parch/environment) fully supporting the weight of the drone, allowing the propellers to be fully switched off, and other methods relied on balancing and would allow for a great decrease in power consumption. These previous experiments with quadrotors featured the addition of landing gear increasing the weight and complexity of the drones. The design of the DFA features an integrated docking system, which when combined with specialized docking ports situated about the environment could allow for the DFA platform to utilize perching without additional on-board equipment.

11.3.11   Conclusions of the literature review

The Distributed Flight Array (DFA) is a very capable research platform well equipped to handle and adapt to a multitude of situations and uses. Applications of DFAs in warehouses do seem feasible as they have many theoretical benefits such as lower costs for expansions and reparability. However in practice, for successful usage in a warehouse, the modules would require many refinements and new features to be suitable for material handling such as improved structural properties to support the mass of a payload, the compatibility with and installation of an on-site, highly accurate, real-time, position tracking system for array navigation, such as a ultra-wideband RTLS, implementation of an automated charging system for the modules to efficiently manage down-time and maintenance, and algorithms and processing to calculate optimal path planning and order picking, along with the assignment of tasks to the distributed flight arrays and choice of appropriate array structure. Much of this technology has been implemented in other similar systems and can be seen in use in many different areas and industries such as charging stations for AVs in warehouses and RTLS for tracking assets and equipment in facilities, but this gap in the development of DFAs leaves much speculation about their true efficiency and usefulness in material handling. However, the DFA still serves as an excellent research platform for distributed estimation and control and would greatly benefit from developments in these areas, such as gaining the ability to carry a payload.

From this background research the most impactful area to develop, and the first step towards being capable of substantial material handling, was identified as the Distributed flight arrays existing docking system design and its ability to support payloads. Permanent magnet-based systems as described in previous research have very low strength, which is sufficient for the purpose of alignment, unloaded docking, and the disengagement process of the docking ports. The strength of the docking mechanism can be improved numerous ways through the use of added actuation for a more secure latching mechanism and better surface contact and friction with grooves and other surface geometry. However, to produce an effective docking solution that adheres to the requirements of the modules (such as low mechanical complexity, weight, and power requirements), the expected loads, strength requirements, and behavior of the forces need to be understood so that a relevant design specification could be established.

11.4 Methodology

To establish the expected force magnitudes and their behavior under different circumstances, an experiment, or model, of the Distributed flight array is required. The model must be able to simulate the modules accurately, be subjected to different forces/support payloads, and to record the forces experienced by the joints. There are two main methods that fulfill these requirements, physical experimentation and computer simulation. To conduct the investigation using physical experimentation would require the construction of a set of DFAs or some analogous testing setup. Although it has the possibility to be an extremely accurate method and an insightful process, building an entire set of DFA modules equipped with the necessary sensors is very resource and time intensive, due to the number of parts required for the large variety of testing. Modeling and experimentation is often an iterative process requiring many testing and development cycles before being ready; the process is unpredictable and can take up large amounts of time. The final production is also likely to contain many imperfections due to the scale of the model and the fabrication/iteration process, which increases the chance to negatively affect the results.

Alternatively, computer simulation is much less resource intensive than physical experimentation, with the only requirements being time and computation power. Digital recreation of the DFA can be achieved with a relatively high level of detail and accuracy in the module systems and in the physics. This also considerably reduces the time invested in the iteration cycles compared to physical experimentations, modifications can be made much quicker, and the modules can be simply duplicated as required by the investigation. However, increase in scale and fidelity is often at the cost of longer computation time. These opposing factors create an optimal balance of computational time and detail, which is achieved by only including the necessary features, simplifications where possible, and reducing the scale of the simulation. Within the method of computer simulation, there are many different choices of software available each with their individual strengths, capabilities, and limitations. One particular method of simulation considered for the project was to be based in MATLAB®, consisting of custom written code to mathematically calculate the forces distributed across a given array under a set load. This method would allow the user to input a custom array setup using a hexagonal coordinate system [13] and display the results as a graphical render of the module array overlaid with a heatmap of shear force intensity. However, this method is extremely complicated, requiring deep understanding of the physics at play, and is unable to simulate the dynamics of the cables supporting the payload without vastly exceeding the project scope.

Alternatively, choosing to use a simulation software with a physics engine greatly decreases the work required to create an accurate simulation. One such program is "Simscape Multibody", a simulation package within "Simulink®", a MATLAB-based graphical programming environment. "Simscape Multibody" is a specialized program within the broader "Simulink" program, which is capable of modeling and simulating a variety of multidomain physical systems such as wireless communications, electric motors, and control systems using block diagrams [14]. "Simscape Multibody" specializes in modeling and simulating mechanical systems in a 3D environment, such as autonomous robots, pulley systems, and vehicle suspension systems. Another popular computer simulation method is Finite Element Analysis (FEA), which uses numerical methods to simulate the effects of forces on deformable bodies that have been subdivided into "finite elements". However, FEA is a very specialized simulation type, which limits its focus to smaller features with high resolution and is not suitable for the multidomain simulation of an entire DFA. FEA would be much more suited to analyzing the structure of specific docking mechanism designs to find weaknesses, failure modes, and their safety factor.

11.5 Methods

11.5.1 Payload simulation

Simulation began by breaking down the DFA design to fit the Simscape format of frames, bodies, and joints by identifying the required level of detail for the simulation. This process involved listing all the DFA characteristics and subsystems and filtering them by which are the most impactful and important to capture within the simulation, then translating those aspects into their Simscape equivalent. The critical areas identified were the shape of the module chassis, approximate weight of the materials, the inertia, the upwards thrust produced by the propellers, and the docking function. The main element of the list was the chassis, which was represented as an "Extruded Solid" body element. This element contained a variety of variables, one of which was the geometry parameter containing information about the form of the body. Here the number of sides, radius, and thickness of the extruded regular polygon could be defined, allowing the creation of a hexagon, with an outer radius of 0.155 m and thickness of 0.11 m, along with the mass of the body 0.297 kg (these values were based upon previous DFA designs). Features such as cut outs were ignored as they would require a much more complicated multi-body assembly made of numerous bodies in the shape of beams and plates to replicate, and instead were reproduced by lowering the material density to account for the increase in volume.

The simulation assumes that the chassis is perfectly rigid, which means that certain effects of the material properties such as bending, plastic deformation, fracturing, and other modes of failure that could potentially alter the DFA performance and ability to perform tasks are unaccounted for. While this knowingly introduces inaccuracies, the complexity, time, and computational resources, the demands of all these factors are all extremely high and due to limitations in scope and software, were left out. However, this does allow the simulation results to be purely based upon the forces applied by the payload, without interference from dampening effects caused by material properties. The final step to complete the creation of the module chassis was to define the frames. In Simscape Multibody frames, there are points in space bound by the geometry of the body with a set orientation in the 3D plane (xyz), and these frames act as sockets allowing other bodies to physically connect to one another, defining their spatial location and orientation. This meant that the individual DFA modules required seven extra frames (eight in total, as all bodies have a frame at their origin), six of which are located at the center of each docking face to allow the modules to connect to one another, and one located on the bottom of the chassis to define the attachment point of a payload tether. The orientation of the docking faces had to be set carefully in an alternating pattern so that an array of connected modules should face in the same direction (Fig. 11.1), resulting in a set of alternating oppositely oriented frames. This completed the assembly of the main component of the simulation, the individual module, which could then be converted to its own "subassembly" and duplicated for each simulation as required. These duplicate modules could then be assembled into any array structure by connecting them together using the docking ports. However, to monitor the force data through each connection, the "weld joint" was used as an intermediate between each of the docking ports.

FIGURE 11.1 A single module with 8 "frames".

Joints in "Simscape Multibody" are a class of blocks that defines certain properties about a physical connection and allows for data such as torque and force to be outputted; in the case of the weld joint, it defined zero degrees of movement/rotation, creating a perfectly rigid connection, imitating an ideal docking mechanism. The data collected was in the form of a physical signal, which was then converted to a "Simulink output signal" and exported to a MATLAB workspace for data processing and plotting. This MATLAB workspace also holds several variables that are referenced by the simulation in Simscape multibody, such as calculated thrust values required to maintain a hover, and the dimensions and mass of the individual modules. This data processing involved separating the data into two variables as the raw signal was an array with separate Y and Z data, and then the two parts of the data could be input into the Pythagorean theorem. The resultant data was of the magnitude of the combined vector, which allows for much easier interpretation of the results. However, this simplification of the data means that it does not display the direction of the shear force, which may be of interest if the strength of the docking joint design is significantly anisotropic.

The next component of the simulation was to create the payload system. This involved creating a body to represent the mass of the payload and the cables that connect it to the module array. Creating the mass is a very simple procedure; however, the cables are much more complicated. In essence the cables in the system describe the freedom of movement of the payload, and because "Simscape Multibody" does not contain a "rope/cable/tether" element, it must be recreated with what is available. Ropes are a complex component to simulate to a high level of accuracy due to their spring-like properties. Although it is not often observable by the naked eye, rope, like every material, is subject to a level of elastic deformation, which plays an important role in their behavior and is therefore able to be modeled as a very stiff damped spring. Because of this, ropes seen in video games are commonly simulated by a series of segments, each connected in a chain via small springs, creating the illusion of a rope that looks visually correct and is able to interact with objects [15]. Recreating this method of rope simulation in Simscape Multibody requires body elements to represent each of the segments, spring elements, to connect the segments and to add the spring properties, and also revolute/ball joint elements to give the joints flexibility. Due to limitations of the software, joints cannot be connected directly to one another; the solution to this is to use telescoping joints. These joints are described as having one prismatic joint and one spherical joint primitive, resulting in an element with rotational freedom in the three-axis and one translational degree of freedom. This combination most closely describes that of a rope out of the joint elements available. However, this method, when combined with accurate spring stiffness and damping coefficient values, causes severe slowdown of the simulation software, slowing further as the number of tethers used increases. This method is therefore inadequate for use in Simscape Multibody. The solution to this is a different method that uses only a single telescoping joint to represent each tether rather than a series of small repeating joints. The parameters of the joint allow for the spring effects to only be activated past a certain extension and so could replicate the constraining effects a tether would have on a payload, while sacrificing the visual element and finer mechanics of a full rope simulation.

There are several parameters within the telescoping joint for the rotational and translational components that describe the internal mechanics, with the prismatic primitive "internal mechanics" and "limits" being the most important to the simulation. The "internal mechanics" allow the equilibrium position of the "spring" to be defined; this is the length of the spring at which the spring force equals zero. The Limits' parameters include the "upper and lower limits" of the joint, which control the range of motion of the prismatic component, the "value", which defines a location through which the joint resists travel, the spring stiffness, which defines the resistance through the joint limit, the damping coefficient, which defines the resistance of the damper through the joint limit, and the transitional region, which defines a region of the joint over which the spring-damper force must be increased to its full value. These parameters were then set to values based on the application in the simulation and on real-world material data. The cable length corresponds to the upper and lower limits of the joint, retraining it to the desired length; this function is also partly shared by the "value" variable that determines at what length to activate the effects of the spring properties, also being the max length of the rope.

For the variables "springs stiffness" and "damping coefficient", research of the material properties of cables and ropes approved for lifting, and material handling was conducted to determine realistic values. However, it was found through further research that due to the nonlinearity and uncertainty, these values are often found using physical testing and are not available as documented material properties. These material properties are instead replaced by metrics to convey a rope's material properties such as "static elongation" (ΔL), which is used by climbing rope manufacturers to describe the percentage that a given rope will stretch by under an 80 kg load. By using climbing ropes as a model for the tethers in the simulation it is possible to estimate realistic spring stiffness K values for the tethers as follows:

$$K = \frac{80 \times 9.81}{\epsilon L}, \quad (11.1)$$

where $\epsilon = \frac{\Delta L}{L}$ is the strain of the rope material, ΔL is the elongation, and L is the length of the rope. These values were sourced from the average of several "static class climbing ropes", which are designed for hauling loads and rescue work as they have a low elongation percentage. For determining a realistic damping coefficient, the simulation uses a target damping ratio. Damping ratio is a parameter that describes the level of damping within a system. Damping ratio values are categorized in the following way: systems with a value of 0 have no damping (undamped), systems between 0 and 1 are considered as underdamped, systems with a damping ratio of 1 are considered as critically damped, and systems with values above 1 are considered over-damped. The damping ratio determines how the system behaves as it returns back to the equilibrium position and the time it takes.

The decision to use a target damping ratio over a fixed damping coefficient is that it is more reflective of how a commercial distributed flight array system would operate as the specification of the material properties of the tethers used would be based on the desired behavior of the payload system and subsequently the damping ratio. For these reasons, a target damping ratio value of 0.65 was selected. An ideal system would be a critically damped system; however, due to inaccuracies, wear, and environmental factors, the system would become either over- or underdamped, in which case underdamping is more desirable as the system will reach the equilibrium position in a much more reasonable amount of time.

With the aim to create a large number of simulations testing a variety of DFA formations, it was important to create a naming system that was capable of accurately describing the structure of the array for organizational and identification purposes. During the development stage of the digital recreation of the DFA, naming was based on the visual similarity to common shapes, such as "tri-" denoting triangular qualities; however, this system is very inadequate as the variety of formations grows. To solve this, a simple naming convention using the number of modules and their status regarding being tethered or not was developed. The convention followed the general form $a + bt$, where the value of a corresponds to the number of modules in the formation not tethered to the payload, the value of b corresponds to the number of modules in the formation that are tethered to the payload, and t is a suffix representing the term "tethered". This naming system, while simple, allows information about the array size and shape

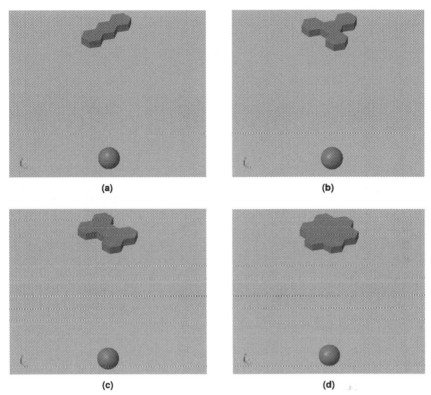

FIGURE 11.2 Various modules with a payload including (a) three arrays, (b) four arrays, (c) five arrays, and (d) seven arrays.

to be presented without visual aid. The system relies on only being applied to relatively small-sized formations and on the assumption that the formation has at least second-order radial symmetry. Fig. 11.2 shows the configuration of various module arrays with a payload. This notation is suitable for the project as all array formations used in the simulations follow the requirements. With the individual modules and the tether system completed, the simulation stage could begin, the first step of which was to determine what formations and variables to test. Three categories were decided upon: Single tether simulations, "N" Tether simulations, and payload configuration simulations. There were several "Single tether simulations", each involving a module array configuration performing a hover with a single central module tethered to the payload. These simulations aimed to identify the effect of increasing the number of docked modules had on joint shear forces at the center of the array. The single tether simulations category also included simulations investigating changes to the tether length. These tests consisted of the same array structure ($3 + 1t$) with different tether lengths, ranging from 0.5 m to 4 m.

These simulations each had two versions, one conducted with a static load and one conducted with a dynamic load, which involved applying a sinusoidal force on the payload in the X and Y directions, causing the payload to move in a circular motion. This type of testing was done to investigate the pattern of redistributed forces on individual joints that occur when attempting to keep the array stable and parallel to the ground over a range of conditions. Additionally, some simulations also used sinusoidal force in a single axis to investigate the effects of bilateral symmetry of tethers compared to radial symmetry and any biases as a product of the array formation. "N-Tether simulations" involved testing multiple different array configurations under both static and dynamic payloads. The arrays featured different numbers of untethered modules and tethered modules. This class of simulations were aimed to investigate the force distributing properties of the tethers across a variety of arrays. The following array configurations were tested: $1 + 2t$, $1 + 3t$, $1 + 4t$, $1 + 6t$, $4 + 3t$, and $3 + 4t$. "Payload configuration tests" featured a variety of simple cable configurations, most of which were alterations to the tether angle. All simulations so far featuring multiple tethers had them originate at the array spaced out and then come to a single point on the payload. Simulations with slight spacing on the payload, parallel tethers, and overspaced tethers were conducted. An additional simulation where the tethers combine into a single tether before attacking to the payload was also carried out. These simulations were all run using dynamic loads. All simulations used the same software and simulation conditions and produced shear force values taken from the central set of docking joints over a simulated period of 10 seconds, producing approximately 370 data points per joint.

11.5.2 Single module flight simulation

Simulation of flight mechanics and control systems began by representing a dimensionally accurate single DFA module through the use of an extruded solid body element with a weight roughly equal to that of a its real work equivalent. An infinite plane block was used in conjunction with a contact force block to represent the ground plane intersecting the world origin, and the module body was connected to the world reference frame via a disconnected weld joint. To control the flight of the module six PID controllers must be used, one for each degree of freedom. The PIDs each take a current signal value, such as the current altitude, and compare it with a desired signal value, such as the desired altitude, to calculate an error value. Therefore sensors must be used to gather information on the module's current location and rotational values with respect to the world reference frame. These values are in the form of Cartesian (xyz) coordinates for locational and a quaternion for rotational values. These values are then routed to their corresponding PID controller, with the rotational value being converted from a quaternion into Euler's angles before being routed. Creating a desired value variable for each PID and connecting them to the corresponding force and torque blocks completes to form the control system. Each PID controller was then isolated and hand tuned to generate an appropriate system response. This method treats the translation of PID output to propeller motor rpm to force output as a black box to simplify the control system and avoid complications with propeller force calculations. This simple control system allows for the simulation a number of basic flight manoeuvres. A controlled hover could be achieved by entering desired values of 0 for all PIDs except for the z-axis controlling PID, which could be given a desired altitude at which the module will hover and maintain. Basic flight paths could be created using multiple signal functions, each controlling the desired signal values, which would also allow for basic docking and perching manoeuvres by navigating and orientating a docking face frame with a togglable weld joint block to a predefined environmental frame operating from sensor signals or predefined signal functions.

11.6 Data analysis

11.6.1 Single tether simulations

The single tether simulations consisted of $3 + 1t$ and $6 + 1t$ formations under a standard 10-kg static load. The aim of the simulation was to investigate the effect the number of supporting modules around a tethered module has on the shear forces, with the hypothesis that by increasing the number of neighboring modules the force is more distributed, reducing the magnitude on anyone joint. The results (Table 11.1) show a decrease in average shear force experienced by the joints of 43% as the number of modules surrounding the tethered module increases from 3 to 6. This value is supported by theoretical workings that show that force exerted by a 10-kg mass ($10 \times 9.81 = 98.1$ N) when divided across both 4 and 7 total supporting bodies should show a 42.86% decrease. The maximum and minimum force values of the static simulation show negligible difference as expected, indicating a stable state with minimal system noise.

These configurations also underwent "Dynamic" loading, where the payload was subject to an increasing cyclic sinusoidal lateral force. The shear force plots (Fig. 11.3) illustrate the fluctuations in force on the individual joints that act as

TABLE 11.1 Results table showing force values acting on the joints of the different single tether arrays from static loading and the % reduction in force between the two arrays.

	$3 + 1t$ static	$6 + 1t$ static	Reduction (%)
Maximum force (N)	24.5	14.2	41.91
Minimum force (N)	24.5	13.8	43.91
Average force (N)	24.5	14.0	42.83

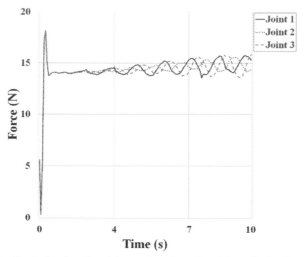

FIGURE 11.3   Graph showing plotted force values from three joints of a $6 + 1t$ array under dynamic loading.

TABLE 11.2 Results table showing force values acting on the joints of a $3 + 1t$ array with various lengths of tether under dynamic loading.

	Length (m)							
	0.5	1	1.5	2	2.5	3	3.5	4
Maximum force (N)	25.5	26.6	33.4	27.4	25.3	25.0	24.9	24.7
Minimum force (N)	23.6	23.8	22.6	23.7	24.1	24.3	24.3	24.3
Average force (N)	24.6	25.0	26.2	25.2	24.7	24.6	24.6	24.5
Variance (N)	0.0526	0.296	4.66	0.66	0.056	0.016	0.0081	0.0062

the "restoring" force to keep the array stable as the payload moves beneath. (All the simulations experienced abnormal force values within the first few moments of being run, most likely due to initializing the physics bodies, which has been disregarded from analysis.)

The "single tether" simulations also included a varying length simulation, where a $3 + 1t$ formation was repeatedly subjected to dynamic loading, each time with the length of tether increasing by 0.5 m (from 0.5 m to 4 m). The results (Table 11.2) show the maximum, average, and variance all following the same trend as the length increases. The shear force rises to a peak of 26.2 N (average) at the 1.5-m length, then falls back down to 24 N (average) as the length increases (Fig. 11.4). This is likely a resonance response brought on by the natural frequency approaching the frequency of the dynamic loading as the length of tether was increase, causing an increase in amplitude.

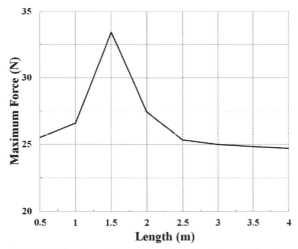

FIGURE 11.4 Graph showing plotted force values from the joints of a 3 + 1t array with various lengths of tether under dynamic loading.

TABLE 11.3 Results table showing force values acting on the joints of different *N*-tether arrays from static loading.

	Static					
	1 + 2t	1 + 3t	4 + 3t	1 + 4t	3 + 4t	1 + 6t
Maximum force (N)	16.9	8.13	5.56	4.97	4.37	2.31
Minimum force (N)	15.8	7.72	5.31	4.50	4.01	1.77
Average force (N)	16.4	8.00	5.48	4.79	4.23	2.12
Variance (N)	0.0258	0.0156	0.0056	0.0202	0.0116	0.0245

11.6.2 N-tether simulations

The "N-Tether" simulations tested a range of simple formations with varying numbers of modules, and tethers ($1+2t, 1+3t, 4+3t, 1+4t, 3+4t$, and $1+6t$). The static simulations involved using a standard 10-kg payload to measure the shear forces across the central joints of the formations. The results shown in Table 11.3 show the recorded maximum, minimum, and average force values, along with the variance between the minimum and maximum values. When the averages are primarily ordered by the number of tethers, the graph (Fig. 11.5) follows a smooth downward curve, with a final dip at the $1 + 6t$ formation. This dip is most likely caused by a lack of data between $3 + 4t$ and $1 + 6t$. A simple symmetrical formation of 5 tethers could potentially smooth out this gap; however, only tether amounts that are a multiple of 2 or 3 could be used due to symmetry requirements. Additionally, the variance across each of the simulations is negligible, as expected of a static load, being caused only by a small amount of noise in the simulation system.

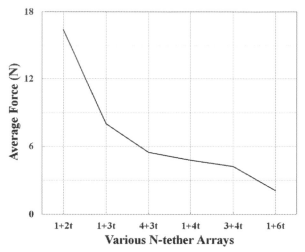

FIGURE 11.5 Graph showing plotted average force values from the joints of different N-tether arrays from static loading.

TABLE 11.4 Results table comparing theoretical force based on the number of modules with simulated force values from the N-Tether simulations.

	Static					
	$1 + 2t$	$1 + 3t$	$1 + 4t$	$4 + 3t$	$3 + 4t$	$1 + 6t$
Theoretical force (N)	32.70	24.53	19.62	14.01	14.01	14.01
Simulated force (N)	16.37	8.00	4.79	5.48	4.23	2.12
Reduction (%)	49.94	67.40	75.57	60.90	69.79	84.90

Arrays simulated with multiple tethers outperform the simple theoretical predictions made using number of modules, as seen from analysis of "Single Tether" simulations, supporting the theory that multiple tethers are able to improve the distribution of shear forces across the array, reducing the magnitude of force translated through the docking joints (Table 11.4).

The results of the dynamic testing, ordered primarily by the number of tethers, show a downward trend of the average forces experienced by the joints, with a sudden dip at $4 + 3t$. If re-ordered primarily by the total number of modules, then the downwards trend smooths out (Fig. 11.6). This could indicate that under dynamic loading, the number of modules, and possibly the specific formation of those modules, has a larger effect than simply the number of tethers or that the effects of the number of tethers and modules are compounding.

Some of the simulations for the "N-Tether" simulations involving dynamic loads also underwent an alternative type of dynamic load test, where instead of both x and y forces being applied periodically, the forces were separated over two tests to investigate any biased responses brought on by the array formations. The results (Figs. 11.7 and 11.8) show that certain formations have reduced

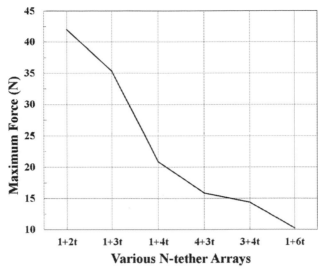

FIGURE 11.6 Graph showing plotted maximum force values from the joints of different N-tether arrays from dynamic loading.

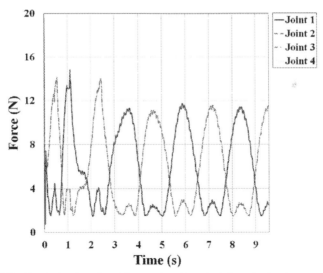

FIGURE 11.7 Graph showing plotted force values from the joints of a $1 + 4t$ array under single axis dynamic loading over time – Mode 1, excitation along the Y-axis.

shear forces when force is applied to the payload in a particular direction, which is observed when the payload is forced to oscillate along the path that follows the circumference of the 2D shape that forms from intersecting spheres of two attachment point's maximum tether length. This path only occurs as a smooth curve along a flat plane in formations with only two tethers. When these condi-

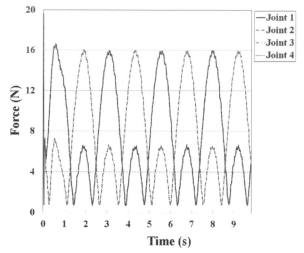

FIGURE 11.8   Graph showing plotted force values from the joints of a $1 + 4t$ array under single axis dynamic loading over time – Mode 2, excitation along the X-axis.

tions are met, the payload will be allowed to oscillate without causing any of the tethers to become slack. This could indicate that certain array formations when executing specific flight manoeuvres with an expected payload response have favorable orientations that reduce shear forces on the docking joints.

When these particular conditions are only partially met, meaning that there are only portions of the circumference of the shape formed from the intersection that are smooth/flat (e.g., cases where there are more than two tethers involved), the points at which the payloads path intersects with the surface of another tethers sphere will cause one or more of the tethers to transition from being slack to tensioned and will cause shock-loading of those particular tethers (Figs. 11.9 and 11.10). Shock-loading of the tethers will cause large spikes in shear force, which can be observed in Fig. 11.10.

Each spike in the graph correlates to a tether transitioning from being slack to tensioned. This effect is much more pronounced depending on the degree of deviation of the payload's oscillation path caused by reaching a tethers maximum length, and with more energetic excitation of the payload, as the dampening effect of the tethers is only able to absorb smaller surges of force.

11.6.3   Payload configuration simulations

The "Payload configuration" simulations tested a number of alternate payload attachment locations for a $1 + 3t$ array in regard to their position on the payload. These include Slightly spaced, where the tether lines would converge roughly to the same point on the payload, "Parallel spacing", where the tether lines would remain parallel to each other (going straight down), and "Over spacing", where

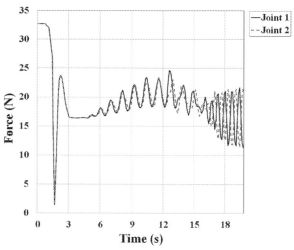

FIGURE 11.9 Graph showing plotted force values from the joints of a $1 + 2t$ array under single axis dynamic loading over time – Mode 1, excitation along the X-axis.

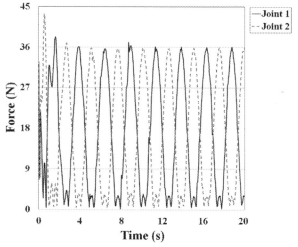

FIGURE 11.10 Graph showing plotted force values from the joints of a $1 + 2t$ array under single axis dynamic loading over time – Mode 2, excitation along the Y-axis.

the lines would diverge outwards (from the center axis of the array and payload) to the attachment points on the payload. Also included was a "3-to-1" tether configuration involving the three tethers combined into a single tether before attaching to the payload. The results of the static loading simulations in Table 11.5 show little change in all metrics across the spacing configurations; however, although the average stayed similar, the 3-to-1 configuration showed significant change to maximum, minimum, and variance. Although it does seem that the

TABLE 11.5 Force values acting on the joints of $1 + 3t$ arrays with different payload attachment configurations under static loading.

	Static configuration			
	Slight	*Parallel*	*Over*	*3-to-1*
Maximum force (N)	8.59	8.29	8.86	52.60
Minimum force (N)	7.63	7.95	7.30	0.006
Average force (N)	8.18	8.18	8.18	12.6

TABLE 11.6 Results table showing force values acting on the joints of $1 + 3t$ arrays with different payload attachment configurations under dynamic loading.

	Static configuration			
	Slight	*Parallel*	*Over*	*3-to-1*
Maximum force (N)	24.0	16.5	15.0	95.5
Minimum force (N)	0.015	1.00	4.56	0.143
Average force (N)	11.2	10.2	10.6	37.1
Variance (N)	32.9	12.1	6.30	508

3-to-1 configuration holds no advantages over the other configurations tested, it is possible that these results could be a product of simulation inconsistency brought on by the tether-combining segments. The dynamic testing (Table 11.6) displays a slight decrease of a few Newtons in the maximum and variance as the spacing increased, whereas the average stayed fairly consistent. The 3-to-1 configuration had increased the maximum and average values, and the variance increased significantly.

11.6.4 Single module flight simulation

The flight simulations successfully demonstrated the usage of PIDs within Simscape Multibody to control the flight of a single drone module. The module, even with a simple control system, was capable of carrying out a series of precise tasks such as accurately following a pre-determined flight trajectory, accurate rotational orienting, and docking with environmental elements in a perching manoeuvre. This was demonstrated by a simulation carried out in which a single co-axial propeller distributed flight array module navigated a small environment by first achieving a controlled hover, then traveling along a pre-defined flight path towards a rendezvous point, at which it then executed a perching manoeuvre by correctly re-orienting and docking with a secondary solid body via enabling a weld joint. Figs. 11.11 and 11.12 demonstrate the time history of pose channels and the flight trajectory, respectively.

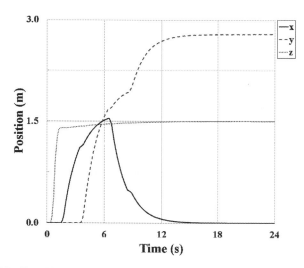

FIGURE 11.11 Time history of pose channels.

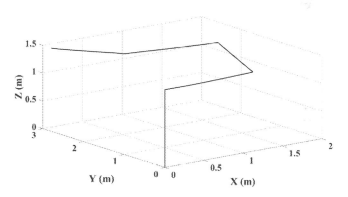

FIGURE 11.12 Flight trajectory.

11.7 Discussions, critical thinking, and reflections

The payload simulations and their results help demonstrate some of the principles that must be followed to effectively operate a DFA payload system such as optimizing the array formation parameters, such as size, shape, and tether distribution, around the type and size of payload. The results also illustrate how using single tether theoretical estimations to calculate shear force of the joints will cause over-estimations in their required strength as multiple tethers when used correctly can effectively distribute force throughout the array formation as seen in the 85% decrease in joint shear force between $6 + 1t$ and $1 + 6t$ (under static loading), as it is illustrated in Tables 11.1 and 11.3. This serves to reinforce the strengths of simulation, which allow for more accurate estimations without requiring more advanced theoretical calculations, as a tool for both the design and operation of a DFA.

The simulations and results help to highlight that the "effectiveness" of a DFA carrying capacity is multivariable problem, and while dominantly dependent on the shear strength of the joints, it is also affected by the nature of the range of intended payloads, the types of flight manoeuvres used, and the tether configuration and its materials. Although this does mean that calculating the exact carrying capacity of any given DFA system based upon its joint strength is not feasible, it does present the possibility of a more complex alternative rating system, which would involve a diagram of multiple plotted curves to describe a systems range of effectiveness with certain given parameters.

Resonance was an effect observed in the single tether-length variation simulations. This caused amplification of the payloads oscillations and subsequently the forces exerted on the docking joints. If unaccounted for resonance has the potential to cause unexpected catastrophic failure, then countermeasures would need to be taken to avoid such events. There are multiple ways this could be executed, with many examples of these theories being applied in other industries, such as in earthquake engineering, and would first involve in modeling the system and simulating the expected excitations such as wind or vibrations from flight and alterations to the tether length could be made to avoid resonance; additionally, dampeners could be incorporated into the array structure or payload attachment system to avoid resonance and absorb energy.

Shock Loading was another phenomenon observed during the "N-Tether Simulations" and is a product of the number and positioning of the payload's tethers under dynamic loading. Much like resonance, this effect can cause much higher loading than expected and result in unexpected failure of the array system. This similarity also means that a lot of the same solutions for resonance also apply to shock-loading, such as additional dampeners and modeling the excitations of the payload to avoid any hazards; however, along with altering the formation layout, the orientation of certain array formations during flight manoeuvres also impact these results, which may allow for a wider range of solutions to avoid the shock loading effect from occurring.

The project has resulted in a significant body of numerical data and has made contributions to the field through exploration of a simulation approach to estimating the forces exerted on the multibody problem of DFAs. The simulation results were of appropriate resolution and precision as they were able to illustrate some of the real-world effects and behaviors such as resonance and shock loading; however, the accuracy of the simulations with respect to real-world models is not verifiable without conducting real-world experiments. The simulation had several limitations, such as within the software, as both computational intensity and increased complexity as a product of "Simscape Multibody" being a graphically based programming language, caused the testing of larger DFA formations to be increasingly laborious and therefore was not used during the project.

The simulation also used several assumptions to reduce complexity and computational time. These include perfect rigidity that removes effects such as bending from the chassis and joints, no air resistance and no weight associated with the tethers, which would have influenced their behavior, and that the arrays maintain a perfectly station hover. Such assumptions could affect some of the behaviors observed in the simulations, such as the resonance and shock-loading. Flexible chassis and more accurate tether simulations that include the weight and air resistance will likely change the exact resonance frequency of the system and would need to be accounted for in any industrial applications. The effect of shock-loading may also be reduced by the increased damping that flexible chassis would add; however, the effects of the chassis bending may cause further complications with flight and would also have to be accounted for in the formation design process and flight planning; the values of the shear forces will also be affected by the changes caused by bending.

The model of the DFA modules in the payload simulations also lacked any real flight control, which would have made adjustments to the propellers to maintain a stationary hover. Such a responsive system would have introduced additional damping as the attitude of the array and corrective flight manoeuvres would absorb some of the vibrations energy. Furthermore, the orientation and stability of the payload during dynamic testing was not considered, and if made, then a requirement would need additional systems and testing to restrict change to an acceptable range. The single module flight simulations demonstrated the capabilities of Simscape Multibody to simulate a simple flight controller using traditional PIDs capable of trivial tasks such as executing a flight path and precise alignment for docking and perching manoeuvres. This research has provided a foundation for more complex flight tasks such as path planning, obstacle avoidance, and payload delivery. The potential for what can be achieved would be further increase by constructing a more complex and precise control system designed to dynamically support flight arrays of larger sizes and various configurations. Combining advancements of the control system with several tools belonging to the "UAV toolbox", an "addons" for Simscape Multibody, could create methods to simulate full environments and synthetic sensor data for researching estimated control algorithms, sensor networks, and computer vision. Simulation of DFAs with a wide arsenal of functions, from grounded driving, controlled flight, docking, perching, and payload carrying allows for a greater depth of research into many aspects of the DFA design and practical capability, as well as functioning as a greater tool for developing distributed control algorithms.

11.8 Conclusions

In conclusion, the projects outcome has managed to fulfill the primary aims of creating an accurate digital model of a Distributed Flight Array (DFA) system complete with a slung payload system, functional implementation of the model within simulation software, and completion of multiple simulations to produce a wealth of useful data. This data was then able to be processed and analyzed to reveal various interesting statistical correlations and behavioral patterns of the DFA system under a variety of load conditions, which illustrate the complex mechanics and niche applications of the various configurations that may allow for unique optimizations in a material handling setting. While providing research into the mechanics of a DFA system and the expected forces and responses to a 10-kg payload under various conditions, the project also serves to document a successful utilization of multibody physics-based simulation software and its benefits over physical testing. While the simulation results cannot be accurately verified until physically reproduced, due to the absence of any applicable statistical verification methods (such as "Mesh independence study" used in Finite Element Analysis), the exact methods of which are predicted to be highly laborious and remain unclear as many aspects of physical testing incur unique inaccuracies and biases due to the measurement equipment and design choices, and are therefore beyond the scope of the project.

The project has shown that DFAs are likely to be highly compatible with a slung payload system based on existing technology given an appropriately designed docking mechanism. This serves to support their potential application in material handling; however, the technology required to fully implement and support such a system is currently uncommon in industry but is poised to become much more prevalent as the world transitions into Industry 4.0. The projects secondary aims of investigating the software ability to support and simulate the control system for a simple single module drone have also been successful. By demonstrating the ability to execute tasks such as maintaining a controlled hover, flight path planning, precise spatial alignment, and a perching manoeuvre through a PID-based flight controller, the simulations have revealed the large potential for much more complex flight control and general DFA functionality research within Simscape Multibody.

11.9 Recommendations for future work

This project research aims to serve as a steppingstone for further research and to bridge the gap between DFAs, material handling technology, and to highlight simulation software capabilities, and therefore there are a large number of topics regarding this research to be further investigated. As the payload simulations in this project handled simple symmetrical array formations, naturally the next step would be to push the software to its limits with larger and more complex non-symmetrical formations to study their unique behaviors. This would also require creating a more robust formation notation system that is able to concisely describe the exact structure of each array, possibly taking inspiration from similar systems like chemical formula notation. The flight and control system simulations used a primitive controller and operated a simple single module DFA, the progression of this would be to develop more accurate and complex control systems with a greater focus on PID tuning and responsiveness/robustness. This control system would also be intended for arrays of larger sizes and be able to adapt to various configuration shapes. Additional features could be integrated with a re-designed control system such as intelligent path planning algorithms that can be used to develop computer vision applications and simulate collision avoidance. These algorithms could also be used to develop robust docking and undocking operations for both grounded and perching applications based on sensor information and evaluate the relationship with array characteristics, such as size and module distribution, with effectiveness and energy efficiency. Combining these possibilities with payload simulations would allow for fully simulated array formation, payload attachment, transportation, and payload offloading routines. This topic could further expand into researching the possibility of mid-flight formation changes to optimize the force distribution in response to required flight manoeuvres and environmental obstacles.

The other natural direction for further research is to begin physical testing of the DFA design either by replicating them completely or by designing a simpler more controlled testing model of which to conduct force testing to investigate the accuracy of the simulation method. This could also include designing a working mechanical docking system and testing variations of the design to maximize its strength capabilities. Physical testing may reveal weaknesses and inaccuracies in the current simulation method and therefore could be used to innovate and improve the simulation method to produce more reliable results. Adjacent to the topic of joint strength and flight control are several other subject matters, such as investigating the aerodynamic effects of the downthrust produced by a DFA on a slung payload and how those forces interact with the payload system. Another aspect of the payload system is the attachment method and investigating the optimal locations and designs for the attachment point on the module chassis as they are currently assumed to be at the center of the bottom face of each module. Furthermore, it is interesting how additional dampers could be implemented into the payload system to reduce the effects of dynamic loading.

References

[1] J. Paulos, B. Caraher, M. Yim, Emulating a fully actuated aerial vehicle using two actuators, in: IEEE International Conference on Robotics and Automation (ICRA), Brisbane, 21–25 May, 2018.

[2] R. Oung, A. Ramezani, R. D'andrea, Feasibility of a distributed flight array, in: Proceedings of the 48th IEEE Conference on Decision and Control (CDC) Held Jointly with 28th Chinese Control Conference, China, 15–18 December, 2009.

[3] R. Oung, F. Bourgault, M. Donovan, R. D'Andrea, The distributed flight array, in: IEEE International Conference on Robotics and Automation, USA, 3–7 May, 2010, pp. 3–7.

[4] R. Oung, R. D'Andrea, The distributed flight array, Mechatronics 21 (2011) 90–917.

[5] R. Oung, R. D'Andrea, The distributed flight array: design, implementation, and analysis of a modular vertical take-off and landing vehicle, The International Journal of Robotics Research 33 (3) (2013) 375–400.

[6] Supermagnete, What is the difference between adhesive force and displacement force/shear force. Available at: https://www.supermagnete.de/eng/faq/What-is-the-difference-between-adhesive-force-and-displacement-force-shear-force, 2021. (Accessed 27 January 2022).

[7] M.R. Mahfouz, C. Zhang, B.C. Merkl, M.J. Kuhn, A.E. Fathy, Investigation of high-accuracy indoor 3-d positioning using UWB technology, IEEE Transactions on Microwave Theory and Techniques 56 (6) (2008) 1316–1330.

[8] A. Benavoli, Distributed estimation in sensor networks. Systems and information technology [online]. Available at: https://people.idsia.ch/~alessio/FullyDecentralizednetworks_FirenzeUn.pdf, 2015. (Accessed 13 March 2022).

[9] R.D. Koster, T. Le-Duc, K.J. Roodbergen, Design and control of warehouse order picking: a literature review, European Journal of Operational Research 182 (2) (2007) 481–501.

[10] A. Madani, M. Karwan, The balancing traveling salesman problem: application to warehouse order picking, Top 29 (2020) 442–469.

[11] R. Oung, M.P. Cruz, R. D'Andrea, A parameterized control methodology for a modular flying vehicle, in: International Conference on Intelligent Robots and Systems, Portugal, 7–12 October, 2012.

[12] K. Hang, L. Ximin, H. Song, J.A. Stork, Perching and resting—a paradigm for UAV maneuvering with modularized landing gears, Science Robotics 4 (28) (2019) 1–10.

[13] A. Patel, Hexagonal grids, Red Blob Games [Blog] October. Available at: https://www.redblobgames.com/grids/hexagons/, 2021. (Accessed 19 January 2022).

[14] K. Mahmood, A.I. Norilmi, Application of multibody simulation tool for dynamical analysis of tethered aerostat, Journal of King Saud University - Engineering Sciences 34 (3) (2022) 209–216.

[15] R. Badea, Simulating a rope, Owlree [Blog] 28 May. Available at: https://www.owlree.blog/posts/simulating-a-rope.html, 2020. (Accessed 8 January 2022).

Chapter 12

Recent advancements in multi-objective pigeon inspired optimization (MPIO) for autonomous unmanned aerial systems

Muhammad Aamir khan[a], Quanmin Zhu[b], Zain Anwar Ali[a], and Muhammad Shafiq[a]

[a]*Department of Electronic Engineering, Sir Syed University of Engineering and Technology, Karachi, Pakistan,* [b]*Department of Engineering Design and Mathematics, University of the West of England, Bristol, United Kingdom*

12.1 Introduction

The evolution in aeronautics and aviation has become popularly necessary and demanding in the modern technology served for human desires. The unmanned aerial vehicle (UAV) is a widely used carrier of this field [1]. Recently, UAVs had many extensive applications, especially in commercial and military fields. Multiple unmanned aerial vehicles developed to achieve search and rescue (SAR) mission to save maximum number of people in a limited time. This approach is proposed as the layered search and rescue (LSAR) technique. So it is also emerging in the areas like plotting scenario, surveillance, and rescue purposes [2–4]. Furthermore, UAV is also feasible in remote sensing and access to vulnerable fields where accurate movement required. The grouped UAVs are more beneficial for the successful implementation by decreasing the mission completion time of different battle scenarios.

The novel optimization method in multiple unmanned aerial vehicles (multi-UAVs) can achieve fast and precise reconfiguration under random attacks as shown in Fig. 12.1. It also suggested that the speedy effect of formation control of multi-UAV resisting the random attacks in a successful manner. This optimization model related to the parameter problem is also feasible to enhance the system resilience by the help of adaptive learning-based pigeon-inspired optimization (ALPIO) algorithm to further enhance the value of resilience to get better results.

Modeling, Identification, and Control for Cyber-Physical Systems Towards Industry 4.0
https://doi.org/10.1016/B978-0-32-395207-1.00024-X

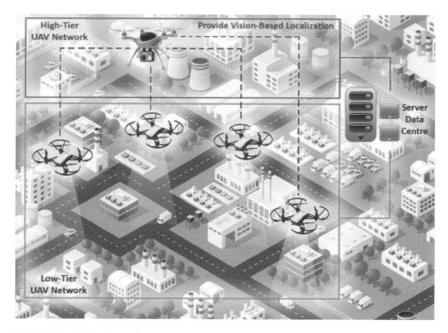

FIGURE 12.1   Two tiers of Unmanned Aerial Vehicles.

Robust optimization is an enormously demanding and intensively researched in academia and applications over the last decade. It has been observed that the proposed algorithms suit very well in the area related to the robustness in the optimization algorithms and yield the novel approach for both current trends of research and open research standards [5,6].

In the real world, optimization problems become challenging in engineering, mathematics, science, and commerce industry. Therefore several scientists have presented multiple optimization techniques and algorithms to address those challenges successfully. As the challenge increases, the algorithms require extensive calculations and computations to solve complex issues. Obviously, optimization techniques that have competent computational power and limited memory are preferable for achieving better results. Scientists [7] have proposed different pigeon-inspired optimization (PIO) techniques in the last decade, which are more precise, accurate, and efficient compared with the classic methods.

Pigeon-inspired optimization technique proved to be a valuable competitor in global optimization problems, and its idea proposed in 2014 by Duan and Qiao [8] is reflected in Fig. 12.2. Recent work modified the PIO algorithm to analyze and evaluate multi-objective optimization issues. In 2015 an emerging swarm intelligence optimization technique was introduced, which follows the behavior of homing pigeons. This algorithm is feasible for solving the problems areas of designing and assembling the brushless direct current motors including

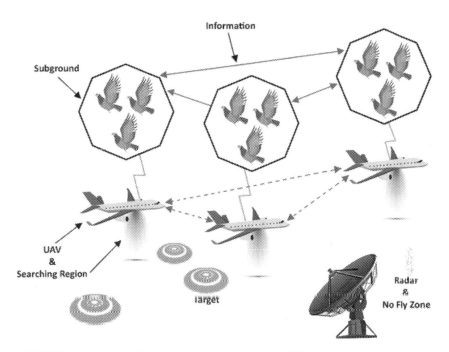

FIGURE 12.2 Inspiration of pigeon-inspired optimization in UAVs.

five design variables, five constraint variables, and two objective variables. In 2016, a model design [20] was proposed based on carrier landing system that works automatically. The proposed pigeon-inspired optimization algorithm was applied to overcome the problems of manual adjustment and converting the system model into efficient design to achieve better optimization. It also prevents the lack of the diversity in the pigeon population.

In 2017 a new technique named as multi-objective genetic algorithm (MOGA) [11] was introduced to achieve oilseed canola production by the combination of environment and economic indices, mixing energy [9]. Pigeon-inspired optimization algorithms need to be combined with some advanced strategies to efficiently solve the complex optimization problems in a better way. The modified PIO algorithm considered the hierarchical learning behavior and presented a distributed flocking control technique to support and guide unmanned aerial vehicles to fly in a stable formation. In 2019 an improved population diversity method [10] was developed for managing the multilevel image thresholding issues by using a cooperative PIO algorithm. In the same year (2019), an analysis of modified PIO technique [11] was proposed for air quality prediction problems by optimizing particle swarm optimization process.

In 2020 a distributed flocking control technique was introduced to coordinate unmanned aerial vehicles fly in a stable formation, which is a modified PIO [12]. Another modified PIO technique was proposed in 2020 to optimize the fuel cost of energy system, efficiency, and power loss in the system [13]. The existing algorithms were gradually improved to produce variety of optimization problems including the aerial field.

In 2021 an emerging algorithm termed multi-objective pigeon-inspired optimization (IMOPIO) was introduced [14] to balance the local and global optimization issues. However, it was observed from the given multiple variants that the PIO algorithm is usually affected by the local convergence due to poor diversity.

In 2022 a novel planning technique named different objective optimization-based multiple UAVs was proposed to address the effects of a single objective function to plan the path for the UAVs [15]. In this method, a non-dominated sorting genetic algorithm (NSGA) was presented to address multi-objective optimization issues by using the fruit fly algorithm (FLO) as the replacement of the genetic algorithm (GA) to further improve the optimization and achieve better efficiency of the system [16].

Many researchers focus on the emerging and demanding optimization algorithm schemes named modified pigeon-inspired optimization to analyze multi-objective optimization problems. Some major analysis is compiled by focusing on various parameters like energy consumption on intelligent algorithm, stable formation in complex environment, model improvements and hardware developments in industrial applications, population convergence and diversity, power loss and fuel cost on different power systems, selection and search ability of proper PIO algorithms. Many researchers employed the recent market trends and developments in the optimization schemes.

The aim and motivation of this survey paper is to compile various parameters in different research fields together in a single forum, which will certainly give easy reference for relevant researchers to extract the best suited and optimized methods for their desired goals considering the comparison of exploitations and explorations of all parameters. To address and overcome the challenges of different restrictions, complexities, and uncertainties, the best and optimized techniques are feasible, which can make the system model more effective and efficient due to the modern and modified pigeon-inspired optimization algorithms to cover all the aspects of multi-objective optimization problems. The main contributions of this survey paper are as follows:

- To introduce the implementation of pigeon-inspired optimization techniques along their technical analysis in terms of efficient energy consumption, power loss and fuel cost of power systems, and population convergence and diversity.
- To analyze the performance and challenges faced by different optimization algorithms and suggest some modifications to improve the efficiency of the system.
- To address the key findings and restrictions of the existing optimization techniques and propose some future recommendations and modifications for better implementation of the system model to achieve network optimization.

The rest of this paper consists of four sections. Section 12.2 presents the state-of-the-art parameters for implementation of multi-objective pigeon-inspired optimization in UAVs. Section 12.3 reviews problem statement and proposes solutions of different aspects of MPIO in UAVs systems. Section 12.4 discusses the recent developments and advancements in MPIO implemented in UAVs. Section 12.5 presents conclusions and future approach for the evaluation of various optimization algorithms and related issues.

12.2 State of the art

The state-of-the-art study in the area of network optimization introduced in 2014 by Duan and Qiao [8] as Pigeon-Inspired Optimization (PIO) algorithm to produce effective navigation system shown in Fig. 12.3 through the mathematical expressions for "Map operator", "Landmark operator", and "Compass operator". This technique is related to the homing behavior of pigeons, which follows the idea of routing to detect any target or returning to its original place. The detection is done by navigation mechanism [9] of homing pigeons and the available data of birds. The navigational map is suitable for using routing information to know the path for a better optimization system. This is applied to the wild birds and migrants in their region.

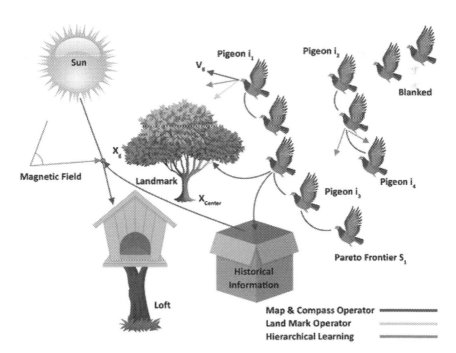

FIGURE 12.3 Pigeon-Inspired Optimization as navigation systems.

Homing pigeons used another method of combining both map and compass system to track their destination. When the sky is clear, the sunlight is used to be a compass. When the sun disappears behind the cloud, pigeons use the magnetic field of Earth to achieve the desired result. The pigeon map uses sensors that can be unknown, but the disorientation of the pigeons in sunny environment suggests the map based on magnetic cues.

By using this optimization algorithm the pigeons use receiver based on magnetic structure to configure and locate the map in their brains to aware the Earth magnetic field. They select and prefer the Sun's elevation to adjust the direction and path of their route. Pigeons fly to their target with very little amount of support magnetic particles and Sun light. They basically depend on the closest landmarks for getting close to their goals. Pigeons fly fast when they trace the landmarks and follow a direct route as the last one. If any pigeon does not find the nearest landmark, then it follows the pigeons that are familiar with the landmark.

The pigeon-inspired optimization was further modified [10] by Industry 4.0 vision integrated with the key technologies and Cyber-Physical-Systems (CPSs), especially in manufacturing sector. However, the responsibilities and decision-making factors of the employees is still an open issue needed to be addressed. So the proposed study solved this challenge by a method to support the configurations and system design of different work by considering the human labor and CPS structure jointly within a comprehensive framework. This approach also covers ordinary production and uncertain parameters like failure detection or maintenance intervention.

12.3 Problem statement and solutions

Pigeon-inspired optimization (PIO) is one of the demanding and novel bio-mimetic swarm intelligence optimization technique introduced in 2014 [8] to provide an efficient algorithm in the fields like energy consumption, power loss, and the fuel cost of the distribution systems. The searching ability of pigeons is unique as compared to other birds in terms of efficiency and accuracy to reach the destination rather than long route traveling. In the telecommunication, the use of pigeons becomes obsolete for communication due to the advancements of emerging technologies dealing with the complex systems that demands more stability and accuracy to solve the convergence and diversity issues. Lists of the research work are briefly introduced for understanding the problems and solutions.

Zhang et al. [17] proposed a hybrid model termed Pigeon-Inspired Optimization (PIO) for solving the issues of reconfiguration for multiple unmanned aerial vehicles, which focuses on achieving the optimum input values including the load factor, thrust, and bank angle to complete the desired task. The proposed model solved the reconfiguration problems for multiple UAVs. The simulation results suggest that the proposed algorithm embedded with PIO is better suited than the existing PSO algorithm.

Dai et al. [18] presented a system design for the implementation of multi-objective pigeon-inspired optimization technique to enhance the capability of search efficiency by the help of the navigation capacity of homing pigeons. In this approach the compass and map operators are capable to record the routing path to obtain accuracy. The landmark operator speeds up the convergence time and improves the diversity of the network as further advancements of this algorithm. The landmark operator used to upgrade the convergence of sub-problems with poor convergence. The advantages of taking two sets of MOPs are shown by comparing six MOEAs and MPIOs.

Qiu et al. [19] proposed a novel swarm intelligence optimization technique, which follows the behavior of homing pigeons. This algorithm is feasible for solving the problems areas of designing and assembling the brushless direct current motors including five design variables, five constraints variables, and two objective variables. This algorithm is feasible for solving multi-objective optimization design parameters. The experimental results show the superiority and validity of the proposed MPIO technique.

Deng et al. [20] introduced a model design based on control parameters for the carrier landing system that works automatically. In this study the pigeon-inspired optimization algorithm is applied to address the problems of manual adjustment and converting the system model into efficient design to achieve better optimization. It also prevents the lack of diversity in the pigeon population. A weighted linear cost function was adopted in time domain to optimize the power compensation system and control parameters in H-dot autopilot. The series of experimental results suggest that the proposed method is feasible and better than the existing methods.

Li et al. [21] proposed a path planning method of many objective optimizations-based multiple UAVs technique to address the effects of path planning considering a single objective function for the UAVs. In this method a non-dominated sorting genetic algorithm (NSGA) was presented to address multi-objective optimization issues by using the fruit fly algorithm (FLO) as the replacement of the genetic algorithm (GA) so that further increase the performance related to optimization and achieve better efficiency of the system. This study focused on multi-objective optimization technique based on multiple UAVs path planning to overcome the limitations of a single objective function. The simulation results show that the improved NSGA-III algorithm is more efficient in the path planning of UAVs than the other methods.

Ruan et al. [22] introduced a multi-objective social learning pigeon-inspired optimization (MSLPIO) for collision avoidance with unmanned aerial vehicle (UAV) formation control. In this approach a learning parameter is added to the compass, map, and nearby landmark operators such that each pigeon adapts the route from the best pigeon to update the process. The blindness of the parameter setting is improved by selecting a dimension-dependent parameter. Simulation results verify the proposed method based on improved convergence performance by comparing multi-objective pigeon-inspired optimization and non-dominated sorting genetic algorithm.

Duan et al. [23] presented a mechanism termed limit-cycle-based mutant multi-objective pigeon-inspired optimization to analyze the behavior affecting the flight of the pigeons. This method is also beneficial to increase the strength of the exploration capacity during the flight process. The proposed algorithm achieved a high accuracy and faster convergence speed to improve population diversity. Comparative analysis of proposed technique along with the other multi-objective approaches is conducted to validate the efficiency, accuracy, and convergence stability.

The above-mentioned analysis of modified pigeon-inspired optimization techniques indicates that the presence of pigeons is not necessary for communication due to the advancements of emerging technologies dealing with the complex systems that demands more stability and accuracy to solve the convergence and diversity issues. The study covers various aspects such as dealing with reconfiguration issues for multiple UAVs, improved navigation capacity of homing pigeons, behavior of homing pigeons, effective utilization of control parameters for carrier landing system, efficient optimization algorithm for multiple UAVs path planning, and limit-cycle-based mechanism to achieve better optimization of the network.

12.4 Advancements of MPIO and its variants

The basic concept of advanced pigeon-inspired optimization (PIO) algorithm is applied on various fields such as scheduling process, long distance traveling, power systems, and effective energy consumption in networks with fast convergence speed and strong robustness. However, the local optimization cannot be denied and considered to be shortcomings of the basic PIO algorithms. To effectively address the complex problems, the PIO need to be modified and combined with some relevant and advanced strategies. Furthermore, the developments and modifications in the concept of PIO algorithm can rely on the latest techniques feasible for both local and global optimization. The comparative analysis of some emerging modified optimization algorithms is gathered in Table 12.1 to analyze the network optimization of various applications in the field of power distribution systems, navigation systems, and long-distance traveling mechanisms. Yearwise analysis of PIO and MPIO is as follows:

In 2014, Ran Hao [24] proposed a novel approach related to UAV assignment model addressing on the effective power consumption applied easily to the intelligence algorithm. This task is completed by the modification in pigeon-inspired optimization (PIO) technique to address mission assignment issues of multiple UAVs.

In 2015, Qiu [19] proposed a hybrid swarm intelligence optimization technique that follows the behavior of homing pigeons. This algorithm is feasible for solving the problems of designing and assembling the brushless direct current motors including five design variables, five constraints variables, and two objective variables.

In 2016, Deng [20] introduced a model design based on control parameters for the carrier landing system that works automatically. In this study, the pigeon-inspired optimization algorithm is applied to overcome the problems of manual adjustment and converting the system model into efficient design to achieve better optimization. It also prevents the lack of diversity of the pigeon population.

In 2018, Huaxin Qiu [25] presented a multi-objective pigeon-inspired optimization method to prevent collision of multiple UAVs among the obstacles. In this method, two types of criteria are addressed to guarantee the flight safety for different objectives. In such a scenario, a modified PIO that follows the UAV distributed flocking control algorithm is presented to achieve the stability of multiple UAVs in terms of flying condition under dynamic and complex environment.

In 2019, Haibin Duan [20] proposed a study to review the analysis and modifications of PIO in the parameters like structure adjustment, component replacement, operation addition, and application expansion. Zhihua Cui [21] addressed the limitation of multi-objective pigeon-inspired optimization for not solving the optimization issues. So this study adapted the selection mechanism termed the balanceable fitness estimation (BFE) approach that combines the convergence and diversity distance to achieve the desired performance.

In 2020, Gonggui Chen [28] proposed a modified pigeon-inspired optimization algorithm (MPIO) to deal with problems regarding multiple contradictory objectives and produce a solution of non-differentiable optimal power flow. In this study the MPIO algorithm is combined with commonly used penalty function method (PFM) to form the MPIO-PFM algorithm to address the parameters of power distribution systems in terms of energy loss, emission, and system fuel cost. Jing Wang [29] presented a modified pigeon-inspired optimization (MPIO) technique that offers emerging and innovative miniature atomic sensors for designing self-shielded uniform magnetic field coils. The proposed method consists of coaxial cylinder surfaces that can provide a highly uniform magnetic field at the center of the coil and attenuate the external field.

In 2021, Xu Zhun [14] proposed an improved pigeon-inspired algorithm to address the manipulations in product demand and uncertainty caused by the demand prediction. In this approach, a dynamic layout facility (DLF) is presented to cope the uncertain product demands by optimizing the cost and considering effective area utilization. Ali [30] presented a hybrid strategy for swarm of multiple unmanned aerial vehicles in terms of formation control. This study provides a three-dimensional function for formation control by using Cauchy mutant (CM) and particle swarm optimization (PSO) to update the fitness function. Dai [18] presented a system design for the implementation of multi-objective pigeon-inspired optimization technique to enhance the capability of search efficiency by the help of navigation capacity of homing pigeons. In this approach, the compass and map operators are capable to record the routing path to obtain

TABLE 12.1 Implementation of PIO and its variants.

S. No	Ref	Title	Technique used	Applied on	Compared	Key Findings	Limitation
1.	[24]	"Multiple UAVs Mission Assignment Based on Modified Pigeon-Inspired Optimization Algorithm"	UAV model focusing energy consumption on intelligent algorithm	MPIO	PIO and DE	MPIO is better for solving multiple UAVs mission then classical algorithms	Sequence of task execution is ignored
2.	[25]	"A multi-objective pigeon-inspired optimization r approach to UAV distributed flocking among obstacles"	multi-objective pigeon-inspired optimization	Unmanned aerial vehicle (UAV)	MPIO and NSGA-II	The technique applied for stable formation to coordinate UAVs to fly in complex environments	Convergence analysis is lacking by using modified MPIO
3.	[26]	"Advancements in pigeon-inspired optimization and its variants"	multi-objective pigeon-inspired optimization	PIO	PIO, PSO, GA, and ABC algorithms	The study is best suited for industrial applications as follows: 1. Model improvement 2. Application intensification 3. Hardware development	The research has two limitations: 1. Theoretical analysis is improper for simultaneous simulation processes. 2. Predominance of PIO has issues
4.	[27]	"A pigeon-inspired optimization algorithm for many-objective optimization problems"	multi-objective pigeon-inspired optimization	Multi-objective model	GrEA and NSGA-III	Diversity and population convergence are addressed by BFE approach. Center to global best positions are extracted	Solution quality can be enhanced by using Operators such as MaPIO, PM and SBX to support non dominated solution.

continued on next page

TABLE 12.1 (continued)

S. No	Ref	Title	Technique used	Applied on	Compared	Key Findings	Limitation
5.	[28]	"Application of modified pigeon-inspired optimization algorithm and constraint-objective sorting rule on multi-objective optimal power flow problem"	Combination of MPIO and PFM	OPF problems	MPIO-COSR and NSGA-II	Optimization on fuel cost and energy loss on different power systems	Reducing the running time can be achieved by using pigeon-inspired algorithm and improving the search efficiency
6.	[29]	"Design of Self-Shielded Uniform Magnetic Field Coil via Modified Pigeon-Inspired Optimization in Miniature Atomic Sensors"	MPIO technique for self-shielded uniform magnetic field coils	Miniature atomic sensors	Traditional design methods	Proposed technique addresses the constraints of optimal parameters of the coil.	The development of atomic sensors can be further enhanced based on SUMF coil structure.
7.	[14]	"An improved pigeon-inspired optimization algorithm for solving dynamic facility layout problem with uncertain demand"	IMOPIO technique	Uncertain products	MOPSO and MOPIC	The proposed algorithm gives better searching ability than the compared algorithms.	Facility re-arrangements process needed to be considered to make model more practical.

accuracy. The landmark operator is used to accelerate the convergence time and improve the diversity of the network as further advancements of this algorithm.

In 2022, Li [21] proposed a multi-objective optimization based multiple UAVs path planning method to address the effects of path planning of single objective function for the UAVs. In this method, a non-dominated sorting genetic algorithm (NSGA) is presented to address multi-objective optimization issues by using fruit fly algorithm (FLO) as the replacement of genetic algorithm (GA) to further increase the performance based on the optimization and achieve better efficiency of the system.

Future Scope: Researchers could focus on designing hybrid optimization model in future that can be beneficial for all the aspects of various fields, especially in the areas like power generation, fuel consumption and optimized cost, long distance traveling in optimum route, and all other fields where the optimization techniques can help enormously to further improve the performance and quality of the systems.

12.5 Conclusions

In the modern communication, optimization algorithms are widely used for the multi-objective purposes by the research scholars. These algorithms are extremely popular and demanding for further enhancements, modifications, and improvements of the system design in terms of the optimized cost and effective consumption of the resources. The basic concept of advanced pigeon-inspired optimization algorithms (PIO) is applied in various fields such as scheduling process, long distance traveling, power systems, and effective energy consumption in networks with fast convergence speed and strong robustness. However, the local optimization cannot be ignored and considered to be a shortcoming of the basic PIO algorithms. The proposed study addresses all the shortcomings of PIO by introducing the implementation of pigeon-inspired optimization techniques along its technical analysis in terms of efficient energy consumption, power loss and fuel cost of power systems, and population convergence and diversity. The proposed work also analyzes the performance and challenges faced by different optimization algorithms and suggests some modifications to improve the efficiency of the system. This work can also accommodate the key findings and restrictions of the existing optimization techniques and proposes some future recommendations and modifications for better implementation of the system model, which tends to achieve network optimization.

References

[1] Huang Yao, Rongjun Qin, Xiaoyu Chen, Unmanned aerial vehicle for remote sensing applications — a review, Remote Sensing 11 (12) (2019) 1443.

[2] Ebtehal Turki Alotaibi, Shahad Saleh Alqefari, Anis Koubaa, Lsar: multi-UAV collaboration for search and rescue missions, IEEE Access 7 (2019) 55817–55832.

[3] Hyunbum Kim, Lynda Mokdad, Jalel Ben-Othman, Designing UAV surveillance frameworks for smart city and extensive ocean with differential perspectives, IEEE Communications Magazine 56 (4) (2018) 98–104.

[4] S. Rhee, T. Kim, Investigation of 1: 1,000 scale map generation by stereo plotting using UAV images, The International Archives of the Photogrammetry, Remote Sensing and Spatial Information Sciences 42 (2017) 319.

[5] Karen Hollebrands, Samet Okumus, Prospective mathematics teachers' processes for solving optimization problems using Cabri 3D, Digital Experiences in Mathematics Education 3 (3) (2017) 206–232.

[6] Marc Goerigk, Anita Schöbel, Algorithm engineering in robust optimization, in: Algorithm Engineering, Springer, Cham, 2016, pp. 245–279.

[7] Randal S. Olson, Nathan Bartley, Ryan J. Urbanowicz, Jason H. Moore, Evaluation of a tree-based pipeline optimization tool for automating data science, in: Proceedings of the Genetic and Evolutionary Computation Conference 2016, 2016, pp. 485–492.

[8] Haibin Duan, Peixin Qiao, Pigeon-inspired optimization: a new swarm intelligence optimizer for air robot path planning, International Journal of Intelligent Computing and Cybernetics (2014).

[9] Wolfgang Wiltschko, Roswitha Wiltschko, Homing pigeons as a model for avian navigation?, Journal of Avian Biology 48 (1) (2017) 66–74.

[10] Paola Fantini, Marta Pinzone, Marco Taisch, Placing the operator at the centre of Industry 4.0 design: modelling and assessing human activities within cyber-physical systems, Computers & Industrial Engineering 139 (2020) 105058.

[11] Seyed Hashem Mousavi-Avval, Shahin Rafiee, Mohammad Sharifi, Soleiman Hosseinpour, Bruno Notarnicola, Giuseppe Tassielli, Pietro A. Renzulli, Application of multi-objective genetic algorithms for optimization of energy, economics and environmental life cycle assessment in oilseed production, Journal of Cleaner Production 140 (2017) 804–815.

[12] Huaxin Qiu, Haibin Duan, A multi-objective pigeon-inspired optimization approach to UAV distributed flocking among obstacles, Information Sciences 509 (2020) 515–529.

[13] Muhammad Shafiq, Zain Anwar Ali, Amber Israr, Eman H. Alkhammash, Myriam Hadjouni, A multi-colony social learning approach for the self-organization of a swarm of UAVs, Drones 6 (5) (2022) 104.

[14] Xu Zhun, Xu Liyun, Ling Xufeng, An improved pigeon-inspired optimization algorithm for solving dynamic facility layout problem with uncertain demand, Procedia CIRP 104 (2021) 1203–1208.

[15] Yun Wang, Guangbin Zhang, Xiaofeng Zhang, Multilevel image thresholding using Tsallis entropy and cooperative pigeon-inspired optimization bionic algorithm, Journal of Bionic Engineering 16 (5) (2019) 954–964.

[16] Feng Jiang, Jiaqi He, Tianhai Tian, A clustering-based ensemble approach with improved pigeon-inspired optimization and extreme learning machine for air quality prediction, Applied Soft Computing 85 (2019) 105827.

[17] Xiaomin Zhang, Haibin Duan, Chen Yang, Pigeon-inspired optimization approach to multiple UAVs formation reconfiguration controller design, in: Proceedings of 2014 IEEE Chinese Guidance, Navigation and Control Conference, IEEE, 2014, pp. 2707–2712.

[18] Cai Dai, A multi-objective pigeon-inspired optimization algorithm based on decomposition, in: International Conference on Advanced Machine Learning Technologies and Applications, Springer, Cham, 2021, pp. 929–936.

[19] HuaXin Qiu, HaiBin Duan, Multi-objective pigeon-inspired optimization for brushless direct current motor parameter design, Science China. Technological Sciences 58 (11) (2015) 1915–1923.

[20] Yimin Deng, Haibin Duan, Control parameter design for automatic carrier landing system via pigeon-inspired optimization, Nonlinear Dynamics 85 (1) (2016) 97–106.

[21] Kun Li, Xinxin Yan, Ying Han, Fawei Ge, Yu Jiang, Many-objective optimization based path planning of multiple UAVs in oilfield inspection, Applied Intelligence (2022) 1–16.

[22] Wan-ying Ruan, Hai-bin Duan, Multi-UAV obstacle avoidance control via multi-objective social learning pigeon-inspired optimization, Frontiers of Information Technology & Electronic Engineering 21 (5) (2020) 740–748.

[23] Haibin Duan, Mengzhen Huo, Yuhui Shi, Limit-cycle-based mutant multiobjective pigeon-inspired optimization, IEEE Transactions on Evolutionary Computation 24 (5) (2020) 948–959.

[24] Ran Hao, Delin Luo, Haibin Duan, Multiple UAVs mission assignment based on modified pigeon-inspired optimization algorithm, in: Proceedings of 2014 IEEE Chinese Guidance, Navigation and Control Conference, IEEE, 2014, pp. 2692–2697.

[25] Zain Anwar Ali, Zhangang Han, Rana Javed Masood, Collective motion and self-organization of a swarm of UAVs: a cluster-based architecture, Sensors 21 (11) (2021) 3820.

[26] Haibin Duan, Huaxin Qiu, Advancements in pigeon-inspired optimization and its variants, Science China. Information Sciences 62 (7) (2019) 1–10.

[27] Zhihua Cui, Jiangjiang Zhang, Yechuang Wang, Yang Cao, Xingjuan Cai, Wensheng Zhang, Jinjun Chen, A pigeon-inspired optimization algorithm for many-objective optimization problems, Science China. Information Sciences 62 (7) (2019) 1–3.

[28] Gonggui Chen, Jie Qian, Zhizhong Zhang, Shuaiyong Li, Application of modified pigeon-inspired optimization algorithm and constraint-objective sorting rule on multi-objective optimal power flow problem, Applied Soft Computing 92 (2020) 106321.

[29] Jing Wang, Xinda Song, Yun Le, Wenfeng Wu, Binquan Zhou, Xiaolin Ning, Design of self-shielded uniform magnetic field coil via modified pigeon-inspired optimization in miniature atomic sensors, IEEE Sensors Journal 21 (1) (2020) 315–324.

[30] Zain Anwar Ali, Han Zhangang, Multi-unmanned aerial vehicle swarm formation control using hybrid strategy, Transactions of the Institute of Measurement and Control 43 (12) (2021) 2689–2701.

Chapter 13

U-model-based dynamic inversion control for quadrotor UAV systems

Ahtisham Aziz Lone[a], Hamidreza Nemati[a], Quanmin Zhu[a], Paolo Mercorelli[b], and Pritesh Narayan[a]

[a]*Department of Engineering Design and Mathematics, University of the West of England, Bristol, United Kingdom,* [b]*Institute for Production Technology and Systems, Leuphana University of Lueneburg, Lueneburg, Germany*

13.1 Introduction

A quadcopter is a special type of non-coaxial multi-rotor aircraft, which has four equally spaced rotors, usually arranged at the corners of a square body. The rotors are directed upwards and are placed in a square formation with equal distance from the center of mass of the quadrotor. Compared with fixed-wing aircraft, quadrotors do not require long runways for take-off and landing; they are also designed to operate with greater agility and rapid manoeuvring and are capable of operating in harsh conditions in the presence of environmental disturbances such as wind gusts.

Due to general advances in the area of electronics (improved sensors, batteries, and actuators etc.), during the last few decades, the performances and capabilities of modern quadcopters has been increasing exponentially. The small size of these unmanned quadcopters makes them ideal for use both in constrained spaces and in open outdoors. These small-size quadcopters can replace full-sized helicopters currently used for tasks like crop inspection, weather and traffic monitoring, power line, building and bridge inspections, aerial photography, and other general surveillance tasks. It is comparatively cost effective, and there is no need to endanger human life as the small quadcopter system is unmanned.

Several different methods have been proposed for regulation and trajectory tracking. By far the most common and widely used one is the Proportional Integral Derivative (PID) controller [9,12], but there are other approaches that have also been studied that use Proportional Integral–Proportional Derivative (PI-PD) controller, Fuzzy PI-PD (FPI-PD) [13] controller as well as Linear Quadratic Regulator (LQR) [11,14] approach and Sliding Mode Controller (SMC) de-

Modeling, Identification, and Control for Cyber-Physical Systems Towards Industry 4.0
https://doi.org/10.1016/B978-0-32-395207-1.00025-1

sign approach. Generally speaking, most of these approaches can be placed in one of the two popular frameworks, the model-based and model-free (data-driven) frameworks. However, there is another less attended third one called the model-independent framework [4,5]. Let us now briefly introduce the three frameworks.

Model-based framework design. Let us consider a general cascade feedback control system shown as shown in Fig. 13.1, where G_p denotes the plant, which could be modeled as a linear transfer function or nonlinear dynamic equation in either polynomial or state-space expression, and G_c represents the classical controller. Let G be the closed-loop transfer function specified in advance according to the requirements of the designer. For a linear plant G_p, the controller could be designed with $G_c = G_p^{-1} \frac{G}{1-G}$. For a nonlinear plant G_p, the controller could be designed with $G_c = f(G - p, G)$, where f is a function to link the plant and closed-loop performance to determine the control in a certain way of plant inversion.

In this approach, the model of the plant G_p is requested in advance, where the model sets include the linear/nonlinear polynomial and state-space expressions. There are many mature approaches available for this design framework as this has been the predominant approach in academic research and industrial applications. Using this framework, it is difficult to design nonlinear plant-based control systems, and it is also difficult to specify the transient responses of nonlinear control systems with this framework [8].

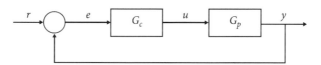

FIGURE 13.1 Classical control framework.

Model-free (data-driven) design. There are various approaches to model-free control system design. Here we describe a few well-known approaches, but in general, all these data-driven approaches use measured data to learn and find proper control in reversion of some expected closed-loop performance, for example, in iterative learning control (ILC) [2] as can be seen in Fig. 13.2. One of the main principles is that it uses iterative learning to improve the controller G_c with repeated reference stimulation to finally achieve $G_c G_p = G_p^{-1} G_p = 1$. This approach considers every possibility for integrating past control information into the next round of control design. There is no need for a clear model structure. This approach is only available in a repeatable control environment under strict conditions. There is no need for the model G_p of the plant in advance, and it is challenging to control nonlinear dynamic plants with this approach. Another representative that falls into this category is PID control tuned by the Ziegler–Nichols approach [1,3]. This approach does not need a model of the plant G_p, even when mild conditions are required for the controlled plants.

It is the most common and easily used trial-and-error approach. However, this approach wastes experimental work to obtain plant models. Almost all engineering plants/processes and input/output measurements are possible to model in principle, although it is sometimes a difficult task [8,17].

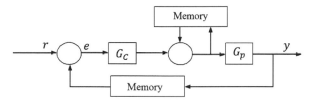

FIGURE 13.2 Iterative learning control (ILC).

Model-independent design. This study refers it to U-model-based control design [4,5]. As shown in Fig. 13.3, G_{cl} is a linear invariant controller, and G_p^{-1} is the dynamic inverse of the plant G_p. In the design procedure, $G_{cl} = \frac{G}{1-G}$ and G_p^{-1} are designed separately to form the resultant controller $G_c = G_{cl}G_p^{-1}$.

The parallel design of the controller and dynamic inversion make the design procedure applicable to linear/nonlinear polynomial/state-space model structures [18]. Transient responses can be specified for nonlinear systems. It is neat in design without waste/repetition if the plant model changes. This approach also complements most existing design approaches. However, this approach is sensitive to model uncertainty and robustness is the paramount issue in designing control systems.

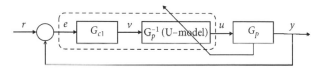

FIGURE 13.3 U-control framework.

Regarding the research status of U-model-based control, the main focus so far has been the study of discrete-time systems. Pole placement control design method [4], U-Smith predictor with input time delay [7], adaptive U-control of total nonlinear dynamic systems [6], U-neural network enhanced control [17], and underactuated coupled nonlinear adaptive control synthesis, using U-model for multivariable unmanned marine robotics [15]. There is very little research on U-model-based control system design for continuous-time systems so far [17]. In this chapter, we develop a general continuous-time (CT) controller for a MIMO UAV control system by using the U-model-based Dynamic Inversion (UM-Dynamic Inversion) approach.

The rest of the chapter is organized as follows. In Section 13.2, we derive the dynamic model of the UAV system. In Section 13.3, we give an overview of the U-Model-based design approach and step-by-step U-model-based dynamic inversion procedure. In Section 13.4, we present the simulation and results of the study. Section 13.5 concludes the study with a brief analysis of results and suggests a possible future work.

13.2 Quadrotor dynamic model

In this study, we use a Parrot Mambo minidrone shown in Fig. 13.4 as a plant. This small and compact minidrone is used because of its cost effectiveness and suitability to test theoretical controllers designed in the MATLAB®/Simulink software environment with the help of the Parrot minidrone support package. Once the controller is tested in the simulation environment, it can be deployed to the hardware using the Simulink Support Package for Parrot minidrones to perform physical experiments on the hardware and verify the simulation results.

FIGURE 13.4 A Parrot Mambo minidrone quadcopter.

The Parrot minidrone has 6 degrees-of-freedom, which means it is able to move translationally in three directions along the x, y, and z axes and rotates in 3 directions with rotations Roll ϕ, Pitch θ, and Yaw ψ, respectively. There are many existing studies describing the nonlinear dynamic model of the quadcopter. Referring to previous studies [9,10], the dynamic model is reviewed and presented in the following section.

In this section, we derive the equations of motion describing the attitude and position of a quadrotor. The quadrotor consists of four propellers in which two of them (r_1; r_3) rotate clockwise and rest of them (r_2; r_4) rotate counterclockwise. The attitude change of the quadrotor results from variations on forces and moments produced by adjusting rotor speeds.

To calculate the dynamic model of the quadrotor, it is assumed that the quadrotor structure and propellers are rigid and symmetric. It is also assumed that the center of mass does not coincide with the origin of the body-fixed frame. Aerodynamic forces and moments are proportional to the square of the rotor speed. Also, the axes of the body-frame coincide with the principle axes of the quadrotor, i.e., the inertia matrix of the quadrotor is diagonal. The Euler angles are also assumed to be small around the hovering position.

The dynamics of the quadrotor is described using the following two reference frames, the Earth-fixed inertial frame $E(X_E Y_E Z_E)$ and the body-fixed frame $B(X_B Y_B Z_B)$, as shown in Fig. 13.5. The attitude of the quadrotor is described by ZYX Euler angle notations, where the Euler angles $\Theta = [\phi \quad \theta \quad \psi]^T$ are respectively known as roll (rotation around x-axis), pitch (rotation around y-axis), and yaw (rotation around z-axis). The attitude angles are bounded as follows: $\phi \in (-\frac{\pi}{2}, \frac{\pi}{2})$, $\theta \in (-\frac{\pi}{2}, \frac{\pi}{2})$, and $\psi \in (-\pi, \pi)$ because the various acrobatic flyings are not admissible. The quadrotor angular velocity in frame B is represented by $\omega^B = [p \quad q \quad r]^T$. The transformation of vectors from the Earth-fixed inertial frame to the body-fixed frame can be expressed using the orthogonal rotation matrix ($R_{\phi\theta\psi}$) in Eq. (13.1). The translational equations of motion are described in the Earth-fixed inertial frame, whereas the rotational ones can be defined in the body-fixed frame. Thus the rotation matrix from the body-fixed frame to the Earth-fixed inertial frame can be calculated as Eq. (13.2). Subsequently, the transformation relations between the angular velocity vector in the Earth-fixed inertial frame and that in the body-fixed frame can be computed by the matrix (M) in Eqs. (13.3) and (13.4).

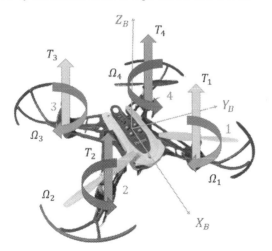

FIGURE 13.5 Coordinate axes for the Mambo minidrone.

$$R_{\phi\theta\psi} = R(\phi)R(\theta)R(\psi)$$

$$= \begin{bmatrix} \cos\psi & \sin\psi & 0 \\ -\sin\psi & \cos\psi & 0 \\ 0 & 0 & 1 \end{bmatrix} \begin{bmatrix} \cos\theta & 0 & -\sin\theta \\ 0 & 1 & 0 \\ \sin\theta & 0 & \cos\theta \end{bmatrix} \begin{bmatrix} 1 & 0 & 0 \\ 0 & \cos\phi & \sin\phi \\ 0 & -\sin\phi & \cos\phi \end{bmatrix}$$

$$= \begin{bmatrix} C\theta C\psi & S\theta C\psi S\phi - S\psi C\phi & S\theta C\psi C\phi + S\psi S\phi \\ C\theta S\psi & S\theta S\psi S\phi + C\psi C\phi & S\theta S\psi C\phi - C\psi S\phi \\ -S\theta & C\theta S\phi & C\theta C\phi \end{bmatrix},$$

$$(13.1)$$

where S and C denote the trigonometric functions sin and cos, respectively.

$$R_{\phi\theta\psi}^{-1} = R_{\phi\theta\psi}^{-T} = \begin{bmatrix} C_\theta C_\psi & S_\phi S_\theta C_\psi - C_\phi S_\psi & C_\phi S_\theta C_\psi + S_\phi S_\psi \\ C_\theta S_\psi & S_\phi S_\theta S_\psi + C_\phi C_\psi & C_\phi S_\theta S_\psi - S_\phi C_\psi \\ -S_\theta & S_\phi C_\theta & C_\phi C_\theta \end{bmatrix}. \quad (13.2)$$

The rotational kinematics is obtained from the transformation of the Euler rates $\dot{\Theta} = \begin{bmatrix} \dot{\phi} & \dot{\theta} & \dot{\psi} \end{bmatrix}^T$ measured in the Earth-fixed inertial frame and angular body rates $\omega^B = \begin{bmatrix} p & q & r \end{bmatrix}^T$ as follows:

$$\omega^B = M\dot{\Theta}$$

$$\begin{bmatrix} p \\ q \\ r \end{bmatrix} = \begin{bmatrix} 1 & 0 & -S_\theta \\ 0 & C_\phi & S_\phi C_\theta \\ 0 & -S_\phi & C_\phi C_\theta \end{bmatrix} \begin{bmatrix} \dot{\phi} \\ \dot{\theta} \\ \dot{\psi} \end{bmatrix} \quad (13.3)$$

and

$$\dot{\Theta} = M^{-1}\omega^B$$

$$\begin{bmatrix} \dot{\phi} \\ \dot{\theta} \\ \dot{\psi} \end{bmatrix} = \begin{bmatrix} 1 & S_\phi T_\theta & C_\phi T_\theta \\ 0 & C_\phi & -S_\phi \\ 0 & \frac{S_\phi}{C_\theta} & \frac{C_\phi}{C_\theta} \end{bmatrix} \begin{bmatrix} p \\ q \\ r \end{bmatrix}, \quad (13.4)$$

where T denotes the trigonometric function tan. Since the Euler angles are assumed to be small around the hovering position, we can easily conclude that $\cos\phi = \cos\theta = 1$ and $\sin\theta = \theta$.

The aerodynamic forces T and moments M produced by the ith propeller can be expressed as

$$T_i = k_{Ti}\Omega_i^2,$$
$$M_i = k_{Mi}\Omega_i^2, \quad (13.5)$$

where Ω_i represents the ith propeller's speed, and k_{Ti} and k_{Mi} are aerodynamic constants, which can be determined experimentally for each rotor. The total moments τ can be derived using the following relationship:

$$\tau_x = \frac{\sqrt{2}}{2}L(T_1 - T_2 - T_3 + T_4),$$

$$\tau_y = \frac{\sqrt{2}}{2}L(-T_1 - T_2 + T_3 + T_4), \quad (13.6)$$

$$\tau_z = (M_1 - M_2 + M_3 - M_4),$$

where L is the distance between the propeller and the center of mass of the quadrotor. Hence the control inputs can be described by combining Eqs. (13.5) and (13.6) in a vector form as

$$
\begin{bmatrix} U_1 \\ U_2 \\ U_3 \\ U_4 \end{bmatrix} = \begin{bmatrix} 1 & 1 & 1 & 1 \\ \frac{\sqrt{2}}{2}L & -\frac{\sqrt{2}}{2}L & -\frac{\sqrt{2}}{2}L & \frac{\sqrt{2}}{2}L \\ -\frac{\sqrt{2}}{2}L & -\frac{\sqrt{2}}{2}L & \frac{\sqrt{2}}{2}L & \frac{\sqrt{2}}{2}L \\ \frac{k_{M1}}{K_{T1}} & \frac{k_{M2}}{K_{T2}} & \frac{k_{M3}}{K_{T3}} & \frac{k_{M4}}{K_{T4}} \end{bmatrix} \begin{bmatrix} T_1 \\ T_2 \\ T_3 \\ T_4 \end{bmatrix}.
\tag{13.7}
$$

Desirably, the translational equations of motion in the Earth-fixed inertial frame can be summarized as

$$
m\ddot{X} = mg + R_{\phi\theta\psi}^T u_1 Z_E,
\tag{13.8}
$$

where g denotes the gravity acceleration, m is the mass of the quadrotor, and $z_E = \begin{bmatrix} 0 & 0 & 1 \end{bmatrix}^T$ is the unit vector expressed in the Earth-fixed inertial frame. Eq. (13.8) can be re-defined as

$$
\begin{bmatrix} \ddot{x} \\ \ddot{y} \\ \ddot{z} \end{bmatrix} = \begin{bmatrix} 0 \\ 0 \\ -g \end{bmatrix} + \begin{bmatrix} C_\phi S_\theta C_\psi + S_\phi S_\psi \\ C_\phi S_\theta S_\psi - S_\phi C_\psi \\ C_\phi C_\theta \end{bmatrix} \frac{U_1}{m}.
\tag{13.9}
$$

The quadrotor rotational dynamics can be derived using the Newton–Euler equations of motion in the body-fixed frame as

$$
J\dot{\omega}^B + \omega^B \times J\omega^B + M_g = \tau,
\tag{13.10}
$$

where J is the inertia matrix, of the quadrotor, M_g is the propeller gyroscopic effect, and $\tau = \begin{bmatrix} \tau_x & \tau_y & \tau_z \end{bmatrix}^T$ presents the total moments acting on the quadrotor in the body frame. The gyroscopic moment resulting from the propeller's kinetic energies can be described as

$$
M_g = \omega^B \times \begin{bmatrix} 0 & 0 & J_r \omega_r \end{bmatrix}^T,
\tag{13.11}
$$

where J_r denotes the propeller's' inertia, and ω_r represents the relative propeller's speed, which can be defined as

$$
\Omega_r = \sum_{i=1}^{4} (-1)^{i+1} \omega_i.
\tag{13.12}
$$

Eventually, the quadrotor attitude dynamics can be written using Eqs. (13.7) and (13.10) in the following form:

$$
\begin{bmatrix} \ddot{\phi} \\ \ddot{\theta} \\ \ddot{\psi} \end{bmatrix} = \begin{bmatrix} \frac{J_{yy}-J_{zz}}{J_{xx}}\dot{\theta}\dot{\psi} \\ \frac{J_{zz}-J_{xx}}{J_{yy}}\dot{\phi}\dot{\psi} \\ \frac{J_{xx}-J_{yy}}{J_{zz}}\dot{\phi}\dot{\theta} \end{bmatrix} - J_r\Omega_r \begin{bmatrix} \frac{\dot{\theta}}{J_{xx}} \\ \frac{\dot{\phi}}{J_{yy}} \\ 0 \end{bmatrix} + \begin{bmatrix} \frac{U_2}{J_{xx}} \\ \frac{U_3}{J_{yy}} \\ \frac{U_4}{J_{zz}} \end{bmatrix}. \qquad (13.13)
$$

Eqs. (13.9) and (13.13) combined together represent the dynamic model of our system.

TABLE 13.1 Values for the Mambo minidrone.

Parameters	Values
m (kg)	0.063
$J_{xx} \times 10^{-5}$ (kg.m^2)	5.83
$J_{yy} \times 10^{-5}$ (kg.m^2)	7.17
$J_{zz} \times 10^{-5}$ (kg.m^2)	10
L (m)	0.108
k_T	0.0107
$k_M \times 10^{-3}$	0.78264

The values of the parameters in Table 13.1 are taken from the Simulink Support Package for Parrot minidrones.

13.3   U-model dynamic inversion-based control system design

A general continuous time U-model with a triplet of $(y(t), \lambda(t), u(t))$ for a Single-Input Single-Output continuous-time polynomial dynamic system with $y(t) \in \mathbb{R}$ as the output and $u(t) \in \mathbb{R}$ as the input at time $t \in \mathbb{R}^+$ is expressed as

$$
\overset{(M)}{y(t)} = \sum_{j=0}^{J} \lambda_{j(Y_{M-1}, U_{N-1}, \Theta)} \left(\overset{(N)}{u(t)} \right)^j, \quad M > N, \qquad (13.14)
$$

where $\overset{(M)}{y(t)}$ and $\overset{(N)}{u(t)}$ are the Mth- and Nth-order derivatives of the output $y(t)$ and input $u(t)$, respectively, $\lambda_j \in R$ is a time-varying parameter absorbing all the other inputs and outputs,

$$Y_{M-1} = \begin{bmatrix} (M-1) \\ y(t), \ldots, y(t) \end{bmatrix} \in R^M, U_{N-1} = \begin{bmatrix} (N-1) \\ u(t), \ldots, u(t) \end{bmatrix} \in R^N,$$

and Θ are the coefficients associated with the input $\left(\dfrac{(N)}{u(t)} \right)^j$.

When using the U-model-based design approach to design a controller according to the performance requirements, the controller is made up of two parts, the general controller G_{cl}, which is designed independently according to performance requirements, and the inverse of the plant G_p^{-1}, as shown in Fig. 13.3. Therefore the main task in the whole process is to calculate the inverse G_p^{-1} of the plant G_p to be controlled and integrate it with the already designed universal controller G_{cl} to form the new controller. The universal controller does not need to be redesigned while designing a new controller and can be reused for different plants if the design requirements are the same; however, the inverse of each plant will be different, and this inverse G_p^{-1} combined with universal controller G_{cl} is the new controller. For a more detailed discussion about the U-model design approach along with examples, we refer to [4,5,16,17].

13.3.1 Design of the universal controller G_{cl}

To proceed with the controller design, we first specify the ideal closed-loop transfer function

$$\frac{Y(s)}{R(s)} = G(s) = \frac{1}{s^2 + 1.4s + 1}, \tag{13.15}$$

which generates the system output response with damping ratio $\zeta = 0.7$, undamped natural frequency $\omega_n = 1$, and zero steady state error to a step reference input [19]. The invariant controller with a unit plant in a feedback control system is determined by taking the inverse of the closed-loop transfer function (13.15) as

$$G_{cl} = \frac{1}{1 - G} = \frac{1}{s(s + 1.4s)}. \tag{13.16}$$

13.3.2 U-model state space realization

The step-by-step procedure for U-model state-space realization for a general Single-Input Single-Output (SISO) continuous-time state space model [16,17] is defined as follows:

$$\begin{cases} \dot{X} = F(x, u), \\ y = H(x). \end{cases} \tag{13.17}$$

The state-space model (13.17) is then expanded into a multi-layer polynomial expression as follows:

$$
\begin{cases}
\dot{x}_1 & = F_1(x_1, x_2, \ldots, x_n), \\
\dot{x}_2 & = F_2(x_1, x_2, \ldots, x_n), \\
& \vdots \\
\dot{x}_n & = F_n(x_1, x_2, \ldots, x_n), \\
y & = H(x_1, x_2, \ldots, x_n).
\end{cases}
\tag{13.18}
$$

The multi-layer polynomial expression (13.18) is converted into a multi-layer polynomial U-model expression:

$$
\begin{cases}
\dot{x}_1 & = \sum_{i=0}^{n} \lambda_{1i} f_{1i}(x_2), \\
\dot{x}_2 & = \sum_{i=0}^{n} \lambda_{2i} f_{2i}(x_2), \\
& \vdots \\
\dot{x}_n & = \sum_{i=0}^{n} \lambda_{ni} f_{ni}(u), \\
y & = \sum_{i=0}^{n} h_i(x_1, x_2, \ldots, x_n).
\end{cases}
\tag{13.19}
$$

Combining this U-model expression (13.19) with the universal controller described in the literature [4,5] in detail, each line of (13.19) can be considered as a separate plant G_{pi}, $i = 1, 2, \ldots, n$, and its inverse G_{pi}^{-1}, which is unique for each plant, is calculated. This inverse combined with the universal controller forms the new unique controller for each plant.

For illustration, consider the state space model

$$
\begin{cases}
\dot{x}_1 & = x_2, \\
\dot{x}_2 & = -x_1 - x_2 + u, \\
y & = x_1,
\end{cases}
\tag{13.20}
$$

$$
\begin{aligned}
\dot{y} &= \dot{x}_1 = x_2, \\
\ddot{y} &= \dot{x}_2 = -x_1 - x_2 + u.
\end{aligned}
\tag{13.21}
$$

The second line of (13.20) directly relates the input and output, and therefore it can be used for dynamic inversion. Then the associated U-model is

$$
\begin{cases}
\ddot{y} & = \lambda_0 + \lambda_1 u, \\
\lambda_0 & = -x_1 - x_2, \\
\lambda_1 & = 1.
\end{cases}
\tag{13.22}
$$

Now we can determine the inverse G_p^{-1} keeping in mind that the eventual goal is to make the plant output y equal to the invariant controller $(G_c l)$ output v. The next step is to integrate the above expression till we get y:

$$y = \frac{\lambda_0}{s^2} + \frac{\lambda_1}{s^2} u. \tag{13.23}$$

To get the second-order derivative of the controller output v, directly modify the universal controller to match the order of y in the expression where the input and output are directly related. For (13.21), this is achieved by multiplying a second-order Laplace operator s^2 with invariant controller G_{cl} so that $\ddot{v} = s^2 v$. Therefore $u = x_1 + x_2 + \ddot{v}$ results in $G_p G_p^{-1} = 1$. Fig. 13.6 shows the setup and result of the simulation.

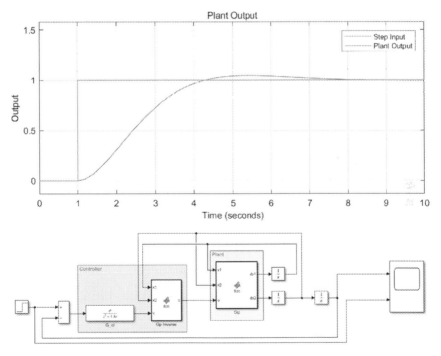

FIGURE 13.6 Example results and simulation setup.

13.4 Simulation study

The quadrotor UAV dynamic model (13.9)–(13.13) is rewritten assuming small Euler angles around hovering position, $\cos(\phi) = \cos(\theta) = 1$ and $\sin(\theta) = \theta$. Also, propeller inertia is assumed to be zero ($J_r = 0$). The equations can now be rewritten for a fully actuated system as follows:

$$\begin{aligned}
\ddot{z} &= -g + C_\phi C_\theta \frac{U_1}{m} & &= -g + \frac{U_1}{m}, \\
\ddot{\phi} &= \frac{J_{yy} - J_{zz}}{J_{xx}} \dot{\theta}\dot{\psi} - J_r \Omega_r \times \frac{\dot{\theta}}{J_{xx}} + \frac{U_2}{J_{xx}} & &= \frac{U_2}{J_{xx}}, \\
\ddot{\theta} &= \frac{J_{zz} - J_{xx}}{J_{yy}} \dot{\phi}\dot{\psi} - J_r \Omega_r \times \frac{\dot{\phi}}{J_{yy}} + \frac{U_3}{J_{yy}} & &= \frac{U_3}{J_{yy}}, \\
\ddot{\psi} &= \frac{J_{xx} - J_{yy}}{J_{zz}} \dot{\phi}\dot{\theta} - J_r \Omega_r \times 0 + \frac{U_4}{J_{zz}} & &= \frac{U_4}{J_{zz}}.
\end{aligned} \tag{13.24}$$

For each line of system (13.24), the associated U-model is derived, and its plant inverse is calculated; this inverse is then paired with the universal controller to give a unique controller for each subsystem.

For altitude Z, the associated U-model is calculated as

$$\begin{cases}
\ddot{z} &= -g + \frac{U_1}{m}, \\
\ddot{z} &= \lambda_0 + \lambda_1 U_1, \\
\lambda_0 &= -g \; ; \lambda_1 = \frac{1}{m}, \\
z &= \frac{\lambda_0}{s^2} + \frac{\lambda_1}{s^2} U_1.
\end{cases} \tag{13.25}$$

For its associated inversion resulting in $G_p G_p^{-1} = 1$, $U_1 = (\ddot{v}_1 + g)m$.

We can see from Fig. 13.7 that the system output has a small overshoot and almost zero steady-state error, which meet the performance requirements of a quadrotor during flight. The controller output can also be seen. The invariant controller G_{cl} is used together with plant inverse G_{p1}^{-1} to form the controller G_{c1} for control of the altitude Z.

For Roll (ϕ), the associated U-model is calculated as

$$\begin{cases}
\ddot{\phi} &= \frac{U_2}{J_{xx}}, \\
\ddot{\phi} &= \lambda_0 + \lambda_1 U_2, \\
\lambda_0 &= 0 \; ; \lambda_1 = \frac{1}{J_{xx}}, \\
\phi &= \frac{\lambda_0}{s^2} + \frac{\lambda_1}{s^2} U_2.
\end{cases} \tag{13.26}$$

For successful inversion resulting in $G_p G_p^{-1} = 1$: $U_2 = \ddot{v}_2 \times J_{xx}$

The plant output and the plant input (controller output) responses for the roll channel are designed using the universal controller G_{cl} together with the plant inverse G_{p2}^{-1} to form the controller G_{c2} for roll (ϕ). The output result for a range of reference values ranging from $\phi \in (\frac{-\pi}{6}, \frac{\pi}{6})$ can be seen in Fig. 13.8.

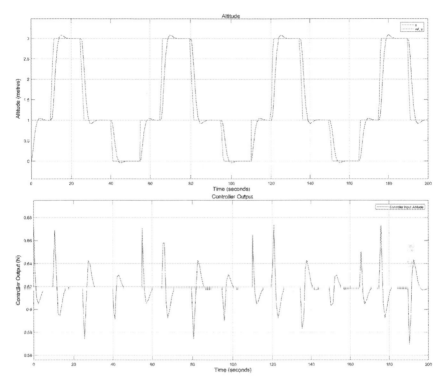

FIGURE 13.7   Plant output with reference values of altitude Z and controller output.

For Pitch (θ), the associated U-model is calculated as

$$
\begin{cases}
\ddot{\theta} & = \dfrac{U_3}{J_{yy}}, \\[2mm]
\ddot{\theta} & = \lambda_0 + \lambda_1 U_3, \\[2mm]
\lambda_0 & = 0 \,;\, \lambda_1 = \dfrac{1}{J_{yy}}, \\[2mm]
\theta & = \dfrac{\lambda_0}{s^2} + \dfrac{\lambda_1}{s^2} U_3.
\end{cases}
\tag{13.27}
$$

For successful inversion resulting in $G_p G_p^{-1} = 1$, $U_3 = \ddot{v}_3 \times J_{yy}$.

The plant output and input (controller output) responses for the pitch are designed using the universal controller G_{cl} together with the plant inverse G_{p3}^{-1} to form the controller G_{c3} for pitch θ. The output result for a range of reference values ranging from $\theta \in (\frac{-\pi}{6}, \frac{\pi}{6})$ can be seen in Fig. 13.9. Comparing Figs. 13.8 and Fig. 13.9, the performance of the controlled object is the same. However, the values for plant input (i.e., controller output) is different, which can be explained by the fact that even though the universal controller G_{cl} is reused here, the plant inverse G_p^{-1} is different, resulting in a new overall controller.

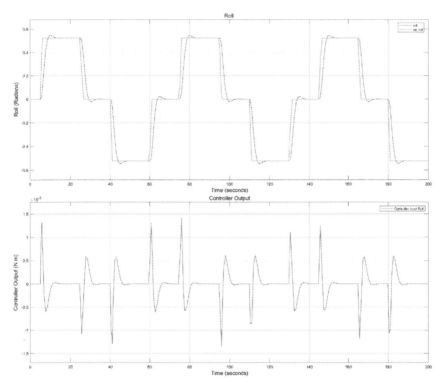

FIGURE 13.8 Plant output with reference values of Roll ϕ and controller output.

For Yaw (ψ), the associated U-model is calculated as

$$
\begin{cases}
\ddot{\psi} &= \frac{U_4}{J_{zz}}, \\
\ddot{\psi} &= \lambda_0 + \lambda_1 U_4, \\
\lambda_0 &= 0 \, ; \, \lambda_1 = \frac{1}{J_{zz}}, \\
\psi &= \frac{\lambda_0}{s^2} + \frac{\lambda_1}{s^2} U_4.
\end{cases}
\tag{13.28}
$$

The plant output and input (controller output) responses for a range of reference values ranging from $\psi \in (\frac{-\pi}{2}, \frac{\pi}{2})$ are designed using the universal controller G_{cl} and are shown in Fig. 13.10. The plant inverse G_{p4}^{-1} is unique, resulting in a unique overall controller, and therefore so is the controller output.

Fig. 13.11 shows the Simulink implementation of the designed control system.

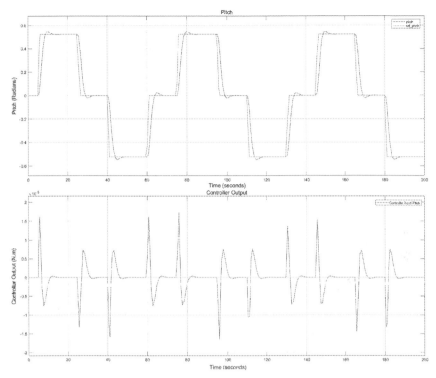

FIGURE 13.9   Plant output with reference values of Pitch θ and controller output.

13.5   Conclusions

This chapter introduces model-independent design approach (U-model-based control) with a working example of a quadcopter UAV. The dynamic model of the system is derived in Section 13.2, followed by a brief introduction to the U-model control approach in Section 13.3, where the basic idea is to find the inverse of the plant G_p^{-1} so that it can be combined with the invariant controller designed independently according to the performance requirements. As the invariant controller G_{cl} is designed independently of the plant G_p, the U-control approach once-off design for all stable nonminimum phase plants, except for the inverse G_p^{-1} of the plant. In Section 13.4 a fully actuated MIMO quadrotor control system is converted into U-Model expression. The system is paired with a universal controller and the inverse of the plant to create a unique controller. The simulations verify that the designed controller provides a good performance with a small overshoot and almost zero steady-state error. The proposed approach not only guarantees the performance requirements that are specified in advance while designing the universal controller, but it also simplifies the controller design process as it can also be extended to other plants with the same performance requirements.

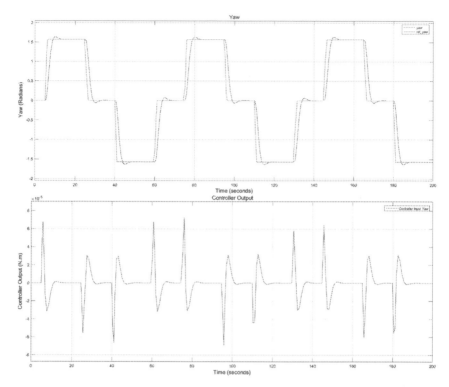

FIGURE 13.10 Plant output with reference values of Yaw ψ and controller output.

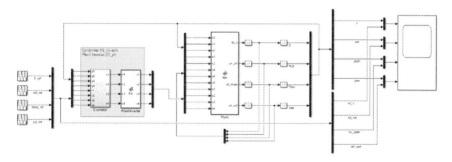

FIGURE 13.11 Simulink implementation of the simulation.

References

[1] K. Ogata, Modern Control Engineering, 4th edition, Prentice Hall, 2002.

[2] J.-X. Xu, Y. Tan, Linear and Nonlinear Iterative Learning Control, vol. 291, Springer, 2003, https://doi.org/10.1007/3-540-44845-4.

[3] M. Fliess, C. Join, Model-free control, International Journal of Control 86 (2013) 2228–2252.

[4] Q.M. Zhu, L. Guo, A pole placement controller for non-linear dynamic plants, Proceedings of the Institution of Mechanical Engineers. Part I, Journal of Systems and Control Engineering 216 (2002) 467–476.

[5] Q.M. Zhu, D. Zhao, J. Zhang, A general U-block model-based design procedure for nonlinear polynomial control systems, International Journal of Systems Science 47 (2016) 3465–3475, https://doi.org/10.1080/00207721.2015.1086930.

[6] Q.M. Zhu, L. Liu, W. Zhang, S. Li, Control of complex nonlinear dynamic rational systems, Complexity 2018 (2018).

[7] X.P. Geng, Q.M. Zhu, T. Liu, J. Na, U-model based predictive control for nonlinear processes with input delay, Journal of Process Control 75 (March, 2019) 156–170, https://doi.org/10.1016/j.jprocont.2018.12.002.

[8] Q. Zhu, W. Zhang, J. Zhang, B. Sun, U-neural network enhanced control of nonlinear dynamic systems, Neurocomputing 352 (2019) 12–21.

[9] J. Li, Y. Li, Dynamic analysis and PID control for a quadrotor, in: 2011 IEEE International Conference on Mechatronics and Automation, 2011, pp. 573–578, https://doi.org/10.1109/ICMA.2011.5985724.

[10] R.V. Jategaonkar, Flight vehicle system identification: a time domain methodology, American Institute of Aeronautics and Astronautics, 2006.

[11] S. Bouabdallah, A. Noth, R. Siegwart, PID vs LQ control techniques applied to an indoor micro quadrotor, in: International Conference on Intelligent Robots and Systems 2004, vol. 3, 2004, pp. 2451–2456.

[12] A. Koszewnik, The parrot UAV controlled by PID controllers, Acta Mechanica et Automatica 8 (2) (2014) 65–69, https://doi.org/10.2478/ama-2014-0011.

[13] M.R. Kaplan, A. Beke, A. Eraslan, T. Kumbasar, Altitude and position control of Parrot Mambo Minidrone with PID and fuzzy PID controllers, https://doi.org/10.23919/ELECO47770.2019.8990445, 2019.

[14] L.M. Argentim, W.C. Rezende, P.E. Santos, R.A. Aguiar, PID, LQR and PID-LQR on a Quadcopter Platform, Retrieved April 18, 2020, https://doi.org/10.1109/ICIEV.2013.6572698, 2013.

[15] N.A.A. Hussain, S.S.A. Ali, M. Ovinis, M. Arshad, U.M. Al-Saggaf, Underactuated coupled nonlinear adaptive control synthesis using U-model for multivariable unmanned marine robotics, IEEE Access 8 (2020) 1851–1965.

[16] Ruobing Li, Quanmin Zhu, Janice Kiely, Weicun Zhang, Algorithms for U-model-based dynamic inversion (UM-dynamic inversion) for continuous time control systems, Complexity 2020 (2020) 3640210, 14 pages, https://doi.org/10.1155/2020/3640210.

[17] Q. Zhu, W. Zhang, J. Na, B. Sun, U-model based control design framework for continuous-time systems, in: 2019 Chinese Control Conference (CCC), 2019, pp. 106–111, https://doi.org/10.23919/ChiCC.2019.8866624.

[18] W. Zhang, Q. Zhu, S. Mobayen, H. Yan, J. Qiu, P. Narayan, U-model and U-control methodology for nonlinear dynamic systems, Complexity 2020 (2020) 1–13, https://doi.org/10.1155/2020/1050254.

[19] K.J. Astrom, B. Wittenmark, Adaptive Control, second edition, Dover Publications, 2008, pp. 92–98.

Chapter 14

Nonlinear control allocation applied on a QTR: the influence of the frequency variation

Murillo Ferreira dos Santos[a], Leonardo de Mello Honório[b], Mathaus Ferreira da Silva[b], Vinícius Ferreira Vidal[b], and Paolo Mercorelli[c]

[a]*Department of Electroelectronics, CEFET-MG, Leopoldina, Brazil,* [b]*Faculty of Engineering, UFJF, Juiz de Fora, Brazil,* [c]*Institute for Production Technology and Systems, Leuphana University of Lueneburg, Lueneburg, Germany*

14.1 Introduction

The applications of Unmanned Aerial Vehicles (UAVs) to perform tasks previously developed by humans increase every day, performing from medical to military tasks [2–5]. Among all possible applications, their topologies became a very important part of the whole project and execution, which are classified as fixed wings (planes for example) and rotary wings (helicopters for example). A quick search on the internet shows some reliable and stable UAVs, such as multicopters [6,7] and fixed wings [8,9].

Aiming to include the fixed-wing and rotary-wing classifications in the same vehicle, hybrid topologies have emerged. Classic examples are the tilt-rotors, which have tilting mechanisms for their propellers, making their maneuvers a partial combination of the two mentioned topologies. These tilt-rotors may or may not have fixed wings, just many propellers may be required. These combinations let them be considered as over-actuated vehicles defined as UAVs with more actuators than the respective Degrees of Freedom (DoF). Also, the over-actuation directly interferes with the system control allocation technique choice, which by definition is responsible for generating signals to the actuators from the control actions resulting from the controllers.

It is important to highlight that the vast majority of robotic systems do not require a complex method of control allocation, both for under- and over-actuated types. In addition, there are still cases where an over-actuated UAV can be simplified as under-actuated [10,11]. However, depending on the aircraft's physical characteristics and design requirements, there is a strict necessity to use a nonlinear and complex method [12].

Modeling, Identification, and Control for Cyber-Physical Systems Towards Industry 4.0
https://doi.org/10.1016/B978-0-32-395207-1.00026-3
311

Recent researches reflect the existence of a wide variety of control allocation methodologies, already consolidated for linear models. Regarding nonlinear systems, there are numerous methodologies that depend specifically on the applied system [13]. Research from the literature remarks some interesting methods: Direct Allocation; Pseudo-Inverse; Linear Programming; and Nonlinear Programming [10].

The Direct Allocation (DA) control technique considers the allocation space unconstrained with a specific pseudo-inverse, in a single iteration, satisfying only the control constraints. Consequently, it aims for an allocated parameter set that preserves the Virtual Control Actions (VCAs) direction [14]. The DA problem is not trivial for cases where the VCA size set is large [10].

Regarding pseudo-inverse methods (such as by redistribution), the first step is to solve the problem without saturation constraints [15]. In the end, if the result satisfies the previously unrestricted actual control actions, then no further steps are required. Otherwise, the optimal free vector with the Real Control Actions (RCAs) is designed in an appropriate configuration to reach the requirements [16].

In contrast to this methodology, Linear Programming (LP) minimizes the weighted error between the desired and estimated VCAs. Thus an optimization problem with geometric/polyhedral constraints is represented. Using defined cost functions, the resulting problem is linearly programmable and can be solved using iterative numerical algorithms such as the simplex method [17].

Taking the nonlinear programming into consideration, there will always be a single optimal solution if all the weights in the cost function are necessarily positive using slack variables [18,19]. Referring to the numerical methods of solving, three algorithms deserve to be highlighted: active-set, interior point, and fixed point methods.

However, an important point needs to be highlighted: as the solution complexity increases, the computational effort becomes a critical point when realtime execution is needed due to the possibility of finding different local minima and numerical sensitivity in the evaluation and validation procedures. Also, the control allocation frequency evaluation becomes a crucial step in the whole UAV design, which means that if the RCAs are not obtained correctly, then the system can become unstable and uncontrollable [12]. Aiming this aspect, it is possible to observe a large gap in the studies reporting these analyses, since incompatible control allocation technique frequencies can lead to instability on the vehicle, therefore generating accidents.

It is in this context that this work is proposed, in which we present a study on the influence of the control allocation execution frequency variation for some simulation flight scenarios analyzed on the vehicle stability. Regarding the considered control allocation, the nonlinear technique applied to the Quadrotor Tilt-Rotor (QTR) developed in [1,12] is used, illustrating a Software In The Loop (SITL) representation.

The main contribution of this work is focused on establishing the minimum and safe frequency of the control allocation task developed in [1,12], where it was assumed to be run in real-time and embedded execution at 400 Hz (the same frequency as the QTR attitude control loop). Then we will study the minimum frequency able to still keep the QTR safely flying with the same control requirements previously set.

This chapter is divided as follows. Section 14.2 presents the UAV kinematics and dynamics modeling, strictly necessary for understanding the control allocation design and the simulation results. Section 14.3 depicts the considered controller topology, such as a brief overview of the Fast Control Allocation Technique proposed by [1,12]. Section 14.4 illustrates the SITL scheme to perform the results. Section 14.5 shows the simulation results, divided into kinematics and dynamics controlled results, and the calculated time threshold values to run the control allocation technique without destabilizing the aircraft. Finally, Section 14.6 presents the conclusions of this work.

14.2 Nonlinear aircraft modeling and control allocation

We present and depict the aircraft kinematics and dynamics modeling, also considering the servomotor tilting angles. For better illustration, Fig. 14.1 shows the UAV and its axis rotations.

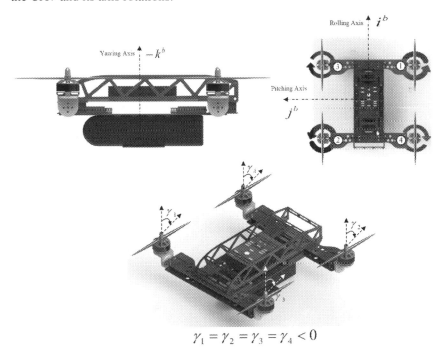

$$\gamma_1 = \gamma_2 = \gamma_3 = \gamma_4 < 0$$

FIGURE 14.1 UAV illustration with respective rotation axes ($\gamma = 0$ degrees is the upward servomotor direction).

The numbers marked in the figure represent the propulsion motor and servomotor number, where 1 is in the front-right part of the QTR $\mathcal{F}^b$, number 2 is in the rear-left, number 3 is in the front-left, and number 4 is in the rear-right. Furthermore, the propulsion motors 1 and 2 have counterclockwise rotation, whereas 3 and 4 rotate clockwise. Also, the QTR RCAs $\boldsymbol{u} = [\delta_1, \delta_2, \delta_3, \delta_4, \gamma_1, \gamma_2, \gamma_3, \gamma_4]$ are the Propulsion Motors Speed (PMS) and tilting angles provided by all the motors and servomotors.

By consequence, considering that each set of motor propulsion force $k_i \delta_i$ and servomotor tilt angle γ_i has its own impact over the resultant virtual control variables [23]:

$$X^b_{p_i} = k_1 \delta_i \sin(\gamma_i), \tag{14.1}$$

$$Z^b_{p_i} = k_1 \delta_i \cos(\gamma_i), \tag{14.2}$$

$$L^b_{p_i} = (\pm k_1 d \cos(\gamma_i) \pm k_2 \sin(\gamma_i)) \delta_i, \tag{14.3}$$

$$M^b_{p_i} = k_1 d \cos(\gamma_i) \delta_i, \tag{14.4}$$

$$N^b_{p_i} = (\pm k_1 d \sin(\gamma_i) \pm k_2 \cos(\gamma_i)) \delta_i, \tag{14.5}$$

where k_1 is the constant of propulsion, characteristic to each propulsion system (motor, electronic speed controller, and propeller), k_2 is the constant of Newton's third law due to the propeller rotations, $d = 0.15\,\text{m}$ is the aircraft arm length, X^p_b and Z^p_b are the forces produced by the propellers in $\mathcal{F}^b$ along the axes $\hat{\boldsymbol{i}}^b$ and $\hat{\boldsymbol{k}}^b$, respectively, and L^p_b, M^p_b, and N^p_b are the torques produced by the propellers in $\mathcal{F}^b$ along the axes $\hat{\boldsymbol{i}}^b$, $\hat{\boldsymbol{j}}^b$, and $\hat{\boldsymbol{k}}^b$.

Then the vehicle position is defined by the vector $\boldsymbol{\eta_1} \in \mathbb{R}^3$ in $\mathcal{F}^I$ (inertial frame), whereas its angles are defined by $\boldsymbol{\eta_2} \in \mathbb{R}^3$ in $\mathcal{F}^v$ (vehicle frame). Moreover, $\boldsymbol{v_1} \in \mathbb{R}^3$ and $\boldsymbol{v_2} \in \mathbb{R}^3$ are the linear and angular velocities, measured in $\mathcal{F}^b$ (body-fixed frame). According to [28], (14.7), and (14.9) present the nomenclature:

$$\boldsymbol{\eta_1} = [p_n \ p_e \ h]^T, \tag{14.6}$$

$$\boldsymbol{\eta_2} = [\phi \ \theta \ \psi]^T, \tag{14.7}$$

$$\boldsymbol{v_1} = [u \ v \ w]^T = [\boldsymbol{v}^b], \tag{14.8}$$

$$\boldsymbol{v_2} = [p \ q \ r]^T = [\boldsymbol{\omega}^b], \tag{14.9}$$

where $\boldsymbol{\eta} = [\boldsymbol{\eta_1} \ \boldsymbol{\eta_2}]^T$, $\boldsymbol{v} = [\boldsymbol{v_1} \ \boldsymbol{v_2}]^T$, $\boldsymbol{v_1} \in \mathbb{R}^3$ is the linear velocity vector, and $\boldsymbol{v_2} \in \mathbb{R}^3$ is the angular velocity vector, both in $\mathcal{F}^b$.

The six DoFs rigid body kinematics is expressed by

$$\dot{\boldsymbol{\eta}} = \boldsymbol{J} \boldsymbol{v}, \tag{14.10}$$

where $\dot{\boldsymbol{\eta}} \in \mathbb{R}^6$ is the velocity vector in $\mathcal{F}^I$, $\boldsymbol{v} \in \mathbb{R}^6$ is the general velocity vector in $\mathcal{F}^b$, and $\boldsymbol{J} \in \mathbb{R}^{6 \times 6}$ is the Jacobian matrix, where the position vector $\eta = \int \dot{\eta}$

is in $\mathcal{F}^I$:

$$
J = \begin{bmatrix} J_1 & 0_{3\times3} \\ 0_{3\times3} & J_2 \end{bmatrix}, \tag{14.11}
$$

$$
J_1 = \begin{bmatrix} c\theta c\psi & s\phi s\theta c\psi - c\theta s\psi & c\phi s\theta c\psi + s\phi s\psi \\ c\theta s\psi & s\phi s\theta s\psi + c\phi c\psi & c\phi s\theta s\psi - s\phi c\psi \\ -s\theta & s\phi c\theta & c\phi c\theta \end{bmatrix}, \tag{14.12}
$$

$$
J_2 = \begin{bmatrix} 1 & s\phi t\theta & c\phi t\theta \\ 0 & c\phi & -s\phi \\ 0 & s\phi/c\theta & c\phi/c\theta \end{bmatrix}. \tag{14.13}
$$

The QTR dynamics can be described by differential equations from the Newton–Euler method:

$$
M^b \dot{v} + C^b(v)v = \tau_p^b + \tau_a^b + \tau_g^b, \tag{14.14}
$$

where $M^b \in \mathbb{R}^{6\times6}$ is the system inertia matrix, $C^b(v) \in \mathbb{R}^{6\times6}$ is the Coriolis-centripetal matrix, both in $\mathcal{F}^b$, and τ_a^b, τ_g^b, $\tau_p^b \in \mathbb{R}^6$ are the aerodynamics, gravitational, and propulsion resultant vectors, respectively, composed of forces and torques, both in $\mathcal{F}^b$.

Eq. (14.14) leads to highlighting the definition of the Control Effectiveness Matrix (CEM) represented by τ_p^b, which relates the effects of all the QTR actuator (RCAs) characteristics directly on the five controlled DoFs from the VCAs. Then the actuator dynamics and their placements on the QTR are expressed in (14.16).

$$
\tau_p^b = M(u), \tag{14.15}
$$

$$
\begin{bmatrix} X_p^b \\ Z_p^b \\ L_p^b \\ M_p^b \\ N_p^b \end{bmatrix} = \begin{bmatrix} k_1\delta_1 s(\gamma_1)+k_1\delta_2 s(\gamma_2)+k_1\delta_3 s(\gamma_3)+k_1\delta_4 s(\gamma_4) \\ -k_1\delta_1 c(\gamma_1) - k_1\delta_2 c(\gamma_2) - k_1\delta_3 c(\gamma_3) - k_1\delta_4 c(\gamma_4) \\ (-k_1 dc(\gamma_1) - k_2 s(\gamma_1))\delta_1 + (k_1 dc(\gamma_2) - k_2 s(\gamma_2))\delta_2 + \\ (k_1 dc(\gamma_3)+k_2 s(\gamma_3))\delta_3 + (-k_1 dc(\gamma_4)+k_2 s(\gamma_4))\delta_4 \\ +k_1 dc(\gamma_1)\delta_1 - k_1 dc(\gamma_2)\delta_2 + k_1 dc(\gamma_3)\delta_3 - k_1 dc(\gamma_4)\delta_4 \\ (-k_1 ds(\gamma_1)+k_2 c(\gamma_1))\,\delta_1 + (k_1 ds(\gamma_2)+k_2 c(\gamma_2))\delta_2 + (k_1 ds(\gamma_3)- \\ k_2 c(\gamma_3))\delta_3 + (-k_1 ds(\gamma_4) - k_2 c(\gamma_4))\delta_4 \end{bmatrix}, \tag{14.16}
$$

where $\tau_p^b \in \mathbb{R}^5$ is the estimated VCA vector, $u \in \mathbb{R}^8$ is the RCAs vector inserted as parameter of the CEM $M \in \mathbb{R}^5$, γ_i is the tilting angle of each servomotor

(0 degrees has upward direction, $-\hat{k}^b$), δ_i is the rotation of each propulsion motor, and i represents the respective servomotor and a propulsion motor number. The propulsion force Y_p^b related to axis $\hat{j}^b$ is not represented because there is no direct force acting on this axis. After the test bench experimental results, $k_1 = 7.0632$ and $k_2 = 0.1413$. For ease of representation, $c\gamma_i = \cos(\gamma_i)$ and $s\gamma_i = \sin(\gamma_i)$.

14.3 P-PID controllers

It is known that UAV control systems are multi-loops, that is, Multiple-Input-Multiple-Output (MIMO). In the vehicle in question, five DoFs are controlled, which are altitude, forward/backward velocity, rolling, pitching, and yawing movements. Four of them are composed of two cascade feedback loops except for the forward/backward velocity, which uses only one feedback loop. Also, all of them use the Successive Loop Closure (SLC) technique to deal with the iteration between the dynamics of altitude and with rolling, pitching, and yawing.

To illustrate all the loops in a simplified form, Fig. 14.2 shows the QTR overall control structure, where the control allocation task is marked with the circle number 3.

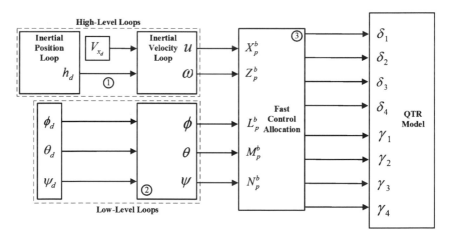

FIGURE 14.2 QTR control structure.

Considering each DoF with cascade loops, the external ones control the position variables through Proportional (P) controllers. The internal ones control their respective velocities, either inertial or angular, using Proportional-Integral-Derivative (PID) controllers. Thus the control action output P will serve as an input for the PID cascade internal controller [20].

A saturation block was also inserted in the integral loop, acting as an anti-windup action of the integrator. For the forward/backward velocity control loop (V_x), only one level was implemented through a PID controller.

Then the MIMO closed control loop equation is

$$\left[\boldsymbol{\eta}_d - \left(\underbrace{\boldsymbol{M}^b \dot{\boldsymbol{v}} + \boldsymbol{C}^b(\boldsymbol{v})\boldsymbol{v} - \boldsymbol{\tau}_a^b - \boldsymbol{\tau}_g^b}_{\text{from the QTR Model}} \right) \right] \boldsymbol{K}^b = \boldsymbol{\tau}_p^b, \qquad (14.17)$$

where $\boldsymbol{\eta}_d \in \mathbb{R}^5$ is the setpoint desired vector, and $\boldsymbol{K}^b \in \mathbb{R}^5$ are the actions from controller technique in $\mathcal{F}^b$.

More details about the QTR controllers and their tuning can be found in [1].

14.3.1 The fast control allocation technique

As presented in (14.16), it is possible to see that the solution of the control allocation procedure has five VCAs (inputs) and eight RCAs (outputs) with defined and non-unique solutions. Also, it is not possible to solve by matrix manipulation, instead more complex and time-consuming algorithms must be used (i.e., such as the primal-dual optimization algorithm [21]). These approaches make the processing costs prohibitive for some onboard hardware.

From this point, the Fast Control Allocation (FCA) technique proposed in [1,12] aims to turn the nonlinear control approach problem presented in (14.16) into a faster linear version by breaking the problem into two interconnected subsets of VCAs and RCAs. The reasons to choose these rules are as follows:

- The nonlinearities were broken into independent problems and solved recursively and iteratively;
- The solution speed and solvability were enhanced with a subsystem coupling all others together;
- Although this approach is not robust as interior point algorithms, it is fast and well established in the literature [22].

Taking this into account, the QTR CEM was broken into two different problems such as

$$\hat{\boldsymbol{\tau}}_a = \boldsymbol{M}_a(\boldsymbol{u}_b)\boldsymbol{u}_a', \qquad (14.18)$$

$$\hat{\boldsymbol{\tau}}_b = \boldsymbol{M}_b(\boldsymbol{u}_a)\boldsymbol{u}_b, \qquad (14.19)$$

where $\boldsymbol{u}_a \in \mathbb{R}^4$, $\boldsymbol{u}_a' \in \mathbb{R}^5$, $\boldsymbol{u}_b \in \mathbb{R}^4$, $\hat{\boldsymbol{\tau}}_a \in \mathbb{R}^2$, $\hat{\boldsymbol{\tau}}_b \in \mathbb{R}^5$, $\boldsymbol{M}_a(\boldsymbol{u}_a) \in \mathbb{R}^{2\times5}$, and $\boldsymbol{M}_b(\boldsymbol{u}_b) \in \mathbb{R}^{5\times4}$. Also, $\boldsymbol{u}_a \cup \boldsymbol{u}_b = \boldsymbol{u}$ and $\boldsymbol{M}_a \subset \boldsymbol{M}_b \subset \boldsymbol{M}$ [12].

In consequence, these subsystems are shown in (14.20)–(14.23):

$$
\overbrace{\begin{bmatrix} X^b \\ N^b \end{bmatrix}}^{\hat{\tau}_a} = M_a(u_b) \overbrace{\begin{bmatrix} \sin(\gamma_1) \\ \sin(\gamma_2) \\ \sin(\gamma_3) \\ \sin(\gamma_4) \\ 1 \end{bmatrix}}^{u'_a}, \tag{14.20}
$$

$$
\overbrace{\begin{bmatrix} k_1\delta_1 & k_1\delta_2 & k_1\delta_3 & k_1\delta_4 & 0 \\ -k_1 d\delta_1 & -k_1 d\delta_2 & k_1 d\delta_3 & k_1 d\delta_4 & k_2(\delta_1 c\gamma_1 + \delta_2 c\gamma_2) - k_2(\delta_3 c\gamma_3 + \delta_4 c\gamma_4) \end{bmatrix}}^{M_a(u_b)}, \tag{14.21}
$$

$$
\overbrace{\begin{bmatrix} X^b \\ Z^b \\ L^b \\ M^b \\ N^b \end{bmatrix}}^{\hat{\tau}_b} = M_b(u_a) \overbrace{\begin{bmatrix} \delta_1 \\ \delta_2 \\ \delta_3 \\ \delta_4 \end{bmatrix}}^{u_b}, \tag{14.22}
$$

$$
\overbrace{\begin{bmatrix} k_1 s\gamma_1, & k_1 s\gamma_2, & k_1 s\gamma_3, & k_1 s\gamma_4 \\ k_1 c\gamma_1, & k_1 c\gamma_2, & k_1 c\gamma_3, & k_1 c\gamma_4 \\ (-k_1 dc\gamma_1 - k_2 s\gamma_1), & (k_1 dc\gamma_2 - k_2 s\gamma_2), & (k_1 dc\gamma_3 + k_2 s\gamma_3), & (-k_1 dc\gamma_4 + k_2 s\gamma_4) \\ k_1 dc\gamma_1, & -k_1 dc\gamma_2, & k_1 dc\gamma_3, & -k_1 dc\gamma_4 \\ (-k_1 ds\gamma_1 + k_2 c\gamma_1), & (k_1 ds\gamma_2 + k_2 c\gamma_2), & (k_1 ds\gamma_3 - k_2 c\gamma_3), & (-k_1 ds\gamma_4 - k_2 c\gamma_4) \end{bmatrix}}^{M_b(u_a)}, \tag{14.23}
$$

where $u'_a = [\sin^{-1}(\gamma_1), \sin^{-}(\gamma_2), \sin^{-}(\gamma_3), \sin^{-}(\gamma_4), 1]^T$. To obtain u_a, it is necessary to take the first four elements of u'_a.

It is important to remark that the chosen subsystem combination allows the QTR to perform maneuvers using the motor's differential rotation speeds and/or tilting its servomotors independently. This means that the VCAs $[X^b, N^b]^T$ are directly actuated by the RCAs u_a. Considering that $[X^b, N^b]^T$ are reached, the remaining VCAs $[Z^b, L^b, M^b]^T$ act on the RCAs u_b.

For feasible constrained solutions, the PMSs and servomotors tilting angle ranges had the normalized values of $[0, 1] = [0, 100\%]$ and $[-1, +1] = [-30, +30]$, respectively.

More details about the control allocation technique are given in [1,12].

14.4 SITL scheme

The SITL simulator allows running the QTR without any hardware device. It is built of the autopilot code using an ordinary C++ compiler, giving to the user an executable to test the aircraft in a virtual environment.

For this matter, the experiments were carried out to emulate the computational effort of the FCA, where all the aircraft dynamics and kinematics responses are from SITL simulator. Therefore the threshold times in the FCA execution directly interfere in the system dynamics due to the respective control loop parallel operation.

To illustrate the procedure, Fig. 14.3 shows the SITL block diagram.

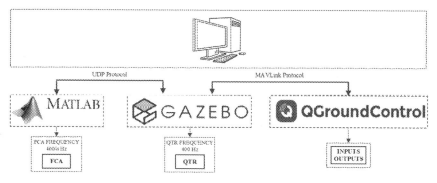

FIGURE 14.3 SITL block diagram illustration.

Fig. 14.3 depicts the software used to run the SITL. The vehicle modeled and presented in Section 14.2 was inserted inside the Gazebo platform, with all the equipment characteristics. It establishes a connection with Matlab® via UDP protocol, where the FCA technique runs at $400/n$ Hz of frequency, making it possible to analyze the minimum frequency allowed to run the control allocation technique without destabilizing the QTR. The communication with QGroundControl is established through MAVLink protocol, where the aircraft inputs and outputs can be manipulated, creating a virtual environment.

To complement the SITL illustration, Fig. 14.4 shows the MIMO control loop of the SITL simulation. Fig. 14.4 shows that the FCA technique can have a different frequency compared to the QTR control board frequency, represented by $400/n$ Hz, where n is the limit number to be obtained for safe and reliable flight conditions.

Note that the three critical DoFs for the QTR stability are roll, pitch, and altitude [23]. Thus, through the Integral of the Squared Error (ISE) index, a threshold was established to define the vehicle instability/fall, which consequently generates a maximum time/minimum frequency of the FCA operation.

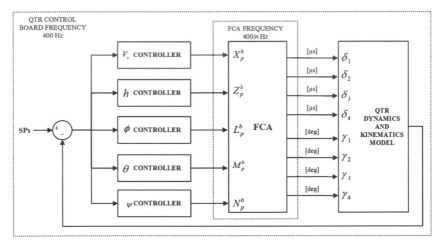

FIGURE 14.4   Illustration of the parts that have different sampling frequency.

These limits are shown in Table 14.1, obtained through experimental curves flying under similar conditions. Then, if bigger controlled responses are obtained, then the delay time at that simulation extrapolated the QTR controllability conditions.

TABLE 14.1 Values of the ISE indexes obtained through experimental tests.	
DoF	ISE
Roll (ϕ)	9×10^3
Pitch (θ)	1×10^1
Backward/forward velocity (V_x)	1×10^1

This index was chosen because a system based on it would have reasonable damping and satisfactory transient response, punishing errors in the quadratic weights during the experiment, regardless of the time they occur [24,25].

14.5   Simulation results

For a better understanding and presentation of the analysis of the frequency control allocation variation effect on the QTR stability, this section is divided into two main parts: the first one depicts the DoF controlled responses as the delay time of the control allocation variation is changed; the second one illustrates some three-dimensional results, aiming to obtain the minimum frequency for the control allocation operation.

14.5.1   Remarks about the control technique

Besides, it is not the purpose of this work either to analyze or to demonstrate the controller topology. For this work, the Successive Loop Closure was used as presented in [23]. However, several other control methodologies could provide similar results [26,27].

14.5.2   Kinematics and dynamics control results

In this section, we present some simulation results that aim to analyze the QTR dynamics responses considering three different delay times that the FCA technique spent to process the VCAs, demanded by the five controllers.

To illustrate the controlled responses, Figs. 14.5 and 14.6 present the vehicle behavior for three different delay times of the FCA operation. Randomly, the SetPoints (SPs) were $h^d = 10$ m at 0 s, $V_x^d = 3$ m/s at 10 s, $\phi^d = 0°$, $\theta^d = 0°$, and $\psi^d = 0°$ at 0 s. Then the red, black, and blue curves (mid gray, dark gray and black curves in print version) represent $n = 1, 2$, and 4, respectively.

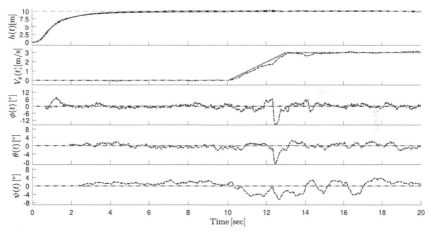

FIGURE 14.5   Controlled responses for altitude, forward/backward velocity, rolling, pitching, and yawing dynamics.

Fig. 14.5 shows that as the time demanded to process the control allocation increases, the vehicle starts to show oscillatory characteristics around the SPs because the RCAs do not match to the VCAs required at that instant loop iteration. For this situation, the minimum frequency of a control allocation acceptable operation would be 100 Hz.

Fig. 14.6 illustrates the RCAs of propulsion systems 1 and 3 on the QTR front part, as shown in Fig. 14.1. The RCAs of systems 2 and 4 showed similar responses. Thus note that as n increases, the RCAs start to show impractical

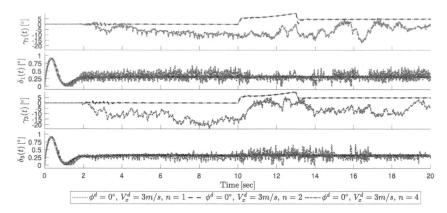

FIGURE 14.6   QTR RCAs – Propulsion motors and servomotors numbers 1 and 3.

amplitudes and oscillations for open field tests, which leads to the conclusion that for these presented SPs, if the allocation control task had a frequency five times lower than the control board frequency, then the vehicle would be unstable or crash to the ground.

14.5.3   Three-dimensional results

This section shows simulation results in three-dimensional graphics for different SPs, close to the SP range considered in the controller tuning, aiming to obtain the minimum frequency for each case. This range was based on the parameters presented by the author in [1,12]. In this way the pitching SP was kept at 0 degrees flying at 10-m altitude. The rolling SP ranged from -10 to 10 degrees and the forward/backward velocity was from -3 to 3 m/s, both requested in 10 seconds of flight after take off.

Figs. 14.7 and 14.8 show the QTR behavior in a three-dimensional perspective and in a top view, respectively.

Note that as the vehicle presents a greater range of roll and forward/reverse velocity, the control allocation technique starts demanding more time to transform the VCAs into RCAs. Although a leveled flight allows a control allocation operation with a minimum frequency of around 67 Hz (six times lower than that of the used system, 400 Hz), the extreme points of the considered ranges do not allow values lower than 200 Hz (two times lower). We can conclude that the control allocation technique should not be performed with frequencies of less than 200 Hz for flights within the control requirements considered in [1,12].

Discontinuities between some operating points in their respective neighborhoods can be explained by the vehicle model nonlinearities. In addition, the insertion of Gaussian noises into the control loops means that the proximity of each operation point cannot have a unique solution.

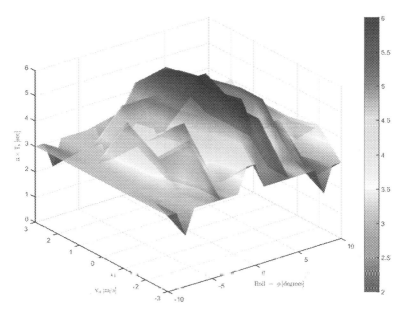

FIGURE 14.7   Three-dimensional graph where $400/n$ Hz is the FCA operation frequency.

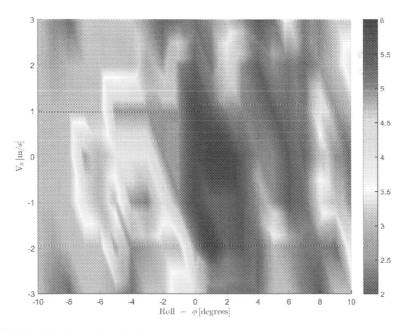

FIGURE 14.8   Three-dimensional top view graph.

14.6 Conclusions

This work can be summarized into two stages: design and control of an over-actuated UAV and development of the SITL simulation considering different delay times for FCA technique operation; and performance of simulation tests.

For the first step, it is possible to conclude that the QTR vehicle was modeled and controlled satisfactorily considering that the FCA technique has the same frequency as the control board, 400 Hz. Therefore it was a crucial step for the FCA development and implementation.

The principal and more important step in this work is the SITL simulation, which allowed it to obtain the minimum frequency acceptable for the control allocation execution, 200 Hz. Lower frequencies than this one would cause instability or crashes in the QTR operation.

As future works, test bench and open field flights are now able to be performed, aiming to validate the simulation tests.

Acknowledgment

The authors thank CEFET-MG, UFJF, Leuphana University of Lüneburg, and DAAD for the financial support.

References

[1] M.F. Santos, Alocação de Controle Desacoplada Rápida em Sistemas de Controle Superatuados, UFJF, 2019.

[2] F.C. Ferreira, M.F. Santos, V.B. Schettino, Computational vision applied to mobile robotics with position control and trajectory planning: study and application, in: 19th International Carpathian Control Conference (ICCC), IEEE, 2018.

[3] M.F. Silva, A.S. Cerqueira, V.F. Vidal, L.M. Honório, M.F. Santos, E.J. Oliveira, Landing area recognition by image applied to an autonomous control landing of VTOL aircraft, in: 18th International Carpathian Control Conference (ICCC), IEEE, 2017, pp. 240–245.

[4] G.M. Prisco, D.J. Rosa, Methods for handling an operator command exceeding a medical device state limitation in a medical robotic system, 2017, US Patent 9, 566, 124.

[5] M.F. Santos, V.S. Pereira, A.C. Ribeiro, M.F. Silva, M.J. Carmo, V.F. Vidal, L.M. Honório, A.S. Cerqueira, E.J. Oliveira, Simulation and comparison between a linear and nonlinear technique applied to altitude control in quadcopters, in: 18th International Carpathian Control Conference (ICCC), IEEE, 2017, pp. 234–239.

[6] Y. Liu, Q. Wang, H. Hu, Y. He, A novel real-time moving target tracking and path planning system for a quadrotor UAV in unknown unstructured outdoor scenes, Transactions on Systems, Man, and Cybernetics: Systems 99 (2018) 1–11, IEEE.

[7] S. Minaeian, J. Liu, Y.J. Son, Vision-based target detection and localization via a team of cooperative UAV and UGVs, Transactions on Systems, Man, and Cybernetics: Systems 46 (7) (2016) 1005–1016, IEEE.

[8] K. Klausen, T.I. Fossen, T.A. Johansen, Autonomous recovery of a fixed-wing UAV using a net suspended by two multirotor UAVs, Journal of Field Robotics 35 (5) (2018) 717–731, Wiley Online Library.

[9] M. Warren, I. Mejias, J. Kok, X. Yang, F. Gonzalez, B. Upcroft, An automated emergency landing system for fixed-wing aircraft: planning and control, Journal of Field Robotics 32 (8) (2015) 1114–1140, Wiley Online Library.

[10] T.A. Johansen, T.I. Fossen, Control allocation – a survey, Automatica 49 (5) (2013) 1087–1103, Elsevier.

[11] M. Saied, H. Shraim, B. Lussier, I. Fantoni, C. Francis, Local controllability and attitude stabilization of multirotor UAVs: validation on a coaxial octorotor, Robotics and Autonomous Systems 91 (2017) 128–138, Elsevier.

[12] M.F. Santos, L.M. Honório, A.P.G.M. Moreira, M.F. Silva, V.F. Vidal, Fast real-time control allocation applied to over-actuated quadrotor tilt-rotor, Journal of Intelligent & Robotic Systems 102 (3) (2021) 1–20, Springer.

[13] W. Gai, J. Liu, J. Zhang, Y. Li, A new closed-loop control allocation method with application to direct force control, International Journal of Control, Automation and Systems 16 (3) (2018) 1355–1366, Springer.

[14] W.C. Durham, Constrained control allocation, Journal of Guidance, Control, and Dynamics 16 (4) (1993) 717–725.

[15] A. Ahani, M.J. Ketabdari, Alternative approach for dynamic-positioning thrust allocation using linear pseudo-inverse model, Applied Ocean Research 90 (2019) 101854, Elsevier.

[16] J. Shi, W. Zhang, G. Li, X. Liu, Research on allocation efficiency of the redistributed pseudo inverse algorithm, Science China. Information Sciences 53 (2) (2010) 271–277, Springer.

[17] M. Bodson, S.A. Frost, Load balancing in control allocation, Journal of Guidance, Control, and Dynamics 34 (2011) 380–387.

[18] D. Simon, O. Härkegård, J. Löfberg, Command governor approach to maneuver limiting in fighter aircraft, Journal of Guidance, Control, and Dynamics 40 (6) (2016) 1514–1527, American Institute of Aeronautics and Astronautics.

[19] E.J. Oliveira, L.W. Oliveira, J.L.R. Pereira, L.M. Honório, I.C.S. Junior, A.L.M. Marcato, An optimal power flow based on safety barrier interior point method, International Journal of Electrical Power & Energy Systems 64 (2015) 977–985, Elsevier.

[20] P. Burggräf, A.R.P. Martínez, H. Roth, J. Wagner, Quadrotors in factory applications: design and implementation of the quadrotors P-PID cascade control system, SN Applied Sciences 1 (7) (2019) 722, Springer.

[21] A.C.Z. Souza, L.M. Honório, G.L. Torres, G. Lambert-Torres, Increasing the loadability of power systems through optimal-local-control actions, IEEE Transactions on Power Systems 19 (1) (2004) 188–194, IEEE.

[22] J.E. Dennis Jr, R.B. Schnabel, Numerical Methods for Unconstrained Optimization and Nonlinear Equations, vol. 16, SIAM, 1966.

[23] R.W. Beard, T.W. McLain, Small Unmanned Aircraft: Theory and Practice, Princeton University Press, 2012.

[24] M.F. Santos, Controle Tolerante a Falhas de um Sistema de Propulsão de Hexacópteros, UFJF, 2014.

[25] A.R. Shahemabadi, S.B.M. Noor, F.S. Taip, Analytical formulation of the integral square error for linear stable feedback control system, in: International Conference on Control System, Computing and Engineering, IEEE, 2013, pp. 157–161.

[26] L.M. Argentim, W.C. Rezende, P.E. Santos, R.A. Aguiar, PID, LQR and LQR-PID on a quadcopter platform, in: International Conference on Informatics, Electronics and Vision (ICIEV), IEEE, 2013, pp. 1–6.

[27] W. Zhao, T. Go, Quadcopter formation flight control combining MPC and robust feedback linearization, Journal of the Franklin Institute 351 (3) (2014) 1335–1355, Elsevier.

[28] T.I. Fossen, Mathematical models for control of aircraft and satellites, Department of Engineering Cybernetics Norwegian University of Science and Technology, 2011.

Part IV

Theoretical and methodological advancements in disturbance rejection and robust control

Paolo Mercorelli[a], Hamidreza Nemati[b], and Quanmin Zhu[b]

[a]Institute for Production Technology and Systems, Leuphana University of Lueneburg, Lueneburg, Germany, [b]Department of Engineering Design and Mathematics, University of the West of England, Bristol, United Kingdom

---

There is a crack in everything, that's how the light gets in.

—Leonard Cohen

Introduction and issues of this part

It is helpful to investigate some fundamental principles from control theory connected with the understanding of robust controls. The idea is to acquire a perspective on with respect to the way theoretical implications are supposed to work. Such theories may be divided into two categories historically at the larger level. The conventional and contemporary methodologies have played an important role in the advancement. The term "conventional control" is always being used at the moment. It refers to principles and practices created before the early 1950s. Modern systems encompass strategies used from that time period. This discussion looks at each of them, and through the chapters of this section, we will present some advancements in the context of disturbance rejection with different uncertainty conditions and different control strategies. In Industry 4.0 process the disturbance rejection represents a key point of the control law. In fact, the systems to be modeled and to be controlled are extremely complex due to their different nature to be integrated in a common frame. Methodology plays very important role in the control theory study; here by methodology we mean how we treat uncertainties of the system to be controlled, how we treat models, and how we model errors. One of the most important classifications of the errors, which can occur in the control loop in terms of robustness, is "small and large errors". For example, if the system uncertainties are small and bounded, and the main characteristic of the system can be modeled, then H_∞ control strategy [1–3] is an appropriate choice to obtain some kinds of optimality of control performances. On the contrary, if the system uncertainties are large (such as structure known, parameters unknown, or slow time varying), then parameter identification-based adaptive control strategy [4–6] is a practical approach. Active disturbance rejection control is another kind of adaptive control strategy with different methodology thinking on models and model errors, which was originally proposed by Professor Jingqing Han [7,8]. In almost all this literature, PID controllers play a crucial role. Advancements of their design taking into account the problem of disturbances rejection is the focus of joined efforts between academic scientists and industrial ones. In fact, PID controllers represent the basic structures of the most important already existing industrial control strategies. The advancement of these in terms of disturbance rejection represent a suitable and challenging issue to improve and guarantee efficient and future reliable control systems.

Emergence of theoretical implications to use control loops

With the emergence of feedback theory, such controls became more intriguing with the passage of time. The system was stabilized by using feedback from all the stakeholders. The creation of the fly ball governor for stabilizing was one of the major advancements. The steam engines in locomotives were an early use of feedback control at that point of time [9]. Towards the onset of the 20th century, this is another example used at the time of feedback for telephone signaling. The issue was signal transmission across large distances, which were one of the major hindrances. It was due to distortion in that respect. The number of repeaters could be added [9]. This is something in series to lines that was limited. It was developed with a feedback device that would decrease distortion via feedback. Even when the additional feedback was reduced, there were some issues. The problem was with respect to the repeater gain. Overall performance was improved during that time [9]. The system equations are then expressed in the frequency domain. It was done using Laplace transformations. It may be adjusted algebraically at the given point of time. A typical control loop is shown in the given time period. The system receives a reference signal at that point of time. This signal reflects the intended value that is needed to be controlled. To infer the plant output, this reference is passed at the given point of time. The signal is forwarded with the help of transfer function. The output is transmitted back into the system as well. This is done through a feedback transfer function. The mechanism is also developed to ensure that the error signal can be calculated [10]. The idea is to retract the feedback [10]. This is necessary to obtain a signal from the reference signal at that point of time. The fourth industrial revolution is something that brought about that change. Often known as Industry 4.0, it is a major improvement in engineering. It is about creating new opportunities for digital production [10]. It is done by increasing flexibility. The operational performance also brings that change. This advancement will also have a significant influence across other areas.

Conclusion

The goal of this part is to show numerous achievements in various sectors. To conduct the study, we must thoroughly research. It must be about the existing literature that will be considered in the following chapters. There is extensive literature on Industry 4.0. The Industry technologies and their relevance in robust mechanisms are also going to be an important part of the whole thing. It employs a variety of digital manufacturing. The information technologies are also needed to be developed. The following chapters, thanks to their general methodological approach, present interesting and general solutions to different problems of disturbance rejection in different control loops. Particular attention is paid to PID controller. In fact, the proportional-integral/proportional-integral-derivative (PI/PID) controllers still dominate the industry and in the context of Industry 4.0 process can play a key point if they will be adapted to new tasks and challenges. Methodological aspects are considered to show advanced and inspiring solutions for different scenarios in the context of present and future issues in Industry 4.0 process.

References

[1] G. Zames, Feedback and complexity, Special plenary lecture addendum, in: Proceedings of the IEEE Conference on Decision and Control, IEEE, 1976.
[2] G. Zames, Optimal sensitivity and feedback: weighted seminorms, approximate inverses, and plant invariant schemes, in: Proceedings of the Allerton Conference, IEEE, 1979, p. 32.
[3] G. Zames, Feedback and optimal sensitivity: model reference transformations, multiplicative seminorms, and approximate inverses, IEEE Transactions on Automatic Control 26 (1981) 301–320.
[4] K.J. Åström, B. Wittenmark, On self-tuning regulators, Automatica 9 (1973) 185–199.
[5] D.W. Clarke, P.J. Gawthrop, Self-tuning controller, Proceedings of the IEEE 122 (9) (1975) 929–934.
[6] G.C. Goodwin, P.J. Ramadge, P.E. Caines, Discrete time stochastic adaptive control, SIAM Journal on Control and Optimization 19 (1981) 829–853.

[7] J.Q. Han, Active Disturbance Rejection Control Technique – the Technique for Estimating and Compensating the Uncertainties, National Defense Industry Press, Beijing, 2008.

[8] J.Q. Han, From PID to active disturbance rejection control, IEEE Transactions on Industrial Electronics 56 (3) (2009) 900–906.

[9] D. Ivanov, S. Sethi, A. Dolgui, B. Sokolov, A survey on control theory applications to operational systems, supply chain management, and Industry 4.0, Annual Reviews in Control 46 (2019) 134–147.

[10] Y. Huang, W. Xue, Active disturbance rejection control: methodology and theoretical analysis, ISA Transactions 53 (4) (2021) 963–976.

Chapter 15

Active disturbance rejection control of systems with large uncertainties

Weicun Zhang[a] and Hui Wang[b]

[a]University of Science and Technology Beijing, School of Automation and Electrical Engineering, Beijing, China, [b]CHINA Metallurgical Group Corporation, Beijing, China

15.1 Introduction

Methodology plays a very important role in the control theory study; by methodology we mean how we treat uncertainties of the system to be controlled and how we treat a model and model errors. For example, if the system uncertainties are small and bounded, and the main characteristic of the system can be modeled, then the H^∞ control strategy [28–30] is an appropriate choice to obtain some kinds of optimality of control performances; if the system uncertainties are large (e.g., the structure known, but the parameters are unknown or slow time varying), then the parameter identification-based adaptive control strategy [31–33] is a practical approach. Active disturbance rejection control (ADRC) is another kind of adaptive control strategy with different methodology thinking on the model and model errors, which was originally proposed by Professor Jingqing Han [1,2]. It is a natural development of PID with a motivation to improve or strengthen PID control performances. There are many further research results after the pioneer work; see, for example, methodology and theoretical analysis of ADRC [3], ADRC with adaptive extended state observer [4], stability analysis of ADRC [5], and disturbance decoupling control with ADRC [6]. According to Han's analysis, there are four shortcomings of PID controller:

1. Set-point is often given as a step function, not appropriate for most dynamics systems because it amounts to asking the output and therefore the control signal to make a sudden jump;
2. PID is often implemented without the D (Differential) part because of the noise sensitivity;
3. The weight sum of the three terms in PID control law, although simple, may not be the best control law based on the current and past of the errors and their rate of change;

https://doi.org/10.1016/B978-0-32-395207-1.00028-7

4. The integral term, although critical to get rid of steady-state error, introduces other problems such as saturation and reduced stability margin due to phase lag.

As a solution to address the above-mentioned shortcomings, ADRC is suitable for many practical control engineering with satisfactory performances, or, in another words, it has advantages over PID control strategy in many applications, such as power plant, fuel cell, diesel engine, machining, etc. [7–12].

Although there are a lot of achievements have been made on the development of ADRC, it still needs further study to cope with large uncertainties of plant parameters. To tackle this problem of ADRC, in this paper, we combine the ADRC with the weighted multiple model adaptive control scheme to design a weighted multiple model ADRC control system with simulation verification.

With the multiple model ADRC control scheme, the large (global) uncertainties rely on multiple model scheme to deal with, whereas the small (local) uncertainties rely on ADRC to deal with. The weighting algorithm plays the role of "soft" switching, and it performs well under disturbances and noises to identify the most appropriate local model and corresponding local controller.

15.2 From PID to ADRC, MPC, and adaptive control

15.2.1 PID as model reference adaptive control

It is well known that PID is the most popular control strategy all over the word up to now. There are probably two reasons for this situation:

1. PID is the most simple controller among all kinds of control strategies;
2. PID is adequate for most application situations, because most control plants can be described by first- or second-order transfer function models (sometimes with time delay) considering reasonable control requirements, and because PID is in fact a model reference adaptive control taking into account the human adjustment of parameters in control engineering fields.

Fig. 15.1 is the block diagram of a PID control system.

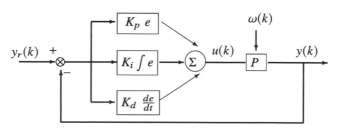

FIGURE 15.1 Block diagram of a PID control system.

As we mentioned in the Introduction, PID has four shortcomings to be addressed if we pursue high performances in relative complex application scenarios. However, in most situations, PID is adequate for reasonable control performance requirements, because PID control systems actually perform like model reference adaptive control, as shown in Fig. 15.2, taking into account the human adjustment of the parameters K_p, K_i, K_d, i.e., there would be a desired step response curve representing the desired closed-loop control system model in the operator's mind. Although the "adaptive" mechanism in PID control system is not as perfect as in traditional adaptive control systems, it indeed exists. In fact, when the PID controller parameters are set up or reset during on-line operation, the plant model is "identified" as in traditional adaptive control systems.

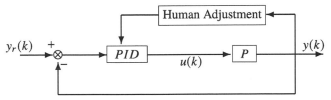

FIGURE 15.2   Block diagram of a PID control system taking into account the human adjustment of parameters.

15.2.2   Brief introduction of ADRC

Generally speaking, ADRC is a feedback liberalization method to cancel the uncertainty by estimating the "total disturbance" and taking the estimated total disturbance into consideration in the control law. Extended State Observer (ESO) is a key point in ADRC, which plays the role of integrator in PID control, but the "error" signal is replaced by "total disturbance" including both internal and external disturbances. Next, we take a second-order uncertain plant as an example to explain the ESO concept and ADRC strategy.

Consider a second-order plant in both ordinary differential equation and state space:

$$\ddot{y} = f(y, \dot{y}, \omega, t) + b_0 u \tag{15.1}$$

or

$$\begin{cases} \dot{x}_1 = x_2, \\ \dot{x}_2 = x_3 + b_0 u, \\ \dot{x}_3 = \dot{f}, \\ y = x_1. \end{cases} \tag{15.2}$$

We first design an ESO to estimate the total disturbance $z_3 = \hat{f}$:

$$\begin{cases} \varepsilon_1 = z_1 - y, \\ \dot{z}_1 = z_2 - \beta_{01}\varepsilon_1, \\ \dot{z}_2 = z_3 - \beta_{02}\varepsilon_1 + b_0 u, \\ \dot{z}_3 = -\beta_{03}\varepsilon_1. \end{cases} \tag{15.3}$$

If the ESO is stable, then we have

$$
\begin{cases}
z_1 \rightarrow x_1, \\
z_2 \rightarrow x_2, \\
z_3 \rightarrow x_3 \rightarrow f.
\end{cases}
\tag{15.4}
$$

Obviously, the ADRC control law includes two parts, one part to cancel the total disturbance f and another part (PD controller) to control the nominal model or ideal model, i.e., $\ddot{y} = b_0 u$. Thus the ADRC strategy should be

$$
u = u_0 - \frac{\hat{f}}{b_0} = u_0 - \frac{z_3}{b_0},
\tag{15.5}
$$

where u_0 is PD controller or PD controller with tracking differentiator (TD) to get desired transient profile (v_1, v_2) to replace y_r and $\dot{y}_r$, respectively. To be specific, we have the following two kinds expressions of u_0.

Without TD, we have that

$$
u_0 = K_p e + K_d \dot{e},
\tag{15.6}
$$

where $e = y_r - y$.

With TD, we have that

$$
u_0 = K_p(v_1 - z_1) + K_d(v_2 - z_2).
\tag{15.7}
$$

For more detail, see the block diagram of an ADRC system shown in Fig. 15.3.

The purpose of tracking differentiator (TD) is to generate v_1 and v_2. There are usually two types of TD, linear and nonlinear. The linear TD is very simple; it can be described by the transfer function $\frac{1}{Ts+1}$ for the first-order TD or by $\frac{1}{(Ts+1)^2}$ for the second-order TD. The nonlinear TD is a little complicated; for more detail, see [1,2].

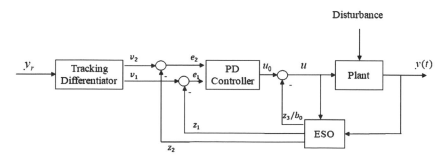

FIGURE 15.3 The block diagram of an ADRC system.

15.2.3 ADRC as reinforced PID control and adaptive control

In this section, we study the differences of control strategies PID, ADRC, MPC, and adaptive control according to their VESs. VES is an equivalent system in the input–output sense to the original control system [13]. It is rather convenient to analyze the stability (or convergence) of complicated systems, such as self-tuning adaptive control systems, multiple model adaptive control systems, ADRC systems, and MPC systems. It was first proposed by Weicun Zhang in the 1990s to get a unified and easy-to-understand stability and convergence theory of self-tuning adaptive control. Next, we give the VES of a general second-order ADRC system (Fig. 15.3), as shown in Fig. 15.4. For simplicity and without loss of generality, we leave out the TD in Fig. 15.4, because TD is out of the closed-loop, although it is very important to the transient performances of the ADRC system. In addition, we suppose that b_0 is known in Fig. 15.4.

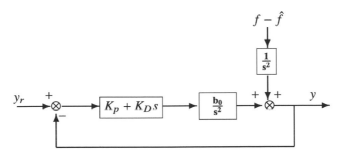

FIGURE 15.4 VES of second-order ADRC system without TD.

The closed-loop transfer function of the second-order ADRC system has the form

$$y(s) = \frac{b_0(K_p + K_D s)}{s^2 + b_0(K_p + K_D s)} y_r(s) + \frac{1}{s^2 + b_0(K_p + K_D s)} (f - \hat{f}). \quad (15.8)$$

From the closed-loop transfer function we see that there are in fact two channels in the ADRC system.

Channel I–Tracking channel:

$$y_1(s) = \frac{b_0(K_p + K_D s)}{s^2 + b_0(K_p + K_D s)} y_r(s), \quad (15.9)$$

which is responsible for tracking the reference input y_r.

Channel II–Disturbance rejection channel:

$$y_2(s) = \frac{1}{s^2 + b_0(K_p + K_D s)}(f - \hat{f}), \tag{15.10}$$

which is responsible for the disturbance rejection, working together with ESO to reject the total disturbance, ESO and control law $\frac{z_3}{b_0}$ to cancel f with $\hat{f}$, and the residual $f - \hat{f}$ is left to Channel II to deal with.

From the VES or from the closed-loop transfer function of the ADRC system we have the following conclusions on its stability and tracking performance (we refer to [13,24,25] for details).

Conclusion 1: If only the Channel I system is stable and tracking, i.e., then the PD controller and model $\frac{b_0}{s^2}$ formulate a stable and tracking closed-loop system; if the Channel II system input signal $f - \hat{f}$ is bounded, then the ADRC system is BIBO (bounded-input bounded-output) stable.

Conclusion 2: If only the Channel I system is stable and tracking, i.e., then the PD controller and model $\frac{b_0}{s^2}$ formulate a stable and tracking closed-loop system; if the Channel II system input signal $f - \hat{f}$ tends to 0, then the ADRC system is stable and tracking.

In fact, for any model-based control system, there is a VES that is rather simple in structure but equivalent in the input–output sense to the original complicated system. Next, we present the VES for deterministic self-tuning adaptive control systems and the VES for MPC system. In this way, we will see the connections among different control systems.

Leaving out the details (on-line parameter identification and online controller design), a deterministic self-tuning (or parameter identification-based) adaptive control system can be described as in Fig. 15.5. The corresponding VES is shown in Fig. 15.6.

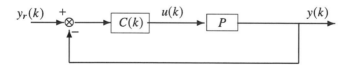

FIGURE 15.5 Simplified block diagram of a self-tuning adaptive control system.

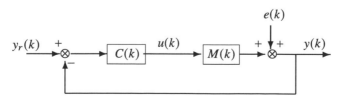

FIGURE 15.6 VES of adaptive control system.

In Fig. 15.6, $e(k)$ is the model output error or identification error. The methodology of adaptive control is to rely on parameter identification to get a precise enough model $M(k) \to M$ to approximate plant P; then the model output error $e(k)$ will tend to zero, i.e., $e(k) \to 0$. If there are unmodeled dynamics, then $e(k) \nrightarrow 0$, and the model output error $e(k)$ should be considered in the control law, like in ADRC or MPC, then the corresponding VES would be changed a little bit as shown in Fig. 15.7.

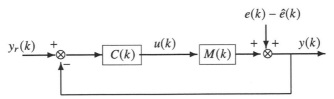

FIGURE 15.7 Another type of VES of adaptive control systems with $e(k)$ considered in control law.

In Fig. 15.7, $\hat{e}(k)$ is the prediction or estimate of $e(k)$.

A natural thinking would be put forward like this: since $e(k)$ could be estimated or predicted and further canceled with a compensation term in control law, why do not we simplify the identified model in Fig. 15.7 to be a fixed model, say a second-order model? The answer is *Yes*, it is possible. Then we have an adaptive control system with $e(k)$ considered in control law but without identification and online controller design, as shown in Fig. 15.8. Along such thinking, ADRC is a special type of adaptive control! Moreover, to some extent, ADRC is more simple and practical than traditional parameter-identification based adaptive control, considering that there are no identification mechanism and online controller design in ADRC systems.

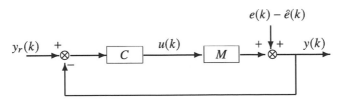

FIGURE 15.8 VES of adaptive control system without identification and online controller design.

Since in most engineering applications, control plant can be described by a second-order transfer function model with reasonable uncertainties, we may select $C = K_p + K_d s$ and $M = \frac{b_0}{s^2}$, in Fig. 15.8. Then we find out that Fig. 15.8 is pretty much like Fig. 15.4!

Now let us consider an MPC system shown in Fig. 15.9. MPC [26] is a form of control strategy in which the current control action is obtained by solving, at each sampling instant, a finite-horizon open-loop optimal control problem using the current state of the plant as the initial state; the optimization yields an optimal control sequence, and the first control in this sequence is applied to the plant [27].

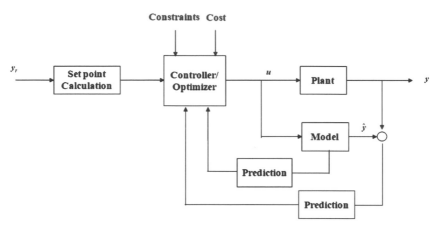

FIGURE 15.9 Block diagram of MPC system.

The VES of MPC is shown in Fig. 15.10, in which the details and characteristics of MPC are also omitted, like in the VESs of ADRC and adaptive control, for simplicity without loss of generality. By "generality" here we mean the stability and tracking performance analysis.

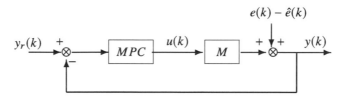

FIGURE 15.10 VES of MPC system.

From the VES of the MPC system we have the following conclusions on its stability and tracking performance without details of proof (we refer to [13,24, 25] for details):

Conclusion 3: If only the MPC system is stable and tracking for the model (on which the output prediction is conducted) without considering the model prediction error, i.e., the MPC controller and model formulate a stable and tracking closed-loop system; and further if the signal $e(k) - \hat{e}(k)$ is bounded, then the MPC system is BIBO (bounded-input bounded-output) stable.

Conclusion 4: If only the MPC system is stable and tracking for the model (on which the output prediction is conducted) without considering the model prediction error, i.e., the MPC controller and model formulate a stable and tracking closed-loop system; and further if the signal $e(k) - \hat{e}(k)$ tends to 0, then the MPC system is stable and tracking.

MPC and ADRC are also very similar from the viewpoint of VES, because MPC and ADRC both take into consideration the model output (prediction) error into their control laws, i.e., error prediction in MPC, and error estimation with ESO in ADRC. However, the difference between MPC and ADRC is that the model considered in MPC is much complicated than the model in ADRC, and as a trade off, the error prediction in MPC is simple, whereas ESO in ADRC is relatively complicated. Of course, MPC is unique in considering controller as an optimizer subject to different kinds of constraints, and the solution is rolling ahead in a short time period; also, it is supposed that the reference input $y_r(k)$ is time varying or even unknown before control system operation, which is quite suitable for control problems like vehicle driving control along an uncertain and unknown road environment.

As a brief summary, from the viewpoint of VES, ADRC, adaptive control, and MPC are similar in methodology, i.e., in treating the model and model errors, such as total disturbance in ADRC, model prediction error in MPC, and model (parameter) identification errors in adaptive control. They also have their own characteristics in some certain details: for example, in MPC the main focus or principle contradiction is the model (nonlinear model and constraints), whereas in ADRC the main focus or principle contradiction is the model errors, i.e., the total disturbance. Also, in adaptive control the main focus or principle contradiction is the model, i.e., model parameter identification. Thus it is reasonable that each control strategy has its own suitable application situation.

15.3 Multiple model ADRC for systems with large uncertainties

Many efforts have been made towards "robust" and "adaptive" characteristics in control field. Among others, the multiple model adaptive control (MMAC) is an important ingredient in the adaptive control family. The thought of multiple models for adaptive estimation originated from Magill [15]. Later on, many scholars, such as Lainiotis [16], Athans et al. [17], and Badr et al. [18], studied MMAC for different purposes of applications. MMAC include switching MMAC (SMMAC) and weighted MMAC (WMMAC) [14]. As one kind of WMMAC, the robust multiple model adaptive control (RMMAC) architecture provides an attractive framework for searching the "holy grail" of robust adaptive control [14,19–23], which inherits the basic structure of classical MMAC with several innovations. In brief, RMMAC differs from CMMAC mainly in the following three aspects:

1. As local controller strategy, output dynamic compensator designed by mixed-μ-synthesis method replaced LQ (linear quadratic) state feedback controller in CMMAC;
2. RMMAC separates control from Kalman filters, which are used for state estimation (also known as multiple model adaptive estimation) and model

identification (also known as weighting algorithm or posterior probability evaluator: PPE);

3. Performance-driven methodology to determine the number of required models.

It is worth pointing out that RMMAC scheme might have poor performance due to either large initial state estimate error or inaccurate knowledge of the disturbance/noise statistics. Additionally, the complexity of RMMAC may hinder its application because every candidate controller requires a Kalman filter and a post posterior evaluation. Consequently, some modified RMMAC scheme have been proposed to leave out Kalman filters.

Inspired by RMMAC, a new weighting algorithm is proposed for WMMAC of discrete-time stochastic plant [24,25], in which "local" controllers could be designed by any possible control strategies such as pole-assignment, ADRC, etc., provided that each "local" controller stabilizes its corresponding "local" model. In addition, the closed-loop stability of the resulting WMMAC system is proved with the help of VES concept and methodology.

The block diagram of a general WMMAC system is shown in Fig. 15.11. Of course, the multiple model ADRC system can also be expressed by Fig. 15.11.

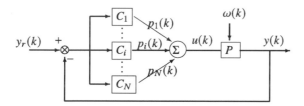

FIGURE 15.11 Block diagram of a multiple model (ADRC) control system.

The basic idea of WMMAC is to design a number of local controllers based on a number of local models and to select the most appropriate local model and local controller according to the performance of local models. Here are the main steps of the design of the local controller.

15.3.1 Model set

We use a number of local models for the uncertainty (structure or parameters change) of the real plant to approximate the dynamic performance. Those local models constitute a multi-model set

$$\Omega = \{M_i | i = 1, 2, \ldots, N\}, \tag{15.11}$$

representing the set of local models for the elements M_i. The selection and the number of elements of the model set affect the accuracy and performance of the entire control system.

15.3.2 Controller set

Design multiple local controllers to obtain the controller set

$$\Phi = \{C_i | i = 1, 2, \ldots, N\}, \tag{15.12}$$

of local controllers according to Ω, where C_i is the local controller designed based on the local model M_i. Each local model and the corresponding local controller constitute a stable closed-loop system. We adopt ADRC to be the local controller for each local model.

Let us consider a second-order time-delay system with parameters uncertainties. The application background is the automatic gauge control (AGC) system in hot rolling mills. The transfer function is

$$y(s)/u(s) = K_1 K_2 e^{-\tau s} / (T_1 s + 1)(T_2 s + 1), \tag{15.13}$$

where the gain K_1 depends on the finished width and thickness, rack outlet temperature, and steel types, τ mainly depends on the distance between the end of the rack to thickness gauge and the strip speed at export, so that K_1 and τ are time-varying parameters, and T_1, T_2, K_2 are all fixed parameters. Time-delay is usually approximated as a first-order inertia link. Because T_2 is too small to be ignored compared with T_1, it is reasonable to select a second-order ADRC. So formula (15.13) can be approximated by

$$\ddot{y} = \bar{f}(y, \dot{y}, d, t) + b \cdot u(t), \tag{15.14}$$

where b is the real gain of the plant. Considering the uncertainties of system parameters, we have the total disturbance

$$f(\cdot) = \bar{f}(\cdot) + (b - b_0)u \tag{15.15}$$

with new disturbance $(b - b_0)u$, where b_0 is not equal to b due to the time-delayed measurements, which is an adjustable parameter in ADRC controller. Formula (15.15) can be rewritten as

$$\ddot{y} = f(y, \dot{y}, d, t) + b_0 u(t). \tag{15.16}$$

The second-order ADRC controller includes the following three parts.

15.3.2.1 Tracking Differentiator (TD)

We need to arrange the transition process $v_1(t)$ reasonably according to the reference signal $v_0(t)$ and the characteristic of the plant, to make the system output $y(t)$ to track $v_0(t)$ rapidly without overshoot, and to give the differential signal $v_2(t)$ at the same time. In this chapter, we consider to remove TD in the design of ADRC because of the slow response of the time-delay plant, so that the output can reach the set value as soon as possible.

15.3.2.2 Extended State Observer (ESO)

We need to build the system state variables and the extended state variables according to the controller output $u(t)$ and the system output $y(t)$, also to generate the estimated signal $z_1(t)$, $z_2(t)$ of $y(t)$ and $\dot{y}(t)$, and the estimated disturbance $z_3(t)$ at the same time. We design the third-order linear ESO (LESO) for the simplified system (15.16):

$$
\begin{cases}
\dot{z}_1 = z_2 - \beta_{01} \cdot \varepsilon, \\
\dot{z}_2 = z_3 - \beta_{02} \cdot \varepsilon + b_0 v, \\
\dot{z}_3 = -\beta_{03} \cdot \varepsilon,
\end{cases}
\tag{15.17}
$$

where the observer error $\varepsilon = z_1 - y$, and β_{0i} and b_0 are adjustable parameters of LESO ($i = 1, 2, 3$). Note that the number of local ESOs is equal to the number of local models (controllers), i.e., for each local model (controller), there is a local ESO.

15.3.2.3 State error feedback

$$
u_{i0} = K_{pi}(v_1 - z_1) + K_{di}(v_2 - z_2),
\tag{15.18}
$$

$$
u_i = u_{i0} - \frac{z_3}{b_{i0}}.
\tag{15.19}
$$

15.3.3 Weighted control strategy

The global control output is the weighted sum of all local outputs of a controller:

$$
u(k) = \sum_{i=1}^{N} p_i(k) u_i(k),
\tag{15.20}
$$

where p_i is the weight of the ith controller, and u_i is the control output of the ith local controller ($i = 1, 2, \ldots, N$.

The weighting algorithm is described as follows:

$$
l_i(0) = \frac{1}{N}, \quad p_i(0) = l_i(0), \quad l_i'(k) = 1 + \frac{1}{k} \sum_{q=1}^{k} e_i^2(q),
$$

$$
l_{\min}(k) = \min_i l_i'(k), \quad l_i(k) = l_i(k-1) \frac{l_{\min}(k)}{l_i'(k)}, \quad p_i(k) = \frac{l_i(k)}{\sum_{i=1}^{N} l_i(k)},
\tag{15.21}
$$

where $e_i(k) = y(k) - y_{mi}(k)$, $y(k)$ is the output of the plant, and $y_{mi}(k)$ is the output of ith local model.

15.4 Simulation verification

Consider the second-order linear time-delay system (15.13) with parameters $K_2 = 0.1576$, $T_1 = 0.5$, $T_2 = 0.01$, $K_1 \in [4.4, 8.2]$, and $\tau \in [0.3, 0.5]$.

1. Select the model set with variation of the switching parameters. There are four local models in the model set with $K_1 = 6.3$ and $\tau = 0.5$ for model 1, $K_1 = 6.3$ and $\tau = 0.3$ for model 2, $K_1 = 5$ and $\tau = 0.4$ for model 3, and $K_1 = 7.7$ and $\tau = 0.5$ for model 4.
2. Design the local ADRC controller and ESO for each local model to constitute the controller set. Each local model (without total disturbance f) and its corresponding local controller formulate a stable closed-loop system. In addition, to verify the effectiveness of the proposed multiple model ADRC scheme, we also design the corresponding PID controllers for comparisons. For model 1, we have the following parameters for ADRC controller, ESO, and PID controller.
   ADRC controller: $K_P = 15.5$, $K_D = 9.7$, $b_0 = 90$;
   ESO: $\beta_{01} = 100$, $\beta_{02} = 4000$, $\beta_{03} = 60000$;
   PID controller: $K_P = 0.58$, $K_I = 0.99$, $K_D = 0$.
   For model 2, we have the following parameters for ADRC controller, ESO, and PID controller.
   ADRC controller: $K_P = 15.3$, $K_D = 8$, $b_0 = 90$;
   ESO: $\beta_{01} = 100$, $\beta_{02} = 4000$, $\beta_{03} = 60000$;
   PID controller: $K_P = 0.7$, $K_I = 1.4$, $K_D = 0.2$.
   For model 3, we have the following parameters for ADRC controller, ESO, and PID controller
   ADRC controller: $K_P = 16$, $K_D = 10.3$, $b_0 = 90$;
   ESO: $\beta_{01} = 100$, $\beta_{02} = 4000$, $\beta_{03} = 60000$;
   PID controller: $K_P = 0.82$, $K_I = 1.28$, $K_D = 0.1$.
   For model 4, we have the following parameters for ADRC controller, ESO, and PID controller.
   ADRC controller: $K_P = 15.7$, $K_D = 9.7$, $b_0 = 90$;
   ESO: $\beta_{01} = 100$, $\beta_{02} = 4000$, $\beta_{03} = 60000$;
   PID controller: $K_P = 0.6$, $K_I = 1$, $K_D = 0.1$.
3. The weighting algorithm is as described in Eq. (15.21).

15.4.0.0.1 Simulation case 1

Set the reference input signal as a step function y_r at the time $t = 0$ with amplitude 1. The parameters of real plant model jump from model 1 to model 2 at $t = 30$ s. The plant and controller outputs are shown in Fig. 15.12. The weight signals are shown in Fig. 15.13, in which the weighting algorithm quickly identified the correct local model and local controller, i.e., from model 1 and controller 1 to model 2 and controller 2. Simulation results verified that the multiple model ADRC achieves better control performances of the system with time-varying parameters, a smaller overshoot, and a shorter settling time than those of PID controller. The output fluctuation of ADRC is relatively smaller than that of PID controller when the real plant model parameter changes. In particular, the output of ADRC controller is much more stable than that of PID controller.

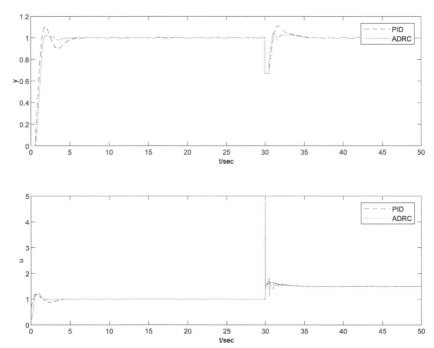

FIGURE 15.12 Outputs of the system and controller: step response.

15.4.0.0.2 Simulation case 2

Set the reference input signal as a step function y_r at the time $t = 0$ with amplitude 1. The parameters of real plant model jump from model 1 to model 3 at $t = 30$ s. The plant and controller outputs are shown in Fig. 15.14. The weights signals are shown in Fig. 15.15, in which the weighting algorithm quickly identified the correct local model and local controller, i.e., from model 1 and controller 1 to model 3 and controller 3. Simulation results verified that the multiple model ADRC achieves better performances, a smaller overshoot, and a shorter settling time than those of PID controller. The output fluctuation of ADRC is relatively smaller than that of PID controller when the real plant model parameter changes. In particular, the output of ADRC controller is much more stable than that of PID controller.

15.4.0.0.3 Simulation case 3

Set the reference input signal as step function y_r at the time $t = 0$ with amplitude 1. The parameters of real plant model jump from model 4 to model 2 at $t = 30$ s.

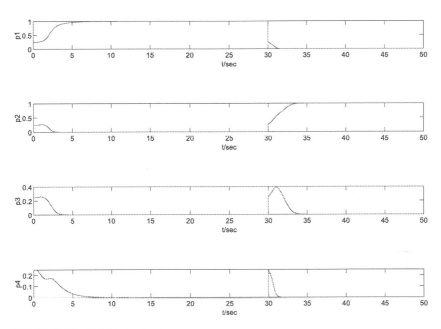

FIGURE 15.13 Weight signals: step response.

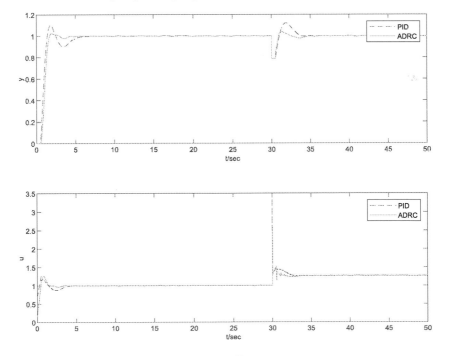

FIGURE 15.14 Outputs of the system and controller.

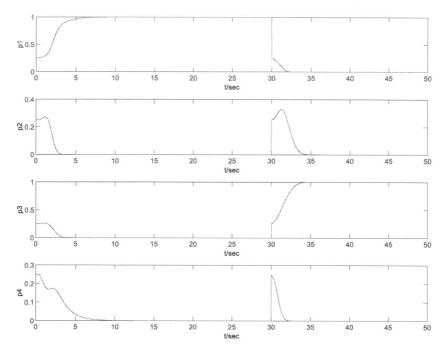

FIGURE 15.15 Weight signals.

The plant and controller outputs are shown in Fig. 15.16. The weights signals are shown in Fig. 15.17, in which the weighting algorithm quickly identified the correct local model and local controller, i.e., from model 4 and controller 4 to model 2 and controller 2. Simulation results verified that the multiple model ADRC can achieve better control performances of the system with time-varying parameters, a smaller overshoot, and a shorter settling time than those of the PID controller. The output fluctuation of the ADRC is relatively smaller than that of the PID controller when the real plant model parameter changes. In particular, the output of ADRC is much more stable than that of the PID controller.

15.4.0.0.4 Simulation case 4

Set the reference input signal as step function y_r at the time $t = 0$ with amplitude 1. The parameters of real plant model jump from model 4 to model 3 at $t = 30$ s. The plant and controller outputs are shown in Fig. 15.18. The weight signals are shown in Fig. 15.19, in which the weighting algorithm quickly identified the correct local model and local controller, i.e., from model 4 and controller 4 to model 3 and controller 3. Simulation results verified that the multiple model ADRC can achieve a better control performances of the system with time-varying parameters, a smaller overshoot, and a shorter settling time than those of PID controller. The output fluctuation of the ADRC is relatively smaller than that of the PID

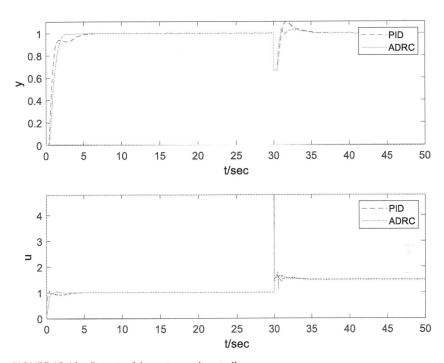

FIGURE 15.16   Outputs of the system and controller.

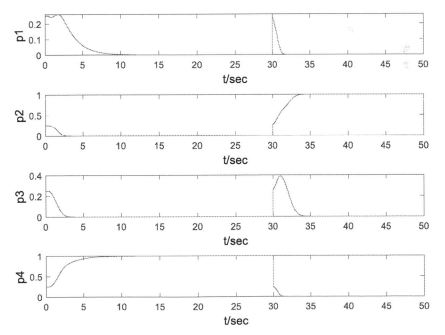

FIGURE 15.17   Weight signals.

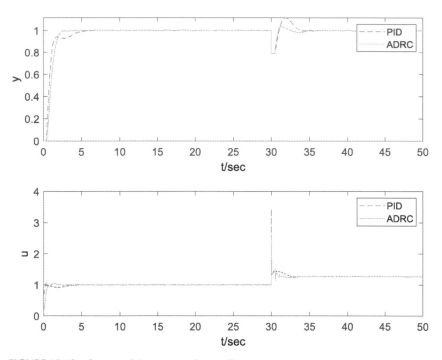

FIGURE 15.18 Outputs of the system and controller.

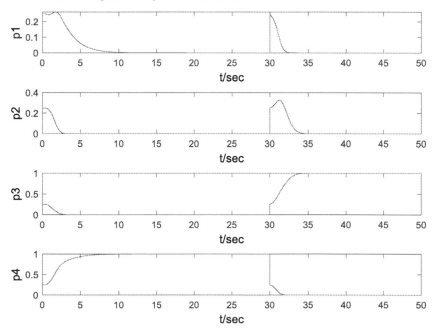

FIGURE 15.19 Weight signals.

controller when the real plant model parameter changes. In particular, the output of the ADRC controller is much more stable than that of the PID controller.

15.5 Conclusions

A multiple model active disturbance rejection control strategy for a class of systems with large uncertainties was proposed with simulation verifications. For the comparison of methodologies, popular control methods PID, model reference adaptive control, self-tuning adaptive control, MPC, and ADRC are discussed according to VES principle. From the viewpoint of VES, ADRC, adaptive control, and MPC are rather similar in methodology how to treat the model and model errors, such as the total disturbance in ADRC, model prediction error in MPC, and model (parameter) identification error in adaptive control. In MPC systems the main focus or principle contradiction is the model (nonlinear model and constraints) with moderate consideration of model errors, whereas in ADRC the main focus or principle contradiction is the model error (total disturbance) estimation. As to adaptive control, the main focus or principle contradiction is the online modeling, i.e., model parameter identification to reduce the model error.

References

[1] J.Q. Han, Active Disturbance Rejection Control Technique - the Technique for Estimating and Compensating the Uncertainties, National Defense Industry Press, Beijing, 2008.

[2] J.Q. Han, From PID to active disturbance rejection control, IEEE Transactions on Industrial Electronics 56 (3) (2009) 900–906.

[3] Y. Huang, W.C. Xue, Active disturbance rejection control: methodology and theoretical analysis, ISA Transactions 53 (2014) 963–976.

[4] W.C. Xue, W.Y. Bai, S. Yang, K. Song, Y. Huang, H. Xie, ADRC with adaptive extended state observer and its application to air-fuel ratio control in gasoline engines, IEEE Transactions on Industrial Electronics 62 (2015) 5847–5857.

[5] Q. Zheng, L.Q. Gao, Z.Q. Gao, On stability analysis of active disturbance rejection control for nonlinear time-varying plants with unknown dynamics, in: Proceedings of the IEEE Conference on Decision and Control, 2007, pp. 3501–3506.

[6] Q. Zheng, Z.Z. Chen, Z.Q. Gao, A practical approach to disturbance decoupling control, Control Engineering Practice 17 (2009) 1016–1025.

[7] Q. Zheng, Z.Q. Gao, On practical applications of active disturbance rejection control, in: Proceedings of the 29th Chinese Control Conference, Beijing, China, 2010, pp. 6095–6100.

[8] L. Sun, D.H. Li, et al., On tuning and practical implementation of active disturbance rejection controller: a case study from a regenerative heater in a 1000 MW power plant, Industrial & Engineering Chemistry Research 55 (23) (2016) 6686–6695.

[9] L. Sun, J. Dong, D.H. Li, et al., A practical multivariable control approach based on inverted decoupling and decentralized active disturbance rejection control, Industrial & Engineering Chemistry Research 55 (7) (2016) 847–866.

[10] W. Xue, X. Zhang, L. Sun, et al., Extended state filter based disturbance and uncertainty mitigation for nonlinear uncertain systems with application to fuel cell temperature control, IEEE Transactions on Industrial Electronics 67 (12) (2020) 10682–10692.

[11] H. Xie, S. Li, K. Song, et al., Model-based decoupling control of VGT and EGR with active disturbance rejection in diesel engines, IFAC Proceedings Volumes 46 (21) (2013) 282–288.

[12] D. Wu, K. Chen, Design and analysis of precision active disturbance rejection control for non-circular turning process, IEEE Transactions on Industrial Electronics 56 (7) (2009) 2746–2753.

[13] Zhang Weicun, On the stability and convergence of self-tuning control-virtual equivalent system approach, International Journal of Control 83 (5) (2010) 879–896.

[14] S. Fekri, M. Athans, A. Pascoal, Issues, progress and new results in robust adaptive control, International Journal of Adaptive Control and Signal Processing 20 (10) (2006) 519–579.

[15] D.T. Magill, Optimal adaptive estimation of sampled stochastic processes, IEEE Transactions on Automatic Control 10 (1965) 434–439.

[16] D.G. Lainiotis, Partitioning: a unifying framework for adaptive systems, I: estimation, II: control, Proceedings of the IEEE 64 (1976) 1126–1143 and 1182–1197.

[17] M. Athans, et al., The stochastic control of the F-8C aircraft using a multiple model adaptive control (MMAC) method – Part I: equilibrium flight, IEEE Transactions on Automatic Control 22 (1977) 768–780.

[18] A. Badr, Z. Binder, D. Rey, Weighted multi-model control, International Journal of Systems Science 23 (1) (1992) 145–149.

[19] M. Kuipers, P. Ioannou, Practical robust adaptive control: benchmark example, in: Proceedings of American Control Conference, Seattle, Washington, USA, 2008, pp. 5168–5173.

[20] M. Kuipers, P. Ioannou, Multiple model adaptive control with mixing, IEEE Transactions on Automatic Control 55 (8) (2010) 1822–1836.

[21] S. Baldi, P. Ioannou, E. Mosca, Multiple model adaptive mixing control: the discrete-time case, IEEE Transactions on Automatic Control 57 (4) (2012) 1040–1045.

[22] N. Sadati, G.A. Dumont, H.R. Feyz Mahdavian, Robust multiple model adaptive control using fuzzy fusion, in: 42nd South Eastern Symposium on System Theory, Tyler, TX, USA, March 7–9, 2010.

[23] Z. Han, K.S. Narendra, New concepts in adaptive control using multiple models, IEEE Transactions on Automatic Control 57 (1) (2012) 78–89.

[24] W. Zhang, Stable weighted multiple model adaptive control: discrete-time stochastic plant, International Journal of Adaptive Control and Signal Processing 27 (7) (2013) 562–581.

[25] W. Zhang, Further results on stable weighted multiple model adaptive control: discrete-time stochastic plant, International Journal of Adaptive Control and Signal Processing 29 (12) (2015) 1497–1514.

[26] J. Richalet, A. Rault, J.L. Testud, J. Papon, Model predictive heuristic control: applications to industrial processes, Automatica 14 (5) (1978) 413–428.

[27] D.Q. Mayne, J.B. Rawlings, C.V. Rao, P.O.M. Scokaert, Constrained model predictive control: stability and optimality, Automatica 36 (6) (2000) 789–814.

[28] G. Zames, Feedback and complexity, special plenary lecture addendum, in: IEEE Conf. Decision Control, IEEE, 1976.

[29] G. Zames, Optimal sensitivity and feedback: weighted seminorms, approximate inverses, and plant invariant schemes, in: Proceedings of the Allerton Conference, IEEE, 1979.

[30] G. Zames, Feedback and optimal sensitivity: model reference transformations, multiplicative seminorms, and approximate inverses, IEEE Transactions on Automatic Control 26 (1981) 301–320.

[31] K.J. Åström, B. Wittenmark, On self-tuning regulators, Automatica 9 (1973) 185–199.

[32] D.W. Clarke, P.J. Gawthrop, Self-tuning controller, Proceedings of the IEEE 122 (9) (1975) 929–934.

[33] G.C. Goodwin, P.J. Ramadge, P.E. Caines, Discrete time stochastic adaptive control, SIAM Journal on Control and Optimization 19 (1981) 829–853.

Chapter 16

Gain scheduling design based on active disturbance rejection control for thermal power plant under full operating conditions

Zhenlong Wu[a], Donghai Li[b], Yali Xue[b], and YangQuan Chen[c]

[a]School of Electrical Engineering, Zhengzhou University, Zhengzhou, China, [b]State Key Lab of Power Systems, Department of Energy and Power Engineering, Tsinghua University, Beijing, China, [c]Mechatronics, Embedded Systems and Automation (MESA) Lab, School of Engineering, University of California, Merced, CA, United States

16.1 Introduction

With the booming growth of the renewable energy such as solar, wind, and tidal power generation in the electricity market, the safety operation of power grid is becoming a challenging issue [1]. The increased impact from these fluctuating energy sources, which have strong intermittency and randomness, significantly affects the operational regime of thermal power plants that have to accelerate the speed of power output responding the automatic generation control (AGC) command and extend the operating range [2]. The frequent and extensive load changes can result in severe thermo-mechanical fatigue, creep, and corrosion, which can cause a lifetime reduction [3]. Besides, strict control requirements for the efficiency and safety put forward great challenges of the management strategies of thermal power plants, especially the daily operational strategies. The proportional-integral (PI) and proportional-integral-derivative (PID) controllers still dominate the industry (more than 98% of all power plant controllers in Guangdong Province, China) and undoubtedly play a key role in current thermal processes [4]. However, facing with strong nonlinear characteristics, strong cross-coupling, wide operating conditions, and strict control requirements, the traditional control strategy cannot obtain the satisfactory control performance [5].

To research dynamic characteristics of thermal power plants, a dynamic power plant model with an innovative level of detail is developed, and the model can be used to optimize startup costs and environmental impact [6]. An experiment-based model of condensate throttling for 1000 MW power units is developed to analyze the dynamics of storage energy [7]. To design better control strategies based on the dynamic model of power plants, some nonlinear dynamic models are built recently. Astrom [8] develops a famous nonlinear dynamic model for natural circulation drum-boilers, which is intended for model-based control focuses. However, the power capacity of this model is small, which is built based on a 160 MW unit in Sweden. Recently, a dynamic model of supercritical once-through boiler units is developed based on the law of conservation of energy and substance, and some necessary assumptions [9]. This dynamic model with three inputs and three outputs is developed for controller design and dynamic analysis. To develop a model suitable for direct energy balance (DEB) coordinated control scheme, a dynamic model is proposed to describe each module by yielding a 6th-order nonlinear model [10]. The accuracy is verified by the field measurements from a 300 MW coal-fired plant. Considering that the DEB control scheme (it will be introduced in Subsection 16.4.1), which is widely used by field engineers in thermal power plants, the analysis and verification of the proposed gain scheduling design based on ADRC are carried out based on this model.

To relieve the adverse effects of the nonlinearity and strong coupling of power plants and obtain the faster power output and smaller pressure fluctuation, many control strategies are proposed to solve the aforementioned control difficulties. Model predictive control (MPC), with distinct advantages in explicitly handling constraints and multivariable couplings, has been applied to the coordinated control system (CCS), which is the most important and crucial loop in any thermal power plant [11,12]. Nonlinear MPC and economic MPC based on nonlinear models are designed and discussed for boiler-turbine systems, respectively. Nonlinear control based on feedback linearization approach is proposed to enhance the disturbance rejection ability and power-tracking rate [13,14]. An optimized nonlinear controller using evolutionary algorithms is designed for boiler-turbine system in [15]. Besides, sliding mode control [16,17], robust control [18], dynamic matrix control [19], self-adaptive PID control [20], and neural network inverse control [21] all show satisfactory control effect in numerical simulations while these aforementioned control strategies were rarely used in practical units due to the following reasons [22]:

1. These control strategies could obtain satisfactory control effect at the cost of a large computation complexity, which results in great implementation difficulty in the distributed control system (DCS) platform.
2. Besides, the accurate mathematical model is the foundation for some model-based control strategies, whereas the accurate model is hard and expensive to build because of the system complexity.

In the past decades, active disturbance rejection control (ADRC) gradually develops into a powerful tool to handle the control difficulties caused by unknown dynamics and external disturbances [23,24]. It offers a new perspective where unmeasured disturbances and unmodeled dynamics can be estimated and compensated in real time by an extended state observer (ESO) [25]. Moreover, it is independent of the accurate mathematical model and can be implemented in DCS easily. The convergence of the tracking differentiator and the stability of ESO are discussed in [26,27], respectively. Based on the theoretical analysis and distinct advantages of ADRC, ADRC has been successfully applied to motion system [28], engines system [29], fan system in server [30], organic Rankine cycle (ORC) system [31], main steam pressure system [2], et al. These systems with ADRC have better tracking performance and stronger disturbance rejection ability than that of other comparative control strategies based on their own experiment platforms. However, the control performance under wide operating conditions is not verified in these experiments.

Recently, ADRC is also proposed for CCS to solve the control difficulties such as the coupling, nonlinearity, et al. In [32] the control strategy based on ADRC is designed for a power plant with a single loop, where a tracking differentiator, ESO, and a nonlinear combination of errors are combined to improve the control performance. ADRC based on the DEB control scheme is designed for CCS to reduce the pressure fluctuation [33]. However, the ability of the ADRC is still limited by faster response speed responding to the AGC command. The faster response speed can result in challenges of the equipment safety. What is worse, thermal power plants have to enlarge their operating conditions to absorb more renewable power into the power grid. Now thermal power plants shift the power output in the range of [50%, 100%] of full load. Thermal power plants will have a larger range of [30%, 100%] of full load to improve the stability of power grid in the future. This can result in stronger nonlinearity of the unit and greater safety pressure. To solve this, a mature method namely gain scheduling can be applied to speed up load response and reduce pressure fluctuations. The whole operating condition can be divided into several typical conditions, and controllers are designed for each condition with the switching law for the design of gain scheduling [34]. The gain scheduling designs combined with adaptive fuzzy PID [35], iterative learning control [36], sliding mode control [37], and MPC [38] have been studied for different systems. These control strategies still have a large computation complexity, which limits the practical application of the gain scheduling.

To deal with control difficulties of thermal power plants and reduce the implementation difficulties of gain scheduling, a gain scheduling design based on ADRC is proposed for thermal power plant under full operating conditions from 30% to 100% of full load. Besides, to the best of authors' knowledge, there are few references to discuss the gain scheduling design based on ADRC, such as the switching method of ADRC parameters and stability analysis for the gain scheduling design based on ADRC. The rest of the paper is organized as follows. A dynamic model of CCS and control objectives are introduced briefly, and the CCS control difficulties are discussed to explain the necessity of the proposed control strategy in Section 16.2. Section 16.3 provides the brief principle of the

ADRC, the tuning method, and stability region of the ADRC. In Section 16.4, the necessary analyses of the gain scheduling design based on ADRC are discussed such as the selection of the scheduling parameter, the switching method for k_p and ω_o, and stability analysis for the gain scheduling design based on ADRC. The superiority of the proposed strategy compared to the regular ADRC and the PI is verified by simulations in Section 16.5. Conclusions are drawn in Section 16.6.

16.2 Problem formulation

16.2.1 Dynamic model of the coordinated control system

The gain scheduling design based on ADRC is discussed for the nonlinear dynamic model in [10] considering that the model is suitable for the DEB control scheme, which is widely used by field engineers in thermal power plants.

This model contains six parts, and the description is as follows:

Dynamic of the flow of coal blowing into the furnace,

$$\dot{q}_f = \frac{1}{22} \left[u_B \left(t - 43 \right) - q_f \right]. \tag{16.1}$$

Dynamic of the steam evaporation amount,

$$\dot{D}_b = \frac{1}{380} \left[2.46 k_c q_f^{1.230} - D_b \right]. \tag{16.2}$$

Dynamic of the boiler pressure,

$$\dot{p}_b = \frac{1}{4057} \left[D_b - 42.51 p_b^{0.956} \sqrt{p_b - p_T} \right]. \tag{16.3}$$

Dynamic of the throttle pressure,

$$\dot{p}_T = \frac{1}{5101} \left[42.51 p_b^{0.956} \sqrt{p_b - p_T} - D_T \right]. \tag{16.4}$$

Dynamic of the governing stage pressure,

$$\dot{p}_1 = \frac{1}{5} \left[0.0083 \mu_t p_T - p_1 \right]. \tag{16.5}$$

Dynamic of the inlet steam mass flow,

$$\dot{D}_T = \frac{1}{5} \left[74.74 p_1 - D_T \right], \tag{16.6}$$

where u_B (t/h) and μ_t (%) are the coal feed and the throttle opening position, respectively.

Besides, the power output N_e can be calculated by

$$N_e = 0.86 D_T^{0.852}. \tag{16.7}$$

Moreover, the normalization coefficient k_c in Eq. (16.2) is introduced to represent the influence of the coal quality, which should be equal to 100% under normal condition. Besides, considering the protection of actuators of boiler and turbine, and the operation rules of boiler and turbine, we can obtain the amplitude limit and rate limit as follows:

$$\begin{cases} 0 \le u_B \le 150, \\ -0.3 \le \dot{u}_B \le 0.3, \\ 0 \le \mu_t \le 100, \\ -0.2 \le \dot{\mu}_t \le 0.2. \end{cases} \tag{16.8}$$

Note that the rate limit listed in Eq. (16.8) is different from that in [10], where $\dot{u}_B$ and $\dot{\mu}_t$ are both ± 0.1. The rate limit in [10] is conservative with the development of material technology and improvement of equipment operation level.

In this model, the coal feed u_B and throttle opening position μ_t are the control inputs. The process outputs are the power output N_e (MW) and throttle pressure p_T (MPa), respectively. Hereby a standard nonlinear model for control can be derived from Eqs. (16.1)–(16.6) as

$$\begin{cases} \dot{x} = f(x,u), \\ y = h(x), \end{cases} \tag{16.9}$$

where $u = [u_B \quad \mu_t]^T$, $x = [q_f \quad D_b \quad p_b \quad p_T \quad p_1 \quad D_T]^T$, and $y = [Ne \quad p_T]^T$.

To better rule in peak-load shaving and play a greater potential, the excepted operating range would be expanded to 30%–100% of the full load. Note that the constant and sliding pressure operation modes both exist in the actual thermal power plant to improve energy efficiency and meet safety requirements. Specifically, the thermal power plant would be on the constant pressure operation mode when the output power locates in the ranges of 30%–40% and 90%–100% to improve energy efficiency and ensure the system security, respectively. In other ranges, the thermal power plant would operate on the sliding pressure operation mode.

Moreover, some typical operating conditions are calculated and listed in Table 16.1 based on the dynamic model in Eqs. (16.1)–(16.6).

Generally, the CCS has some fundamental control requirements to respond the AGC command quickly and satisfy the safety and economic requirements, which are listed as follows [39]:

1) The power output should be adjusted timely as required by AGC command. In China, the tracking rate is about 1.5%–2% of full load per minute,

TABLE 16.1 Fiver typical operating conditions.

Operating condition	N_e (MW)	p_T (MPa)	u_B (t/h)	μ_t (%)
A (30%)	90.0	13.82	40.7	27.4
B (40%)	120.6	13.82	53.8	38.6
C (65%)	195.3	14.81	85.2	63.4
D (90%)	270.0	16.09	116.1	85.4
E (100%)	300.1	16.09	128.4	96.7

which means that it requires to generate 4.5–6 MW more power in one minute for a 300 MW thermal power plant.

2) The reverse change of the throttle pressure when regulating the power output should be limited to a safety bound on sliding pressure operation mode, e.g., ±0.4 MPa from the initial pressure. Besides, the biggest deviation also should be limited to a safe bound on constant pressure operation mode, e.g., ±0.4 MPa from the initial pressure. Note that the system requires a manual intervention when the deviation is bigger than ±0.4 MPa to ensure the system security. The fewer the manual interventions, the better the control performance for the control strategy.

16.2.2 Control difficulties

The CCS of thermal power plant has many difficulties such as the strong nonlinearity, strong coupling, and the wide change of operating conditions. These control difficulties are analyzed in this subsection.

To measure the system nonlinearity of the established physics-informed model, Vinnicombe gap metric is introduced with the definition as in [40],

$$v_g = \max\{\vec{\delta}(P_1, P_2), \vec{\delta}(P_2, P_1)\}, \tag{16.10}$$

where P_1 and P_2 are the two transfer function matrices linearized around two arbitrary operating conditions, and $\vec{\delta}(P_1, P_2)$ is the directed gap of two linear systems P_1 and P_2. Let $P_1 = N_1 M_1^{-1}$ and $P_2 = N_2 M_2^{-1}$ denote the normalized coprime factorizations of P_1 and P_2. The directed gap can be calculated by

$$\vec{\delta}(P_1, P_2) = \inf_{Q \in H_\infty} \left\| \begin{bmatrix} M_1 \\ N_1 \end{bmatrix} - \begin{bmatrix} M_2 \\ N_2 \end{bmatrix} Q \right\|_\infty, \tag{16.11}$$

where Q is a matrix parameter of finite H_∞ norm. The detailed theory of the gap is referred to in [40]. Besides, the gap for any two linear systems is bounded as

$$0 \leq \vec{\delta}(P_1, P_2) \leq 1. \tag{16.12}$$

When P_1 is the linearized system at the nominal operating condition of a nonlinear system and P_2 is the linearized system at another operating condition of the same nonlinear system, $\bar{\delta}(P_1, P_2)$ can be regarded as an indicator of system nonlinearity. The closer the value is to one, the stronger the nonlinearity of the system, and vice versa. The nominal condition is chosen as the full load, E (100%) shown in Table 16.1, where $N_e = 300.1$ MW and $p_T = 16.09$ MPa, and the nominal transfer matrix model can be obtained by the model linearization. The distance measure between the nominal model and other models that are linearized with different N_e and p_T is depicted in Fig. 16.1.

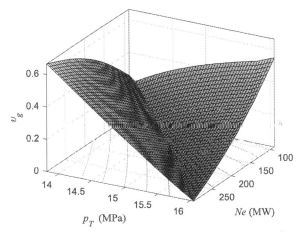

FIGURE 16.1 Distance measure from the nominal model linearized at (300.1 MW, 16.09 MPa) to the models linearized with different N_e and p_T.

Obviously, a large v_g exists when N_e and p_T are far from the nominal model, which means that a strong nonlinearity exists for the model in Eqs. (16.1)–(16.6) or Eqs. (16.9). Besides, an obvious valley in the 3-D figure of distance measure can be seen in Fig. 16.1, showing that least amount of nonlinearity. This indicates that the nonlinearity can be somewhat avoided if the set-points of N_e and p_T are changing in proportion along the valley line. This valley is fully used to reduce the nonlinearity of the system in engineering practice called sliding pressure operation mode. Note that the sliding pressure operation mode is not always reasonable in some operating conditions and is not applicable when the output power is very small or large because of the economical and safe considerations.

The linearized models at typical operating conditions in Table 16.1 can be obtained by the linearization method, and their open-loop responses are presented in Fig. 16.2. The dynamic characteristics of the CCS system vary greatly due to the wide operating conditions, especially the dynamic characteristics of the pressure loop as shown in Fig. 16.2(c) and (d).

Besides, the strong coupling is also seen in Fig. 16.2 because each control input has obvious influence on both outputs. To explain this further, relative gain array (RGA) is applied to measure the coupling between main steam pressure

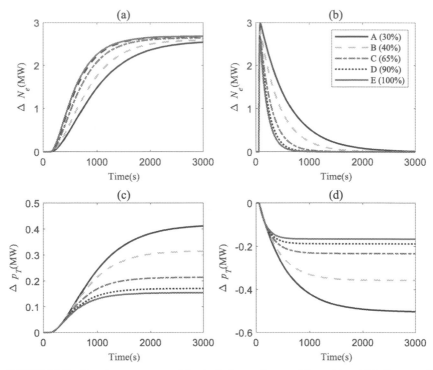

FIGURE 16.2   Open-loop responses of five typical operating conditions. (a) and (c): the step of u_B; (b) and (d): the step of μ_t.

and power loops by transforming the CCS model into a general transfer matrix G by a model linearization method [40]. The definition of RGA is depicted as

$$\text{RGA}(G) \triangleq G \times (G^{-1})^T = \begin{bmatrix} \lambda_{11}(s) & \lambda_{12}(s) \\ \lambda_{21}(s) & \lambda_{22}(s) \end{bmatrix}, \qquad (16.13)$$

where $\times$ and G denote the Schur product and the linear TITO system under operating conditions, respectively, and λ_{ij} means the ratio between the gains from u_j to y_i when the other loop is open and closed, respectively. Studies show that the controller can be designed to the pair of variables u_j and y_i when $|\lambda_{ij}(jw)|$ is close to 1 in all frequency ranges.

Fig. 16.3 shows the RGA distribution under different operating conditions. We can learn that the paring rule has a variation with increasing frequency, which means that there has a strong coupling for the CCS system. y_1 (N_e) and y_2 (p_T) are dominated by u_1 (μ_B) and u_2 (μ_T) in the low-frequency range, respectively. Similarly, y_1 and y_2 are dominated by u_2 and u_1 in the high-frequency range, respectively.

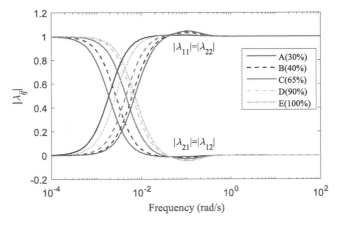

FIGURE 16.3   RGA distribution under different operating conditions.

16.3   Active disturbance rejection control

The ADRC gets more and more attention and successful applications, and stands out among control algorithms owing to the following merits:

1. The ADRC is largely independent of precise mathematical models, which are hard or expensive to obtain for industrial process because of the system complexity.
2. The robustness of the ADRC is stronger than the conventional PI/PID controller and can ensure good control performance when the system is far from the nominal condition.
3. The control law of the ADRC is simple, consisting of some basic algebraic computation, and can be easily implemented via existing function blocks in DCS platform.
4. The ADRC is conceived from the perspective of disturbance rejection, which can balance the tradeoff relationship between tracking and disturbance rejection well, especially for CCS whose two loops both have the tasks of tracking and disturbance rejection.

In this section, we briefly introduce the principle of the first-order ADRC. Besides, we also discuss the parameter tuning and stability region of the ADRC.

16.3.1   Brief principle of ADRC

To reduce the difficulty of implementation, the first- and second-order ADRC controllers are the most widely used ADRCs in practice despite the order mismatch between the system order and the ADRC order. Moreover, the capacity of the low-order ADRC to control high-order systems has been theoretically proved [41] and verified in superheated steam temperature system [42] and fractional-order system [43].

Here the first-order ADRC is discussed, and its structure is illustrated in Fig. 16.4, r, y, d, and u are the reference input, the system output, the external disturbance, and the control signal, respectively, z_2 is the output of the ESO. Besides, k_p and b_0 are the parameters of control law.

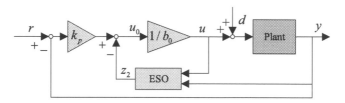

FIGURE 16.4   The structure of the first-order ADRC.

Generally, a system can be arranged into a first-order form

$$\dot{y} = g(t, y, \ddot{y}, \ldots, d, \omega) + bu, \tag{16.14}$$

where g is the synthesis function of time-variant (t), high-order ($\ddot{y}$), external disturbance (d), dynamic uncertainties (ω), etc. in the system, and b is the input gain whose value may be unknown for the plant.

Define $f = g + (b - b_0)u$, where b_0 is an estimate of the input gain b. Then Eq. (16.14) can be transformed into

$$\dot{y} = f + b_0 u, \tag{16.15}$$

where f is called the total disturbance, which includes the external disturbances and the unknown internal dynamics of the system [23]. Let $y = x_1$, and let f be an extended state z_2. The state space representation of Eq. (16.15) can be depicted as

$$\begin{cases} \begin{bmatrix} \dot{x}_1 \\ \dot{x}_2 \end{bmatrix} = \begin{bmatrix} 0 & 1 \\ 0 & 0 \end{bmatrix} \begin{bmatrix} x_1 \\ x_2 \end{bmatrix} + \begin{bmatrix} b_0 \\ 0 \end{bmatrix} u + \begin{bmatrix} 0 \\ 1 \end{bmatrix} \dot{f}, \\ y = \begin{bmatrix} 1 & 0 \end{bmatrix} \begin{bmatrix} x_1 \\ x_2 \end{bmatrix}. \end{cases} \tag{16.16}$$

Then the ESO is designed for system (16.16) as

$$\begin{bmatrix} \dot{z}_1 \\ \dot{z}_2 \end{bmatrix} = \begin{bmatrix} -\beta_1 & 1 \\ -\beta_2 & 0 \end{bmatrix} \begin{bmatrix} z_1 \\ z_2 \end{bmatrix} + \begin{bmatrix} b_0 & \beta_1 \\ 0 & \beta_1 \end{bmatrix} \begin{bmatrix} u \\ y \end{bmatrix}. \tag{16.17}$$

The state z_2 can track f well when β_1 and β_2 are tuned appropriately. The estimated total disturbance can be compensated in real time,

$$u = \frac{u_0 - z_2}{b_0}. \tag{16.18}$$

The plant after the compensation is depicted as

$$\dot{y} = f + b_0 \frac{u_0 - z_2}{b_0} \approx f + \frac{u_0 - f}{b_0} = u_0. \tag{16.19}$$

Therefore the controlled plant becomes an integral process, and the control law shown in Fig. 16.4 is depicted by

$$u_0 = k_p (r - y), \tag{16.20}$$

which is a proportional controller. The desirable closed-loop transfer function can be obtained based on Eqs. (16.19) and (16.20) as

$$G_{lc}(s) = \frac{y(s)}{r(s)} = \frac{k_p}{s + k_p}. \tag{16.21}$$

Note that k_p is the controller bandwidth with practical physical meaning. Moreover, β_1 and β_2 can be tuned by the bandwidth-parameterization method proposed in [44],

$$\begin{cases} \beta_1 = 2\omega_o, \\ \beta_2 = \omega_o^2, \end{cases} \tag{16.22}$$

where ω_o is called the observer bandwidth.

16.3.2 Tuning method and stability region of ADRC

Based on the introduction of the ADRC, we know that there are three parameters to be tuned: the observer bandwidth ω_o, the controller bandwidth k_p, and b_0. The following rules may serve as guidance to tune the parameters:

1. A large k_p or a small b_0 leads to a quick response, which in turn requires increased control force while the stability margin would be small. Besides, the response would have a large overshoot, and the fluctuation would increase due to the strong control intervention with a too large k_p or a too small b_0. Note that the value of b_0 should meet $b/b_0 \in (0, 2)$ to ensure the stability and convergence of the ESO [45].
2. The ability of the observation and compensation of the total disturbance would increase with the increased ω_o, and so does noise sensitivity of the ESO. So ω_o should be gradually augmented to a proper value that could ensure a good estimation of the ESO.

Based on the discussion of the influence of parameters on control performance, a tuning procedure can be summarized as follows:

1. b_0 should be selected firstly, and the value of b_0 should meet $b/b_0 \in (0, 2)$ to ensure the stability and convergence of the ESO. A large b_0 is recommended to avoid the non-convergence when the real gain b is not exactly known.

2. Then k_p can be selected based on the desirable closed-loop dynamic shown in Eq. (16.21), and ω_o can be gradually augmented to a proper value to enhance the observation ability of the ESO.
3. If the ADRC with these parameters can obtain the satisfactory control performance, then the tuning procedure can stop. Otherwise, repeat steps 1–2.

The flow chart is accordingly given in Fig. 16.5 to guide the tuning in engineering practice. Note that the bandwidth-parameterization method works well and the presented tuning procedure involves trial and error tests.

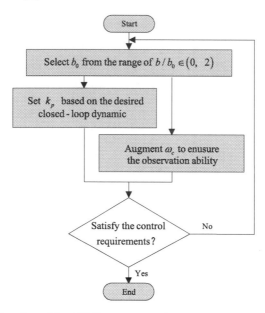

FIGURE 16.5 Flow chart of the ADRC parameters tuning.

The first-order ADRC can be can be transformed into an equivalent two-degrees-of-freedom (TDOF) control structure as shown in Fig. 16.6 [46], where the feedback controller and feedforward controller are depicted by

$$G_c(s) = \frac{k_p s^2 + \left(\omega_o^2 + 2k_p\omega_o\right)s + k_p\omega_o^2}{(s + 2\omega_o)\,b_0 s} \qquad (16.23)$$

and

$$G_f(s) = \frac{k_p(s + \omega_o)^2}{k_p s^2 + \left(2k_p\omega_o + \omega_o^2\right)s + k_p\omega_o^2}. \qquad (16.24)$$

To simplify the stability region analysis of the ADRC, the plant transfer function can be depicted by

$$G_p(i\omega) = r(\omega)\,e^{i\vartheta(\omega)} = a(\omega) + ib(\omega). \qquad (16.25)$$

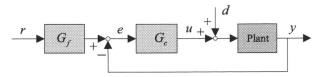

FIGURE 16.6 Equivalent structure of ADRC.

The characteristic equation for the closed-loop system is represented as

$$1 + G_l(s) = 0, \tag{16.26}$$

where $G_l(s) = G_p(s) G_c(s)$ is called the loop transfer function.

By substituting Eqs. (16.23) and (16.25) into Eq. (16.26) and separating the real and imaginary parts we obtain

$$\begin{cases} k_p \left(\omega_o^2 - \omega^2 \right) a(\omega) - \left(\omega_o^2 + 2k_p \omega_o \right) b(\omega) \omega - b_0 \omega^2 - 0, \\ k_p \left(\omega_o^2 - \omega^2 \right) b(\omega) + \left(\omega_o^2 + 2k_p \omega_o \right) a(\omega) \omega + 2b_0 \omega_o \omega = 0. \end{cases} \tag{16.27}$$

By solving Eq. (16.27) with fixed b_0 we can obtain the boundary of stability region. Considering that k_p and ω_o should be greater than zero to ensure the system stability, these boundaries form the stability region of the ADRC.

Now the procedure of the stability region calculation is summarized as follows:

1. The plant in Eqs. (16.25) is known, and b_0 should be fixed firstly according to the discussion above.
2. The boundary of the ADRC can be calculated by solving the expression in Eq. (16.27) with ω varying from zero to $+\infty$. Note that the upper value of ω can be set sufficiently large. We can obtain the stability region of the ADRC with fixed b_0.
3. The whole stability regions of the ADRC can be obtained by repeating the calculation for a set of b_0-values.

Consider a first-order plus dead time (FOPDT) plant defined as

$$G_p(s) = \frac{1.2}{4s + 1} e^{-1.5s}, \tag{16.28}$$

whose real gain b is 0.3. By applying the calculation procedure discussed above we can obtain the stability regions of the ADRC with gridded b_0 from 0.15 to 5 as shown in Fig. 16.7. We see that a larger b_0 means a larger parameter selection region, and this explains the reason why a large b_0 is recommended. The discussion about the stability region of the ADRC offers a parameter selection region and a foundation of the asymptotic stabilization analysis for the smooth switching.

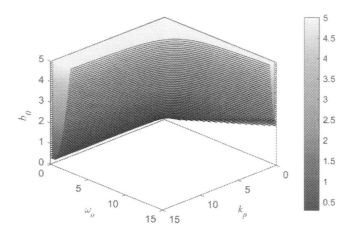

FIGURE 16.7 The stability regions of ADRC with gridded b_0.

16.4 Gain scheduling design based on ADRC

In this section, the principle of DEB is introduced firstly to understand the control structure widely used in thermal power plants, especially in giant-scale thermal power plants. Then some necessary discussions of the division of operating conditions and the selection of scheduling parameters are presented. A linear switching method is developed based on the foundations discussed above. In the end the stability analysis based the Kharitonov theorem and the quantitative calculation are carried out to offer the stability region of the parameter tuning ensuring the asymptotic stabilization of the closed-loop system under full operating conditions.

16.4.1 Brief principle of DEB

The control structure of DEB is shown in Fig. 16.8, where the controlled variable of the throttle pressure loop is the energy heat signal defined by

$$Q_m = p_1 + C_b \frac{dp_b}{dt}, \tag{16.29}$$

which replaces the throttle pressure p_T and $C_b = 120$. Correspondingly, the set-point of the throttle pressure loop becomes the energy demanding signal,

$$r_{Q_m} = r_{p_t} \frac{p_1}{p_T}, \tag{16.30}$$

where r_{p_t} is the set-point of p_T coming from the AGC command by a lookup table.

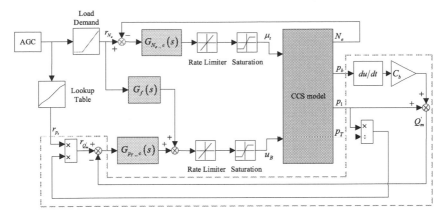

FIGURE 16.8 The DEB control structure for CCS system with actuator constraints.

When the system is at steady state, we have $Q_m = r_{Q_m}$ and $d p_b/dt = 0$, and we have the equation

$$p_1 = p_r \frac{p_1}{p_T}, \tag{16.31}$$

which means that the throttle pressure p_T can converge to its set-point p_r when the system is at steady state. The advantage of the DEB control structure is that it can greatly reduce the coupling effect from the throttle pressure loop to the power output [32].

Based on the control structure of DEB, the output of throttle pressure loop becomes the energy heat signal Q_m, which replaces the throttle pressure p_T, and the linearized model can be depicted by

$$\begin{bmatrix} N_e \\ Q_m \end{bmatrix} = \begin{bmatrix} g_{11}(s) & g_{12}(s) \\ g'_{21}(s) & g'_{22}(s) \end{bmatrix} \begin{bmatrix} u_B \\ \mu_t \end{bmatrix}. \tag{16.32}$$

Similarly, y in Eq. (16.9) changes to $y = [Ne \quad Q_m]^T$, and Q_m can be calculated by

$$Q_m = p_1 + \frac{C_b}{4057} \left[D_b - 42.51 p_b^{0.956} \sqrt{p_b - p_T} \right]. \tag{16.33}$$

Note that the throttle opening position μ_t is selected to control the power output N_e because it is useful for quick tracking of the AGC command, which is the first priority for any thermal power plant. Correspondingly, the coal feed u_B is selected to control the energy heat signal. To accelerate the response of throttle pressure loop, a feedforward PD controller $G_f(s)$ from the load demand is added to the coal feed u_B. $G_{N_e\_c}(s)$ and $G_{pT\_c}(s)$ are the feedback controllers of the power output loop and throttle pressure loop, respectively.

16.4.2 Division of operating conditions and selection of scheduling parameter

Firstly, we discuss the division of operating conditions for CCS, which is the foundation of gain scheduling design. The following factors should be considered,

1. The more scheduling ranges means the more meticulous control, which can ensure the better control performance.
2. The adverse influence of more scheduling ranges should not be neglected, such as the exponentially increased design workload and the implementation difficulty in DCS platform.

Therefore the number of scheduling ranges is a tradeoff between the control performance and the cost of design and implementation. The scheduling ranges are selected as shown in Fig. 16.9. The scheduling ranges are from F to G and from H to I, respectively. Besides, the steady state parameters of the selected operating conditions are listed in Table 16.2, and the scheduling range is about 10 MW. To a great extent, this tradeoff can reduce the design working load and the implementation difficulty, and balance the control requirements of different dynamic characteristics.

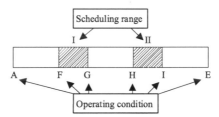

FIGURE 16.9 The scheduling ranges and the operation conditions selected under full operating conditions.

TABLE 16.2 The operating conditions for the gain scheduling design.

Operating condition	N_e (MW)	p_T (MPa)	u_B (t/h)	μ_t (%)
A (30%)	90.0	13.82	40.7	27.4
F (50%)	150.1	14.31	66.3	48.2
G (56.7%)	169.9	14.58	74.6	54.7
H (73.3%)	219.8	15.33	95.4	70.4
I (80%)	240.1	15.63	103.8	76.6
E (100%)	300.1	16.09	128.4	96.7

Besides, the load demand is selected as the scheduling parameter owing to the following reasons:

1. The load is a controlled variable, and it is measurable.

2. The load is a key parameter for CCS, and the change of dynamic characteristics depends on the load change largely. Its variation naturally represents various operating points of the CCS, especially when the power plant is operating under the constant pressure operation mode.

3. The most important factor is that all coefficients of linearized models are monotonous with the monotonous change of load. Take $g_{12}(s)$ in Eq. (16.32) as an example, and denote the coefficients of the numerators and denominators as $\boldsymbol{n} = [n_2 \quad n_1]$ and $\boldsymbol{d} = [d_4 \quad d_3 \quad d_2 \quad d_1]$, respectively. We can obtain the monotonous trends of them at the different operating conditions shown in Fig. 16.10. Note that other coefficients in Eq. (16.32) all have the monotonous trends with the monotonous change of the load.

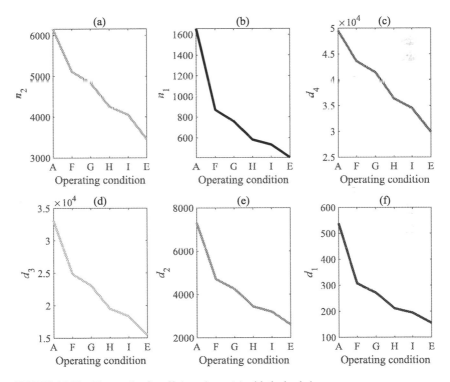

FIGURE 16.10 The trends of coefficients in $g_{12}(s)$ with the load change.

Finally, the ADRC parameters, which should be changed to response to the scheduling parameter, need to be discussed. Although the ADRC has three parameters, the observer bandwidth ω_o, the controller bandwidth k_p, and the estimated input gain b_0, the parameters involved the gain scheduling design are selected as ω_o and k_p except b_0. The reasons of this decision are described qualitatively as follows:

1. As the discussion in Subsection 16.3.2, a larger k_p means a stronger control force, and the changing k_p can reasonably adjust the dynamic characteristics of closed-loop system. Besides, ω_o can reflect the observation and compensation ability of the total disturbance, so a changing ω_o in the gain scheduling design is also necessary.
2. b_0 is the estimate of the real input gain b, and b_0 is often larger than the real value. The changing b_0 would result in a large overshoot because that b_0 is the denominator of Eq. (16.18), whose output is the control signal.

16.4.3 Gain scheduling design based on ADRC under full operating conditions

To simplify the analysis, consider a first-order system with uncertain parameters,

$$G(s) = \frac{\alpha_1(\theta)}{\lambda_1(\theta)s + 1},$$ (16.34)

where $\alpha_1(\theta)$ and $\theta_1(\theta)$ are bounded real numbers with scopes obtained according to the operating conditions of the system, and θ is the scheduling parameter as discussed in Subsection 16.4.2. The division of operating conditions can be rearranged as shown in Fig. 16.11 based on Fig. 16.9. The division of operating conditions is a continuous range as shown in Fig. 16.11, and the two adjacent operating ranges have a small overlap range, which is the scheduling range as shown in the hatched section in Fig. 16.11. For example, the range $[\theta_{i+1} \quad \theta_{i+1+\Delta}]$ is the overlap range between the operating ranges $[\theta_i \quad \theta_{i+1+\Delta}]$ and $[\theta_{i+1} \quad \theta_{i+2+\Delta}]$.

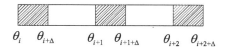

$\theta_i \quad \theta_{i+\Delta} \qquad \theta_{i+1} \quad \theta_{i+1+\Delta} \qquad \theta_{i+2} \quad \theta_{i+2+\Delta}$

FIGURE 16.11 The rearranged diagram of the scheduling ranges.

For each operating range, we can calculate the intervals of all coefficients in Eq. (16.34). The intervals for the operating range $[\theta_{i+1} \quad \theta_{i+1+\Delta}]$ can be obtained as $\alpha_1 \in [\alpha_{1\min} \quad \alpha_{1\max}]$ and $\lambda_1 \in [\lambda_{1\min} \quad \lambda_{1\max}]$, and we can obtain the function family in the operating range $[\theta_{i+1} \quad \theta_{i+1+\Delta}]$,

$$G(s) = \frac{\alpha_1}{\lambda_1 s + 1},$$ (16.35)

where $\alpha_1 \in [\alpha_{1\min} \quad \alpha_{1\max}]$ and $\lambda_1 \in [\lambda_{1\min} \quad \lambda_{1\max}]$.

Design the ADRC controller $\{k_{p\_i}, \omega_{o\_i}, b_0\}$ for the operating range $[\theta_i \quad \theta_{i+1+\Delta}]$ and the ADRC controller $\{k_{p\_i+1}, \omega_{o\_i+1}, b_0\}$ for the operating range $[\theta_{i+1} \quad \theta_{i+2+\Delta}]$. Note that these two ADRC controllers both ensure the stability for the system in the range $[\theta_{i+1} \quad \theta_{i+1+\Delta}]$ with fixed b_0. Based on the feedback controller $G_c(s)$ in Eq. (16.23) and the controlled plant in Eq. (16.35), the eigenpolynomial can be depicted as $T_i(s)$ and $T_{i+1}(s)$ with specific expressions (A.1)–(A.2) in Appendix A. Therefore we have inequalities (A.3)–(A.10) listed in Appendix A by applying the Routh stability criterion.

For the scheduling range $[\theta_{i+1} \quad \theta_{i+1+\Delta}]$, we design the ADRC controller based on the gain scheduling, and the ESO is depicted by

$$
\begin{bmatrix} \dot{z}_1 \\ \dot{z}_2 \end{bmatrix} = \begin{bmatrix} -2\left[\varphi\omega_{o\_i} + (1-\varphi)\omega_{o\_i+1}\right] & 1 \\ -\left[\varphi\omega_{o\_i} + (1-\varphi)\omega_{o\_i+1}\right]^2 & 0 \end{bmatrix} \begin{bmatrix} z_1 \\ z_2 \end{bmatrix}
$$
$$
+ \begin{bmatrix} b_0 & \left[\varphi\omega_{o\_i} + (1-\varphi)\omega_{o\_i+1}\right] \\ 0 & \left[\varphi\omega_{o\_i} + (1-\varphi)\omega_{o\_i+1}\right] \end{bmatrix} \begin{bmatrix} u \\ y \end{bmatrix}.
$$
(16.36)

The control law is depicted by

$$
u_0 = \left[\varphi k_{p\_i} + (1-\varphi)k_{p\_i+1}\right](r - y).
$$
(16.37)

Based on the TDOF control structure of ADRC, the feedback controller can be depicted as $G_{c*}(s)$ with specific expression (A.11), and we have $\varphi \in [0 \quad 1]$. Based on the feedback controller $G_{c*}(s)$ and the system in Eq. (16.35), we can obtain the eigenpolynomial $T_{i*}(s)$ depicted in Eq. (A.12) in Appendix A. By the transformation of inequalities (A.3)–(A.10) we can obtain inequality (A.13), which means that the controller in Eq. (A.11) can ensure the convergence of the closed-loop system [48].

Therefore we have the scheduling method of ADRC parameters for the scheduling range $[\theta_{i+1} \quad \theta_{i+1+\Delta}]$,

$$
k_p = \begin{cases} k_{p\_i}, & \theta \in (\theta_{i+\Delta}, \theta_{i+1}), \\ \frac{\theta_{i+1+\Delta}-\theta}{\theta_{i+1+\Delta}-\theta_{i+1}}k_{p\_i} + \frac{\theta-\theta_{i+1}}{\theta_{i+1+\Delta}-\theta_{i+1}}k_{p\_i+1}, & \theta \in [\theta_{i+1}, \theta_{i+1+\Delta}], \\ k_{p\_i+1}, & \theta \in (\theta_{i+1+\Delta}, \theta_{i+2}), \end{cases}
$$
(16.38)

$$
\omega_o = \begin{cases} \omega_{o\_i}, & \theta \in (\theta_{i+\Delta}, \theta_{i+1}), \\ \frac{\theta_{i+1+\Delta}-\theta}{\theta_{i+1+\Delta}-\theta_{i+1}}\omega_{o\_i} + \frac{\theta-\theta_{i+1}}{\theta_{i+1+\Delta}-\theta_{i+1}}\omega_{o\_i+1}, & \theta \in [\theta_{i+1}, \theta_{i+1+\Delta}], \\ \omega_{o\_i+1}, & \theta \in (\theta_{i+1+\Delta}, \theta_{i+2}). \end{cases}
$$
(16.39)

Note that the analysis about the scheduling method is based on a first-order system; when the controlled plant is a plant with time delay, the analysis difficulty increases exponentially. Fortunately, considering controlled plant in Eq. (16.32) can be approximated as a first-order plus time delay (FOPTD) system depicted by

$$
G_{eq}(s) = \frac{k}{Ts+1}e^{-Ls}.
$$
(16.40)

The equivalent plant is a lag-dominant process considering $L/(T+L)$ is about 0.15 ($\pm$ 0.05) under all operating conditions, and the influence of the time delay is weak and can be dealt with a first-order system. Therefore we can conclude that the proposed gain scheduling method works well for the system in Eq. (16.32) roughly. Besides, the effectiveness of the scheduling method for Eq. (16.32) can be verified in the next section. The proposed gain scheduling method is a simple yet effective linear switching method and can greatly reduce the implementation complexity in DCS. The scheduling method can ensure the stability of the closed-loop system when the system changes slowly [47].

16.4.4 Stability analysis based on the Kharitonov theorem

In this subsection, we discuss the qualitative stability analysis based on the Kharitonov theorem and then is carry out the quantitative calculation of the stability regions of ADRC parameters to provide the parameter regions. The stability analysis based on the Kharitonov theorem offers us a theoretical method to analyze the stabilizability of the proposed gain scheduling design based on ADRC for the interval systems.

The Kharitonov theorem offers a sufficient condition for the judgement whether the ADRC controller can stabilize the entire interval function family [47]. Take $g'_{22}(s)$ in Eq. (16.32) as an example, which can be rearranged with uncertain parameters as

$$g'_{22}(s) = \frac{n_3 s^3 + n_2 s^2 + n_1 s + n_0}{d_4 s^4 + d_3 s^3 + d_2 s^2 + d_1 s + 1},\tag{16.41}$$

where $n_i \in [n_{i\min} \quad n_{i\max}]$ ($i = 0, 1, 2, 3$) and $d_j \in \left[d_{j\min} \quad d_{j\max}\right]$ ($i = 1, 2, 3, 4$).

Based on the feedback controller $G_c(s)$ in Eq. (16.23) and the controlled plant in Eq. (16.41), we can obtain the eigenpolynomial $T(s)$ depicted as Eq. (B.1) in Appendix B.

We can obtain the corresponding four Kharitonov functions (Eqs. (B.2)–(B.5) in Appendix B, where the expressions of $A_{0\_\min}$, $A_{0\_\max}$, etc. in Kharitonov functions are listed in Appendix B.

If ADRC parameters can ensure that all Kharitonov functions in Eqs. (B.2)–(B.5) are Hurwitz, then we can say that these ADRC parameters $\{k_p, \omega_o, b_0\}$ can stabilize the entire interval function family.

Note that the application of the Kharitonov theorem to calculate the stability regions of ADRC parameters is difficult because the coefficients of Kharitonov functions are not independent and influence each other when one of them changes.

To simplify the calculation of stability region, the discussion about the stability region of ADRC provides a method to calculate the stability region that can stabilize the entire interval function family.

Consider that the throttle opening position μ_t is selected to control the power output and take $g_{12}(s)$ in Eq. (16.36) as an example to show the effectiveness of the proposed method. The operating ranges from A to G and from F to I are selected as the design range, and the scheduling range is from F to G as shown in Fig. 16.9. Based on the linearized model $g_{12}(s)$ in Eq. (16.32) at the selected operation conditions, we can obtain the stability regions of ADRC with fixed $b_0 = 1$ at the typical operation conditions (A, F, G, H, I) as shown in Fig. 16.12. We know that the typical operation conditions (A, F, G, H, I) have the similar stability regions even though operation conditions vary greatly. This verifies that the varying ADRC parameters in the scheduling ranges can locate in the stability region except that they locate near the stability boundaries where the ADRC parameters cannot ensure the satisfactory performance. Fig. 16.12 can also explain the rationality of the proposed gain scheduling design based on ADRC roughly.

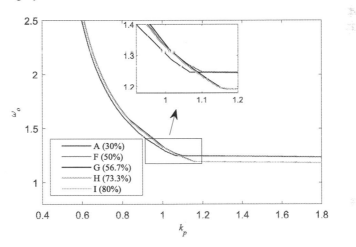

FIGURE 16.12   The stability regions of ADRC at typical operation conditions.

16.5   Simulations validations

In this section, we carry out simulations of the gain scheduling design based on ADRC with and without physical constraints of actuators such as amplitude limiting and rate limiting to verify the superiority of the proposed control strategy under full operating conditions.

Based on the proposed control strategy discussed in the previous section, the control structure of DEB combined with the gain scheduling design based on ADRC is shown in Fig. 16.13, where the blue part (mid gray part in the print version) is the new content for the proposed control strategy. The proposed control strategy does not change the original structure and can greatly reduce the implementation difficulty in DCS by adding two parameter tables.

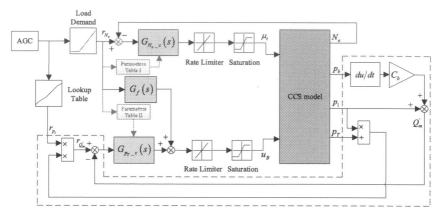

FIGURE 16.13 The DEB control structure with the gain scheduling design based on ADRC.

To design Table I of the parameters of the power output loop and Table II of the throttle pressure loop, the parameters of the regular ADRC under the operating conditions A, E, and G are tuned based on the tuning procedure shown in Fig. 16.5. Besides, the tuned parameters should be checked to ensure that all they locate in the stability region. Then the scheduling methods in Eqs. (16.38) and (16.39) are applied to the gain scheduling design based on ADRC. We can obtain Tables I and II of the parameters as shown in Fig. 16.14. Besides, the feedforward PD controller is $G_f(s) = 0.3 + 110s/(1 + 20s)$, and the fixed b_0 is equal to one for two control loops. The ADRC and the PI $(G_{PI}(s) = k_{p\_pi} + k_{i\_pi}/s)$ control strategies are comparative control strategies and are tuned by the multi-objective parameter optimization. Their parameters are listed in Table 16.3.

TABLE 16.3 Parameters of the comparative controllers. The superscripts a and b denote the controllers of PI and ADRC, respectively.

Controllers	Parameters of different controllers
$G^a_{N_{e\_c}}$	$k_{p\_pi} = 1.95, k_{i\_pi} = 0.11$
$G^a_{P_{T\_c}}$	$k_{p\_pi} = 69.1, k_{i\_pi} = 0.025, G_f(s) = 0.35 + \frac{7.84s}{52s+1}$
$G^b_{N_{e\_c}}$	$k_{p\_pi} = 1.95, \omega_o = 0.3, b_0 = 1$
$G^b_{P_{T\_c}}$	$k_{p\_pi} = 70, \omega_o = 0.002, b_0 = 1, G_f(s) = 0.3 + \frac{117s}{11.7s+1}$

To better compare the control performance with different control strategies quantitatively, the integrated absolute error (IAE) and the input variation (TV) are recorded and defined as

$$IAE_i = \int_0^\infty |r_i(t) - y_i(t)| dt, i = 1, 2, \qquad (16.42)$$

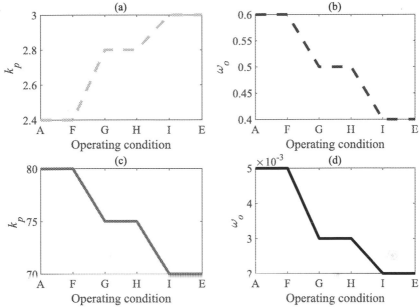

FIGURE 16.14 The parameters of the ADRC for the gain scheduling design at different operating conditions: (a), (b) the parameters of ADRC for the power output loop; (c), (d) the parameters of ADRC for the throttle pressure loop.

$$TV_i = \sum_{j=0}^{n-1} |u_i(j+1) - u_i(j)|, i = 1, 2, \tag{16.43}$$

where $i = 1, 2$ for the power output loop and throttle pressure loop, respectively.

Firstly, we set the load tracking rate as 1.5% of full load per minute. Note that the simulation contains the full operating conditions from 30% (90 MW) to 100% (300 MW), where the system can experience the constant pressure operation and sliding pressure operation modes. The simulation results are shown in Figs. 16.15–16.18.

Without physical constraints of actuators, the gain scheduling design based on ADRC ("The Proposed" in figures) has the best control performance under full operating conditions and the PI ("PI" in figures) has the largest reverse change under sliding pressure operation mode and the largest deviation under constant pressure operation mode. Besides, the control signal of the proposed control strategy in Fig. 16.16 is flatter than that of the regular ADRC ("ADRC" in figures) and the PI. Similar conclusions can be obtained considering physical constraints of actuators as shown in Figs. 16.17–16.18. To show the results clearly, the local enlarged drawings of Figs. 16.17–16.18 from 35000 s to 37000 s are shown in Figs. 16.19–16.20, and the discussions above can be verified. Moreover, all control performance indices of these control strategies are

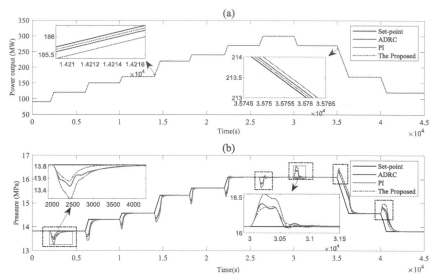

FIGURE 16.15 The output responses of 1.5% load tracking rate without physical constraints of actuators: (a) power output loop; (b) throttle pressure loop.

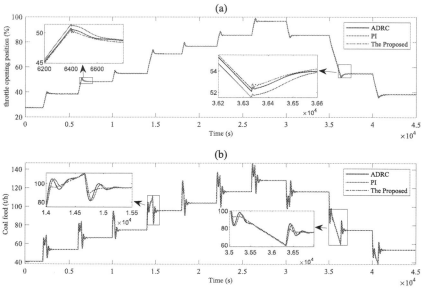

FIGURE 16.16 The control signals of 1.5% load tracking rate without physical constraints of actuators: (a) power output loop; (b) throttle pressure loop.

recorded in Table 16.4. IAEs of the proposed control strategy are about 50.6% and 82.8% of IAE of the regular ADRC in the power output loop and throttle pressure loop, respectively. Besides, IAEs of the proposed control strategy are

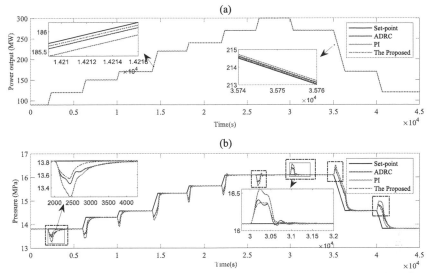

FIGURE 16.17 The output responses of 1.5% load tracking rate with physical constraints of actuators: (a) power output loop; (b) throttle pressure loop.

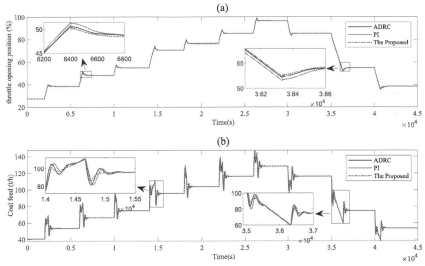

FIGURE 16.18 The control signals of 1.5% load tracking rate with physical constraints of actuators: (a) power output loop: (b) throttle pressure loop.

about 24.1% and 62.4% of IAE of the PI in the power output loop and throttle pressure loop, respectively. Therefore the gain scheduling design based on ADRC can greatly improve the control performance of power output and throttle pressure loops.

During the process of load rising (0–28000 s), the overshoots of the power output loop with the proposed control strategy are 0.33%, 0.40%, 0.50%, 0.26%, 0.60%, 0.40%, and 0.37%, respectively. This means that the overshoot is relevant to the starting point caused by the nonlinearity of the CCS, and it corresponds to the real power plants.

Note that the closed-loop system with the PI has some severe operating conditions (the reverse change and the biggest deviation is larger than ~ 0.4 MPa) as shown in black dashed boxes of Figs. 16.15(b) and 16.17(b). These may result in the irreversible damage for the main steam pipes and bring the necessary manual interventions.

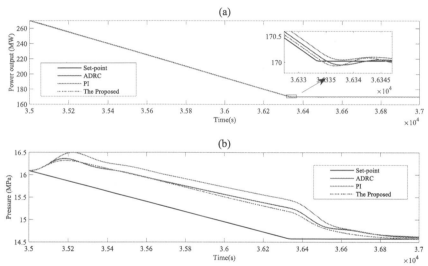

FIGURE 16.19 The local enlarged drawing of Fig. 16.17 from 35000 s to 37000 s: (a) power output loop; (b) throttle pressure loop.

TABLE 16.4 The control performance indices of different control strategies under 1.5% load tracking rate. The superscripts c and d denote the controllers of PI and ADRC, respectively.

Controllers	IAE_1	IAE_2	TV_1	TV_2
PI^c	1669.8	4379.8	192.9	998.7
$ADRC^c$	794.5	3301.5	180.1	1093.7
The Proposed^c	402.0	2734.9	186.1	557.6
PI^d	1674.3	4395.3	193.4	1034.1
$ADRC^d$	794.5	3301.5	180.1	1093.7
The Proposed^d	402.6	2737.2	186.7	681.0

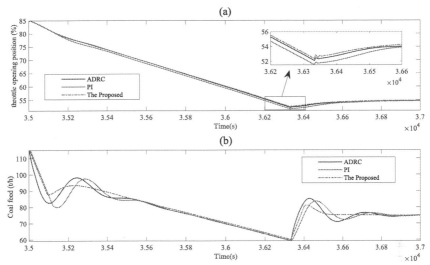

FIGURE 16.20 The local enlarged drawing of Fig. 16.18 from 35000 s to 37000 s: (a): power output loop (b): throttle pressure loop.

To integrate more renewable energy into the power grid, the thermal power plant has to accelerate the power output responding the AGC command, and the upper limit of the load tracking rate should be taken into consideration.

The simulation with the load tracking rate, 2% of full load per minute, is also carried out. All control performance indices of these control strategies with 2% load tracking rate are recorded in Table 16.5. Note that the PI cannot ensure the convergence and stability of two loops when physical constraints of actuators are considered as presented in Table 16.5. IAEs of the proposed control strategy are no more than 52.5% and 86.5% of IAE of the regular ADRC in the power output loop and throttle pressure loop, respectively. Moreover, the closed-loop system with regular ADRC also has some severe operating conditions (the reverse change and the biggest deviation larger than ±0.4 MPa), where the regular ADRC can result in irreversible damage for main steam pipes and bring the necessary manual interventions. However, the proposed control strategy still has no severe operating conditions, which means that the proposed control strategy still has approving control performance when the system has larger load tracking rate under full operating conditions.

Generally, the proposed control strategy can improve the control quality significantly, whereas the PI is not competent: the task for the large load tracking rate and the regular ADRC has poor control quality.

Now the concern goes to the issue of coal quality variation. Since k_c in Eq. (16.2) is the index of the coal quality, to imitate the step and periodic disturbance of coal quality variation, the test can be done by increasing k_c by 20% at 500 s and oscillating k_c at 5000 s with the period of 628 s under the typical operating condition D (90%) in Table 16.1. Simulation results are shown in

TABLE 16.5 The control performance indices of different control strategies under 2% load tracking rate. The superscripts c and d denote the controllers of PI and ADRC, respectively.

Controllers	IAE_1	IAE_2	TV_1	TV_2
PI^c	19515.2	44500.0	211.8	1214.9
$ADRC^c$	8366.0	33233.0	194.4	1334.5
The Proposed^c	4209.3	27627.1	203.2	693.2
PI^d	3.1902×10^6	7.9742×10^5	1758.0	5183.9
$ADRC^d$	8366.0	33233.5	194.4	1334.6
The Proposed^d	4389.4	28751.3	220.3	1084.5

Figs. 16.21–16.22. Note that the test is carried out with physical constraints of actuators in Eq. (16.8).

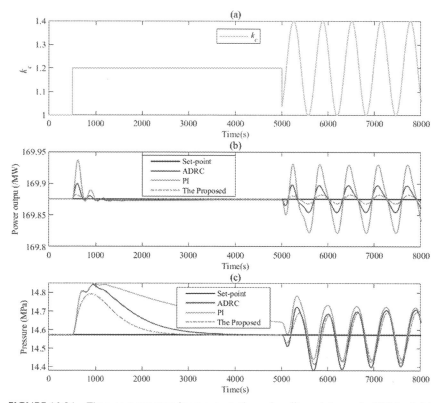

FIGURE 16.21 The output responses in response to the coal quality variation under 90% load: (a) the disturbance of coal quality variation; (b) power output loop; (c) throttle pressure loop.

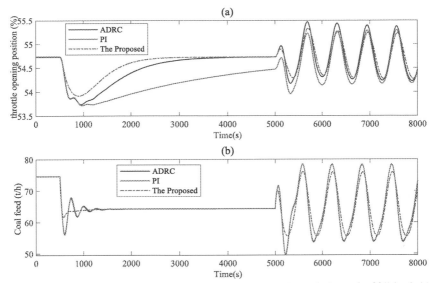

FIGURE 16.22 The control signals in response to the coal quality variation under 90% load: (a) power output loop; (b) throttle pressure loop.

The proposed control strategy can obtain best disturbance rejection no matter which type of coal quality variation as shown in Figs. 16.21–16.22. Moreover, the disturbance rejection indices of these control strategies are recorded in Table 16.6, and the indices verify the superiority of the proposed control strategy in disturbance rejection. The disturbance rejection indices under the operating condition G (56.7%) are also recorded in Table 16.6, and we can learn that the proposed control strategy still works best under other operating conditions.

TABLE 16.6 The indices with the coal quality variation under the operating conditions D and G. The superscripts e and f denote the operating conditions of 90% and 56.7%, respectively.

Controllers	IAE_1	IAE_2	TV_1	TV_2
PI^e	2934.8	13281.4	33.2	614.0
$ADRC^e$	1026.5	8073.9	30.1	533.8
The Proposed^e	418.2	7382.9	25.2	377.1
PI^f	1096.3	11125.2	12.3	294.1
$ADRC^f$	430.6	6096.8	12.6	291.0
The Proposed^f	147.1	4375.0	10.3	213.8

Based on simulation results of the tracking performance with different load tracking rates and disturbance performance with the coal quality variation, the proposed control strategy can obtain the best control performance compared to the regular ADRC and PI, which are tuned by the multi-objective parameter optimization. The successful comparison indicates a promising future of the gain scheduling design based on ADRC for thermal power plants with the increasing demand on integrating more renewable energy into the power grid.

16.6 Conclusions

The renewable energy, which has strong intermittency and randomness, plays an increasingly significant role in energy supply. To integrate more renewable energy into the grid, the thermal power plant has to accelerate the speed of power output responding the AGC command and enlarge its operating range. This puts forward great challenges on the safe operation and the control of thermal power plant. To this end, the gain scheduling design based on ADRC is proposed for the thermal power plant under full operating conditions from 30% to 100% of full load considering the constant and sliding pressure operation modes. The main work can be summarized as follows: 1) The necessity and importance of the proposed control strategies is discussed by analyzing the control difficulties of the CCS. 2) The gain scheduling design based on ADRC under full operating conditions about the scheduling parameter selection and division of operating conditions is discussed. 3) A linear switching method of the observer bandwidth ω_o and controller bandwidth k_p is derived, the stability analysis based on the Kharitonov theorem is carried out qualitatively, and the stability regions with different operating conditions are discussed quantitatively. 4) Simulations of power tracking with different load tracking rates and disturbance rejection with the coal quality variation are carried out to verify the superiority of the proposed control strategy under full operating conditions compared to the regular ADRC and PI. The theoretical analysis and successful comparison indicate a promising potential for application to thermal power plant.

Appendix A

$$T_i\,(s) = b_0\lambda_1 s^3 + \left(k_{p\_i}\alpha_1 + b_0 + b_0\omega_{o\_i}\lambda_1\right)s^2$$
$$+ \left(b_0\omega_{o\_i} + \omega_{o\_i}^2\alpha_1 + 2k_{p\_i}\omega_{o\_i}\alpha_1\right)s + k_{p\_i}\omega_{o\_i}^2\alpha_1, \tag{A.1}$$

$$T_{i+1}\,(s) = b_0\lambda_1 s^3 + \left(k_{p\_i+1}\alpha_1 + b_0 + b_0\omega_{o\_i+1}\lambda_1\right)s^2$$
$$+ \left(b_0\omega_{o\_i+1} + \omega_{o\_i+1}^2\alpha_1 + 2k_{p\_i+1}\omega_{o\_i+1}\alpha_1\right)s + k_{p\_i+1}\omega_{o\_i+1}^2\alpha_1, \tag{A.2}$$

$$b_0\lambda_1 > 0, \tag{A.3}$$

$$k_{p\_i}\alpha_1 + b_0 + b_0\omega_{o\_i}\lambda_1 > 0, \tag{A.4}$$

$$b_0\omega_{o\_i} + \omega_{o\_i}^2\alpha_1 + 2k_{p\_i}\omega_{o\_i}\alpha_1 > 0, \tag{A.5}$$

$$k_{p\_i}\omega_{o\_i}^2\alpha_1 > 0, \tag{A.6}$$

$$\left(k_{p\_i}\alpha_1+b_0+b_0\omega_{o\_i}\lambda_1\right)k_{p\_i}\omega_{o\_i}^2\alpha_1 > \left(b_0\omega_{o\_i}+\omega_{o\_i}^2\alpha_1+2k_{p\_i}\omega_{o\_i}\alpha_1\right)b_0\lambda_1, \tag{A.7}$$

$$k_{p\_i+1}\alpha_1 + b_0 + b_0\omega_{o\_i+1}\lambda_1 > 0, \tag{A.8}$$

$$b_0\omega_{o\_i+1} + \omega_{o\_i+1}^2\alpha_1 + 2k_{p\_i+1}\omega_{o\_i+1}\alpha_1 > 0, \tag{A.9}$$

$$k_{p\_i+1}\omega_{o\_i+1}^2\alpha_1 > 0, \tag{A.10}$$

$$\left(k_{p\_i+1}\alpha_1 + b_0 + b_0\omega_{o\_i+1}\lambda_1\right)k_{p\_i+1}\omega_{o\_i+1}^2\alpha_1$$
$$> \left(b_0\omega_{o\_i+1} + \omega_{o\_i+1}^2\alpha_1 + 2k_{p\_i+1}\omega_{o\_i+1}\alpha_1\right)b_0\lambda_1, \tag{A.11}$$

$$G_{c*}(s) = \frac{\left[\varphi k_{p\_i} + (1-\varphi)k_{p\_i+1}\right]s^2}{\left[s+2\varphi\omega_{o\_i}+2(1-\varphi)\omega_{o\_i+1}\right]b_0 s}$$
$$+ \frac{\left\{\left[\varphi\omega_{o\_i}+(1-\varphi)\omega_{o\_i+1}\right]^2+2\left[\varphi k_{p\_i}+(1-\varphi)k_{p\_i+1}\right]\left[\varphi\omega_{o\_i}+(1-\varphi)\omega_{o\_i+1}\right]\right\}s}{\left[s+2\varphi\omega_{o\_i}+2(1-\varphi)\omega_{o\_i+1}\right]b_0 s}$$
$$+ \frac{\left[\varphi k_{p\_i}+(1-\varphi)k_{p\_i+1}\right]\left[\varphi\omega_{o\_i}+(1-\varphi)\omega_{o\_i+1}\right]^2}{\left[s+2\varphi\omega_{o\_i}+2(1-\varphi)\omega_{o\_i+1}\right]b_0 s}, \tag{A.12}$$

$$T_{i*}(s) = b_0\lambda_1 s^3$$
$$+ \left\{\left[\varphi k_{p\_i}+(1-\varphi)k_{p\_i+1}\right]\alpha_1+b_0+b_0\left[\varphi\omega_{o\_i}+(1-\varphi)\omega_{o\_i+1}\right]\lambda_1\right\}s^2$$
$$+ \left\{b_0\left[\varphi\omega_{o\_i}+(1-\varphi)\omega_{o\_i+1}\right]+\left[\varphi\omega_{o\_i}+(1-\varphi)\omega_{o\_i+1}\right]^2\alpha_1\right.$$
$$\left.+2\left[\varphi k_{p\_i}+(1-\varphi)k_{p\_i+1}\right]\left[\varphi\omega_{o\_i}+(1-\varphi)\omega_{o\_i+1}\right]\alpha_1\right\}s$$
$$+ \left[\varphi k_{p\_i}+(1-\varphi)k_{p\_i+1}\right]\left[\varphi\omega_{o\_i}+(1-\varphi)\omega_{o\_i+1}\right]^2\alpha_1, \tag{A.13}$$

$$\left\{\left[\varphi k_{p\_i}+(1-\varphi)k_{p\_i+1}\right]\alpha_1+b_0+b_0\left[\varphi\omega_{o\_i}+(1-\varphi)\omega_{o\_i+1}\right]\lambda_1\right\}$$
$$\times \left[\varphi k_{p\_i}+(1-\varphi)k_{p\_i+1}\right]\left[\varphi\omega_{o\_i}+(1-\varphi)\omega_{o\_i+1}\right]^2\alpha_1$$
$$> \left\{b_0\left[\varphi\omega_{o\_i}+(1-\varphi)\omega_{o\_i+1}\right]+\left[\varphi\omega_{o\_i}+(1-\varphi)\omega_{o\_i+1}\right]^2\alpha_1\right.$$
$$\left.+2\left[\varphi k_{p\_i}+(1-\varphi)k_{p\_i+1}\right]\left[\varphi\omega_{o\_i}+(1-\varphi)\omega_{o\_i+1}\right]\alpha_1\right\}b_0\lambda_1. \tag{A.14}$$

Appendix B

$$T(s) = k_p\omega_o^2 n_0 + \left(k_p\omega_o^2 n_1 + 2k_p\omega_o n_0 + 2\omega_o b_0 + \omega_o^2 n_0\right)s$$
$$+ \left(b_0 + 2\omega_o b_0 d_1 + \omega_o^2 n_1 + 2k_p\omega_o n_1 + k_p\omega_o^2 n_2 + k_p n_0\right)s^2$$
$$+ \left(b_o d_1 + 2\omega_o b_o d_2 + k_p n_1 + \omega_o^2 n_2 + 2k_p\omega_o n_2 + k_p\omega_o^2 n_3\right)s^3 \quad \text{(B.1)}$$
$$+ \left(b_o d_2 + 2\omega_o b_o d_3 + k_p n_2 + \omega_o^2 n_3 + 2k_p\omega_o n_3\right)s^4$$
$$+ \left(b_o d_3 + 2\omega_o b_o d_4 + k_p n_3\right)s^5 + b_o d_4 s^6,$$

$$T_1(s) = A_{0\_min} + A_{1\_min}s + A_{2\_max}s^2 + A_{3\_max}s^3 + A_{4\_min}s^4$$
$$+ A_{5\_min}s^5 + A_{6\_max}s^6, \tag{B.2}$$

$$T_2(s) = A_{0\_max} + A_{1\_min}s + A_{2\_min}s^2 + A_{3\_max}s^3 + A_{4\_max}s^4$$
$$+ A_{5\_min}s^5 + A_{6\_min}s^6, \tag{B.3}$$

$$T_3(s) = A_{0\_max} + A_{1\_max}s + A_{2\_min}s^2 + A_{3\_min}s^3 + A_{4\_max}s^4$$
$$+ A_{5\_max}s^5 + A_{6\_min}s^6, \tag{B.4}$$

$$T_4(s) = A_{0\_min} + A_{1\_max}s + A_{2\_max}s^2 + A_{3\_min}s^3 + A_{4\_min}s^4$$
$$+ A_{5\_max}s^5 + A_{6\_max}s^6. \tag{B.5}$$

The expressions of A_{0_min}, A_{0_max}, etc.:

$$A_{0\_min} = k_p \omega_o^2 n_{0\min},$$

$$A_{0\_max} = k_p \omega_o^2 n_{0\max},$$

$$A_{1\_min} = k_p \omega_o^2 n_{1\min} + 2k_p \omega_o n_{0\min} + 2\omega_o b_o + \omega_o^2 n_{0\min},$$

$$A_{1\_max} = k_p \omega_o^2 n_{1\max} + 2k_p \omega_o n_{0\max} + 2\omega_o b_o + \omega_o^2 n_{0\max},$$

$$A_{2\_min} = b_0 + 2\omega_o b_o d_{1\min} + \omega_o^2 n_{1\min} + 2k_p \omega_o n_{1\min} + k_p \omega_o^2 n_{2\min} + k_p n_{0\min},$$

$$A_{2\_max} = b_0 + 2\omega_o b_o d_{1\max} + \omega_o^2 n_{1\max} + 2k_p \omega_o n_{1\max} + k_p \omega_o^2 n_{2\max} + k_p n_{0\max},$$

$$A_{3\_min} = b_o d_{1\min} + 2\omega_o b_o d_{2\min} + k_p n_{1\min} + \omega_o^2 n_{2\min} + 2k_p \omega_o n_{2\min}$$
$$+ k_p \omega_o^2 n_{3\min},$$

$$A_{3\_max} = b_o d_{1\max} + 2\omega_o b_o d_{2\max} + k_p n_{1\max} + \omega_o^2 n_{2\max} + 2k_p \omega_o n_{2\max}$$
$$+ k_p \omega_o^2 n_{3\max},$$

$$A_{4\_min} = b_o d_{2\min} + 2\omega_o b_o d_{3\min} + k_p n_{2\min} + \omega_o^2 n_{3\min} + 2k_p \omega_o n_{3\min},$$

$$A_{4\_max} = b_o d_{2\max} + 2\omega_o b_o d_{3\max} + k_p n_{2\max} + \omega_o^2 n_{3\max} + 2k_p \omega_o n_{3\max},$$

$$A_{5\_min} = b_o d_{3\min} + 2\omega_o b_o d_{4\min} + k_p n_{3\min},$$

$$A_{5\_max} = b_o d_{3\max} + 2\omega_o b_o d_{4\max} + k_p n_{3\max},$$

$$A_{6\_min} = b_o d_{4\min},$$

$$A_{6\_max} = b_o d_{4\max}.$$

References

[1] N. Alamoodi, P. Daoutidis, Nonlinear decoupling control with deadtime compensation for multirange operation of steam power plants, IEEE Transactions on Control Systems Technology 24 (1) (2016) 341–348.

[2] Z. Wu, T. He, Y. Liu, D. Li, Y.Q. Chen, Physics-informed energy-balanced modeling and active disturbance rejection control for circulating fluidized bed units, Control Engineering Practice 116 (2021) 104934.

[3] A. Benato, S. Bracco, A. Stoppato, A. Mirandola, LTE: a procedure to predict power plants dynamic behaviour and components lifetime reduction during transient operation, Applied Energy 162 (2016) 880–891.

[4] L. Sun, D. Li, K.Y. Lee, Optimal disturbance rejection for PI controller with constraints on relative delay margin, ISA Transactions 63 (2016) 103–111.

[5] X. Wu, J. Shen, Y. Li, K.Y. Lee, Steam power plant configuration, design, and control, Wiley Interdisciplinary Reviews: Energy and Environment 4 (6) (2015) 537–563.

[6] M. Hübel, S. Meinke, M. Andrén, C. Wedding, J. Nocke, C. Gierow, E. Hassel, J. Funkquistc, Modelling and simulation of a coal-fired power plant for start-up optimization, Applied Energy 208 (2017) 319–331.

[7] D. Long, W. Wang, C. Yao, J. Liu, An experiment-based model of condensate throttling and its utilization in load control of 1000 MW power units, Energy 133 (2017) 941–954.

[8] K. Astrom, R. Bell, Drum-boiler dynamics, Automatica 36 (3) (2000) 363–378.

[9] J. Liu, S. Yan, D. Zeng, Y. Hu, L. You, A dynamic model used for controller design of a coal fired once-through boiler-turbine unit, Energy 93 (3) (2015) 2069–2078.

[10] L. Sun, D. Li, K.Y. Lee, Y. Xue, Control-oriented modeling and analysis of direct energy balance in coal-fired boiler-turbine unit, Control Engineering Practice 55 (3) (2016) 38–55.

[11] M. Lawrynczuk, Nonlinear predictive control of a boiler-turbine unit: a state-space approach with successive on-line model linearisation and quadratic optimisation, ISA Transactions 67 (3) (2017) 476–495.

[12] X. Liu, J. Cui, Economic model predictive control of boiler-turbine system, Journal of Process Control 66 (2018) 59–67.

[13] H. Moradi, A. Alasty, G. Vossoughi, Nonlinear dynamics and control of bifurcation to regulate the performance of a boiler–turbine unit, Energy Conversion and Management 68 (3) (2013) 105–113.

[14] H. Zhou, C. Chen, J. Lai, X. Lu, Q. Deng, X. Gao, Z. Lei, Affine nonlinear control for an ultra-supercritical coal fired once-through boiler-turbine unit, Energy 153 (3) (2018) 638–649.

[15] T. Piraisoodi, W. Iruthayarajan, A. Kadhar, An optimized nonlinear controller design for boiler–turbine system using evolutionary algorithms, IETE Journal of Research 64 (4) (2018) 451–462.

[16] M. Ataei, R. Hooshmand, S. Samani, A coordinated MIMO control design for a power plant using improved sliding mode controller, ISA Transactions 53 (2) (2014) 415–422.

[17] S. Ghabraei, H. Moradi, G. Vossoughi, Multivariable robust adaptive sliding mode control of an industrial boiler–turbine in the presence of modeling imprecisions and external disturbances: a comparison with type-I servo controller, ISA Transactions 58 (2014) 398–408.

[18] V. Iannino, V. Colla, M. Innocenti, A. Signorini, Design of a H_∞ robust controller with μ-analysis for steam turbine power generation applications, Energies 10 (7) (2017) 1026.

[19] U. Moon, Y. Lee, K.Y. Lee, Practical dynamic matrix control for thermal power plant coordinated control, Control Engineering Practice 71 (2018) 154–163.

[20] L. Yu, J. Lim, S. Fei, An improved single neuron self-adaptive PID control scheme of superheated steam temperature control system, International Journal of System Control and Information Processing 2 (1) (2017) 1–13.

[21] L. Ma, Z. Wang, K.Y. Lee, Neural network inverse control for the coordinated system of a 600 MW supercritical boiler unit, IFAC Proceedings Volumes 47 (3) (2014) 999–1004.

[22] W. Wang, J. Liu, D. Zeng, Y. Niu, C. Cui, An improved coordinated control strategy for boiler-turbine units supplemented by cold source flow adjustment, Energy 88 (2015) 927–934.

[23] J. Han, From PID to active disturbance rejection control, IEEE Transactions on Industrial Electronics 56 (3) (2009) 900–906.

[24] R. Madoński, P. Herman, Survey on methods of increasing the efficiency of extended state disturbance observers, ISA Transactions 56 (2015) 18–27.

[25] Y. Zhang, D. Li, Z. Gao, Q. Zheng, On oscillation reduction in feedback control for processes with an uncertain dead time and internal–external disturbances, ISA Transactions 59 (2015) 29–38.

[26] Y. Huang, W. Xue, Active disturbance rejection control: methodology and theoretical analysis, ISA Transactions 53 (4) (2014) 963–976.

[27] H. Feng, B. Guo, Active disturbance rejection control: old and new results, Annual Reviews in Control 44 (2017) 238–248.

[28] Y. Yuan, Z. Wang, Y. Yu, L. Guo, H. Yang, Active disturbance rejection control for a pneumatic motion platform subject to actuator saturation: an extended state observer approach, Automatica 107 (2019) 353–361.

[29] K. Song, T. Hao, H. Xie, Disturbance rejection control of air–fuel ratio with transport-delay in engines, Control Engineering Practice 79 (2018) 36–49.

[30] Q. Zheng, Z. Ping, S. Soares, Y. Hu, Z. Gao, An optimized active disturbance rejection approach to fan control in server, Control Engineering Practice 79 (2018) 154–169.

[31] R. Shi, T. He, J. Peng, Y. Zhang, W. Zhuge, System design and control for waste heat recovery of automotive engines based on organic Rankine cycle, Energy 102 (2016) 276–286.

[32] Y. Xia, B. Liu, M. Fu, Active disturbance rejection control for power plant with a single loop, Asian Journal of Control 14 (1) (2012) 239–250.

[33] L. Sun, Q. Hua, D. Li, L. Pan, Y. Xue, K.Y. Lee, Direct energy balance based active disturbance rejection control for coal-fired power plant, ISA Transactions 70 (2017) 486–493.

[34] W. Rugh, J. Shamma, Research on gain scheduling, Automatica 36 (10) (2020) 1401–1425.

[35] A. Dounis, P. Kofinas, C. Alafodimos, D. Tseles, Adaptive fuzzy gain scheduling PID controller for maximum power point tracking of photovoltaic system, Renewable Energy 60 (2013) 202–214.

[36] D. Huang, J. Xu, V. Venkataramanan, T. Huynhet, High-performance tracking of piezoelectric positioning stage using current-cycle iterative learning control with gain scheduling, IEEE Transactions on Industrial Electronics 61 (2) (2013) 1085–1098.

[37] Y. Yang, Y. Yan, Attitude regulation for unmanned quadrotors using adaptive fuzzy gain-scheduling sliding mode control, Aerospace Science and Technology 54 (2016) 208–217.

[38] A. Gallego, G. Merello, M. Berenguel, E. Camacho, Gain-scheduling model predictive control of a Fresnel collector field, Control Engineering Practice 82 (2019) 1–13.

[39] X. Wu, J. Shen, Y. Li, K.Y. Lee, Fuzzy modeling and stable model predictive tracking control of large-scale power plants, Journal of Process Control 24 (10) (2014) 1609–1626.

[40] W. Tan, H. Marquez, T. Chen, J. Liu, Analysis and control of a nonlinear boiler-turbine unit, Journal of Process Control 15 (8) (2005) 883–891.

[41] C. Zhao, D. Li, Control design for the SISO system with the unknown order and the unknown relative degree, ISA Transactions 53 (4) (2014) 858–872.

[42] Z. Wu, T. He, D. Li, Y. Xue, L. Sun, L.M. Sun, Superheated steam temperature control based on modified active disturbance rejection control, Control Engineering Practice 83 (2019) 83–97.

[43] M. Li, D. Li, J. Wang, C. Zhao, Active disturbance rejection control for fractional-order system, ISA Transactions 52 (3) (2013) 365–374.

[44] Z. Gao, Scaling and bandwidth-parameterization based controller tuning, in: Proceedings of the American Control Conference (ACC), Denver, CO, USA, June 4–6, vol. 6, 2006, pp. 4989–4996.

[45] W. Xue, Y. Huang, Performance analysis of active disturbance rejection tracking control for a class of uncertain LTI systems, ISA Transactions 58 (2015) 133–154.

[46] Z. Wu, T. He, D. Li, Y. Xue, The calculation of stability and robustness regions for active disturbance rejection controller and its engineering application, Control Theory & Applications 35 (11) (2018) 1635–1647.

[47] M. Pakmehr, N. Fitzgerald, E. Feron, J. Shamma, A. Behbahani, Gain scheduled control of gas turbine engines: stability and verification, Journal of Engineering for Gas Turbines and Power 136 (3) (2014) 031201.

[48] H. Chapellat, S. Bhattacharyya, A generalization of Kharitonov's theorem; robust stability of interval plants, IEEE Transactions on Automatic Control 34 (3) (1989) 306–311.

Chapter 17

Active disturbance rejection control of large-scale coal fired plant process for flexible operation

He Ting[a], Zhenlong Wu[b], Donghai Li[c], and Wang Jihong[d]

[a]*Jinan University, International Energy College, Zhuhai, China,* [b]*School of Electrical Engineering, Zhengzhou University, Zhengzhou, China,* [c]*State Key Lab of Power Systems, Department of Energy and Power Engineering, Tsinghua University, Beijing, China,* [d]*University of Warwick, School of Engineering, Coventry, United Kingdom*

17.1 Introduction

Under the rising awareness of global warming and energy crisis, the world has been in a new era of energy revolution. In the meantime, China promises to reach carbon dioxide peak by 2030 and strives to reach carbon neutrality by 2060 [1]. This "double carbon" action plan requires increase share of clean and low carbon energy in energy consumption. However, the growing integration of intermittent and volatile renewable energy, such as solar and wind, has bring risk to the grid safety. In September 2021, abrupt decline of the wind power exacerbated the electricity supply gap, and China Liaoning Province experienced several emergent power cuts to prevent further grid collapse. Since these incidents, the Chinese government has refocused on the fundamental supporting role of coal-fired power plants for the grid safety.

In China, coal-fired power plants still provide 60.75% of total power in 2020 [2]. As the basic power supply, the coal-fired power plants are required to operate in a wider load range to integrate more renewable power into the grid. This poses operation challenges on the plant flexibility, which means that the power plants have to run within the load range from 100% to 50%, even 20%, and varying load more frequently according to the ACG commands. Besides, the increasing demand for safe and economic operation has also put an emphasize on advanced and intelligent control technology.

Proportional-Integral-Derivative (PID) control has been playing a dominant role in industrial processes for more than half a century. According to a survey [3], in Guangdong Province, China, about 98% feedback control loops of

Modeling, Identification, and Control for Cyber-Physical Systems Towards Industry 4.0
https://doi.org/10.1016/B978-0-32-395207-1.00030-5

the coal-fired power unit apply PID controllers. However, PID controllers are found inadequate in dealing with uncertainties and disturbances under frequent and wide-range load varying [4], especially when certain control loops are more easily influenced by the load change, such as the coordinate control system, steam temperature, and combustion system. To enhance the control performance under flexible operation, many advanced control methods, such as model predictive control, adaptive control, disturbance observer-based control, have been configured in the power units [5]. Even so, requirements on a precise process model and high computation power have made these advanced control methods difficult to implement and vulnerable to fluctuations.

In the recent decades an emerging control technology named as active disturbance rejection control (ADRC) has demonstrated its strong robustness and anti-disturbance ability in field practices [6,7]. ADRC inherits the error-based feedback control from PID and the state-space-based observer design from modern control theory [8]. The core concept of ADRC is to estimate and offset the total disturbance via the extended state observer (ESO). Hence ADRC is able to achieve high disturbance rejection performance while keeps simplicity in structures.

ADRC has been extensively investigated and applied by many researchers in different countries and industrial sectors. For example, the theoretical analysis of ADRC with regard to the proof of convergence and stability has been performed [9,10]. Performance and properties of ADRC have been analyzed in frequency domain [11]. Improved ADRC has been proposed to address the challenges caused by large time delays [12], non-minimum phase [13], and multi-variable coupling [14,15]. ADRC were initially studied via simulations and experiments and then extended to industrial applications with the range from motion control [16] and electronic and mechanical systems [17] to process control [18].

With regard to parameter tuning of ADRC, quantitative tuning rules are essential for improving field applications. Most of the tuning work is performed manually, although the process of trial and error is tedious and, in some cases, frustrating. There are successful attempts using heuristic algorithms to optimize parameters, but the tuning process is time consuming and not convenient enough for industrial sectors to adopt. An important progress was made by Gao [19] in 2006. The bandwidth parameterization has greatly simplified the tuning process by reducing six tuning parameters to three. The stable region of second-order ADRC parameters is graphically presented, and the tuning process is refined in the sight of closed-loop desired dynamics [20]. However, those studies on parameter tuning give little attention to high-order processes with no consideration to the sensitivity constraint. In fact, many industrial processes are of high order, such as the superheated steam temperature, main steam pressure, and combustion system including the air-smoke system. Distributed parameter systems, which are common in power plant system, are inherently high-order. In addition, delay-dominated processes can be approximated to high-order processes. Therefore the main motivation of this study is to propose an effective and straightforward ADRC parameter tuning method that can satisfy the designed robustness level.

The chapter is organized as follows. Section 17.2 formulates the problem. The design process of the quantitative ADRC tuning method is developed in Section 17.3. Section 17.4 initially demonstrates the feasibility of the proposed method via water tank experiment and power plant simulator. After that, the ADRC control strategy and the proposed tuning method are implemented in an actual power plant in Section 17.5. Conclusions are drawn in Section 17.6.

17.2 Problem formulation

For derivation of tuning rules, the process is assumed to be in the form of high-order, $K/(Ts+1)^n$. The model parameters K, T, and n will be incorporated in the derivation of the tuning rule.

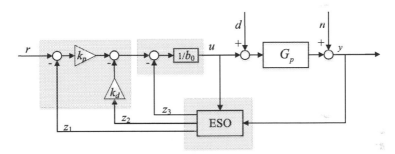

FIGURE 17.1 The schematic diagram of second-order linear ADRC.

The second-order ADRC is used as an example to show how the low-order ADRC controls a high-order process. The tuning rules of first-order ADRC can be easily derived based the method introduced. The schematic diagram is shown in Fig. 17.1, from which the second-order ADRC is formulated by the state feedback control, the extended state observer (ESO), and the real time disturbance compensation.

For the second-order ADRC, the process can be formulated into the canonical form of two cascaded integrators with the external disturbance d, noise w, and nonlinear high-order dynamics lumped in the total disturbance f:

$$\begin{cases} \dot{x}_1 = x_2, \\ \dot{x}_2 = f\left(x_1, x_2, \ldots, x_1^{(n)}, d, w\right) + b_0 u, \\ \dot{x}_3 = \dot{f}, \\ y = x_1, \dot{y} = x_2, f = x_3. \end{cases} \tag{17.1}$$

The ESO is designed based on the above canonical form, so the mathematical expression of the ESO is presented in (17.2), where β_1, β_2, and β_3 are the observer gains, and b_0 is the input gain.

$$\begin{cases} \dot{z}_1 = z_2 + \beta_1 (y - z_1) \\ \dot{z}_2 = z_3 + \beta_2 (y - z_1) + b_0 u \\ \dot{z}_3 = \beta_3 (y - z_1) \end{cases} \tag{17.2}$$

The ESO states z_1 and z_2 are used for feedback control, and k_p and k_d are the feedback control parameters:

$$u_0 = k_p (r - z_1) - k_d z_2. \tag{17.3}$$

The state z_3 is the estimated total disturbance, and it is compensated in real time by

$$u = (u_0 - z_3) / b_0. \tag{17.4}$$

The second-order ADRC has six parameters k_p, k_d, b_0, β_1, β_2, and β_3. The bandwidth-parameterization [19] has greatly simplified the tuning process. It makes k_p and k_d as functions of the desired closed-loop bandwidth ω_c, and β_1, β_2, and β_3 as functions of ESO bandwidth ω_o:

$$k_p = \omega_c^2, k_d = 2\omega_c, \beta_1 = 3\omega_o, \beta_2 = 3\omega_o^2, \beta_3 = \omega_o^3 \tag{17.5}$$

It leaves three parameters ω_c, ω_o, and b_0 to tune. However, it is still not easy to find three proper values for ω_c, ω_o, and b_0. Therefore there is a demand for developing an efficient ADRC tuning rule that can reduce the workload of manual tuning.

17.3 Design procedure

17.3.1 Sensitivity constraint

In process control design, the models used for controller design are often imprecise, and the process parameters and dynamics change with time and also with operating conditions. Therefore it is desired that the control system should be insensitive to the process variations and disturbances. The maximum sensitivity M_s and maximum complementary sensitivity M_t are typical measures of sensitivity to process variation [21]. This paper uses the maximum sensitivity M_s constraint to develop a parameter tuning method for ADRC controlled high-order processes. The M_s is defined as

$$M_s = \max_{\omega} |1/ (1 + G_l(i\omega))|, \quad \omega \in (-\infty, +\infty), \tag{17.6}$$

where $G_l(i\omega)$ is the frequency characteristic of the open-loop transfer function.

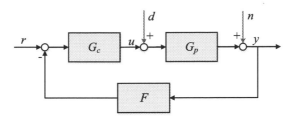

FIGURE 17.2   2-DOF configuration of ADRC.

The open-loop transfer function $G_l(s)$ can be deducted by looking at the two-degree-of-freedom (2-DOF) structure of the second-order ADRC control system.

In Fig. 17.2 the transfer functions of three blocks are

$$G_p = \frac{K}{(Ts+1)^n}, \quad G_c = \frac{k_p\left(s^3+\beta_1 s^2+\beta_2 s+\beta_3\right)}{b_0\left[s^3+(\beta_1+k_d)s^2+\left(\beta_2+k_d\beta_1+k_p\right)s\right]},$$

$$F = \frac{\left(k_p\beta_1+k_d\beta_2+\beta_3\right)s^2+\left(k_p\beta_2+k_d\beta_3\right)s+k_p\beta_3}{k_p\left(s^3+\beta_1 s^2+\beta_2 s+\beta_3\right)}. \tag{17.7}$$

Then the open-loop transfer function $G_l(s)$ and closed-loop transfer function $G_{cl}(s)$ can be obtained:

$$G_l(s) = G_p G_c F = \frac{\left(k_p\beta_1+k_d\beta_2+\beta_3\right)s^2+\left(k_p\beta_2+k_d\beta_3\right)s+k_p\beta_3}{b_0\left[s^3+(\beta_1+k_d)s^2+\left(\beta_2+k_d\beta_1+k_p\right)s\right]}\frac{K}{(Ts+1)^n},$$

$$\tag{17.8}$$

$$G_{cl}(s) = \frac{G_p G_c}{1+G_p G_c F} = \frac{k_p C(s) K}{b_0 A(s)(Ts+1)^n + K B(s)},$$

$$A(s) = s^3 + (\beta_1+k_d)s^2 + \left(\beta_2+k_d\beta_1+k_p\right)s, \tag{17.9}$$

$$B(s) = \left(k_p\beta_1+k_d\beta_2+\beta_3\right)s^2 + \left(k_p\beta_2+k_d\beta_3\right)s + k_p\beta_3,$$

$$C(s) = s^3 + \beta_1 s^2 + \beta_2 s + \beta_3.$$

Thus the frequency characteristic of the open-loop transfer function $G_l(i\omega)$ is

$$G_l(i\omega) = G_c(i\omega)G_p(i\omega)F(i\omega)$$

$$= \frac{k_p\beta_3-\left(k_p\beta_1+k_d\beta_2+\beta_3\right)\omega^2+\left(k_p\beta_2+k_d\beta_3\right)\omega i}{-(\beta_1+k_d)\omega^2+\left[\left(\beta_2+k_d\beta_1+k_p\right)\omega-\omega^3\right]i}\frac{K}{b_0(T\omega i+1)^n}. \tag{17.10}$$

Fig. 17.3 shows the Nyquist diagram of $G_l(i\omega)$. Combined with Eq. (17.6), the definition of the maximum sensitivity M_s can be graphically interpreted: M_s equals the inverse of the shortest distance between the Nyquist curve of $G_l(i\omega)$ and the critical point $(-1, 0i)$.

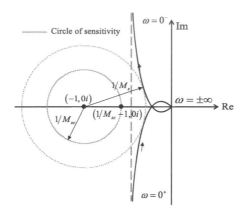

FIGURE 17.3 Graphical interpretation of maximum sensitivity M_s.

Given a certain maximum sensitivity constraint M_{sc}, a circle of sensitivity, centered at $(-1, 0i)$ with radius $1/M_{sc}$, can be constructed. Then the actual value of the maximum sensitivity M_s is guaranteed to be not higher than M_{sc}, provided that the Nyquist curve of $G_l(i\omega)$ does not enter the circle of sensitivity. This important principle will be used in the next subsection to derive tuning rules of ADRC parameters.

17.3.2 Derivation of second-order ADRC tuning rule

As it is mentioned in Section 17.2, ω_c is called the desired closed-loop bandwidth and also the desired closed-loop system poles. When the observer gains are properly chosen so that the ESO states z_1, z_2, and z_3 track y, $\dot{y}$ and f well, combining Eqs. (17.3) and (17.4), the second equation in (17.1) can be rewritten as

$$\ddot{y} \approx f + b_0 \frac{\left[k_p(r - y) - k_d \dot{y}\right] - f}{b_0} = k_p(r - y) - k_d \dot{y}. \qquad (17.11)$$

Therefore the transfer function from the reference r to the output y can be approximated to

$$G_{yr}(s) = \frac{R(s)}{Y(s)} \approx \frac{k_p}{s^2 + k_d s + k_p} = \frac{\omega_c^2}{(s + \omega_c)^2}. \qquad (17.12)$$

The choice of ω_c mainly influences the set-point tracking performance. In [20], it is recommended that $\omega_c = 10/t_s^*$, where t_s^* is the desired settling time. The choice of the desired settling time t_s^* can incorporate the information of the process model. In this study, the controlled process $K/(Ts + 1)^n$ can be seen as n first-order processes $1/(Ts + 1)$ being cascaded connecting together. Since the parameters n and T influence the transient time of the process most, it is natural to assume that t_s^* is proportional to nT. Therefore ω_c is decided as

$$\omega_c = 10/knT, \tag{17.13}$$

where k is the desired settling time factor, and it is the only tuning parameter in this proposed tuning method.

The bandwidth ω_o is the ESO pole. In general, a larger ω_o value can speed up the total disturbance being estimated and rejected, so a large ω_o is usually preferred. However, in practice, the sampling rate limits the upper bound of ω_o. In general, it is recommended that

$$\omega_o = 10\omega_c. \tag{17.14}$$

The tuning equation of ω_c is derived from the perspective of the desired set-point tracking, and the choice of ω_o considers the disturbance rejection speed. Now the design of b_0 will consider the closed-loop system robustness, that is, the sensitivity constraint will be applied in determination of b_0.

Using the bandwidth-parameterization (17.5) and Eq. (17.14), the frequency characteristics of the open-loop system transfer function $G_l(i\omega)$, Eq. (17.10), can be rewritten as

$$G_l(i\omega) = \frac{\left(\omega_c^2/\omega - 1.63\omega\right) + 2.3\omega_c i}{-32\omega_c\omega + \left(361\omega_c^2 - \omega^2\right)i} \times \frac{1}{(1+T\omega i)^n} \frac{10^3 \omega_c^3 K}{b_0}. \tag{17.15}$$

When come to derive PID control parameters under the sensitivity constraints, previous studies [21,22] have made effort on solving the nonlinear equations related to the definition of maximum sensitivity and the open-loop transfer function. It is possible to yield explicit solutions for PID design, although derivation operation and numerical calculation are often involved, making it not easy for control engineers to execute the solving procedure. For the ADRC control system studied in this work, it is very difficult to directly solve the sensitivity constraint from Eqs. (17.6) and (17.15). It is highly nonlinear because of the absolute operation. In addition, it is at least of the fifth degree, which means that it is almost impossible to obtain an explicit solution.

Instead of strictly solving the sensitivity constraint, an alternative asymptote constraint is proposed to determine b_0 in this study. The asymptote is vertical to the real axis in the Nyquist plot of $G_l(i\omega)$ (see Fig. 17.3). Considering the relationship between the asymptote and the circle of sensitivity, an asymptote condition is further proposed.

Provided that the vertical asymptote of Nyquist curve $G_l(i\omega)$ is located at the right side of the circle of sensitivity, and the Nyquist curve of $G_l(i\omega)$ does not enter into the circle of sensitivity, then the actual maximum sensitivity M_{sc} is guaranteed to be smaller than M_{sc}.

The asymptote function of Nyquist curve $G_l(i\omega)$ needs to be found. Let x and y denote the real and imagine axes, respectively, so $G_l(i\omega) = x(\omega) + y(\omega)i$.

The line $x = a$ is a vertical asymptote of the plot of $G_l(i\omega)$ when there exists $\omega = \omega^*$, so that

$$\begin{cases} \lim_{\omega=\omega^*} x(\omega) = \lim_{\omega=\omega^*} \text{Re}\,[G_l(i\omega)] = a, \\ \lim_{\omega=\omega^*} y(\omega) = \lim_{\omega=\omega^*} \text{Im}\,[G_l(i\omega)] = \pm\infty. \end{cases} \qquad (17.16)$$

As observed from Fig. 17.3, the imaginary coordinate of the Nyquist curve tends to infinity as $\omega \to \pm 0$, so the limiting values of the real and imaginary parts of $G_l(i\omega)$ are concerned as $\omega \to \pm 0$. Let

$$(1 + T\omega i)^n = p_1 + p_2 i. \qquad (17.17)$$

Then

$$\begin{aligned} p_1 &= 1 + C_n^2 (T\omega)^2 (-1)^1 + C_n^4 (T\omega)^4 (-1)^2 + \cdots, \\ p_2 &= nT\omega + C_n^3 (T\omega)^3 (-1)^1 + C_n^5 (T\omega)^5 (-1)^2 + \cdots. \end{aligned} \qquad (17.18)$$

Therefore we have

$$\begin{aligned} G_l(i\omega) &= \frac{\left[\left(798.3\omega_c^3 + 49.86\omega_c\omega^2\right) - \left(361\omega_c^4/\omega - 515.83\omega_c^2\omega + 1.63\omega^3\right)i\right]}{\left[(-32\omega_c\omega)^2 + \left(361\omega_c^2 - \omega^2\right)^2\right]} \\ &\quad \times \frac{(p_1 - p_2 i)\, 10^3 \omega_c^3 K}{\left(p_1^2 + p_2^2\right) b_0}. \end{aligned} \qquad (17.19)$$

Then

$$\begin{aligned} \text{Re}\,[G_l(i\omega)] &= \frac{\left(798.3\omega_c^3 + 49.86\omega_c\omega^2\right) p_1 - \left(361\omega_c^4/\omega - 515.83\omega_c^2\omega + 1.63\omega^3\right) p_2}{\left[(-32\omega_c\omega)^2 + \left(361\omega_c^2 - \omega^2\right)^2\right]} \\ &\quad \times \frac{10^3 \omega_c^3 K}{\left(p_1^2 + p_2^2\right) b_0} \end{aligned} \qquad (17.20)$$

$$\begin{aligned} \text{Im}\,[G_l(i\omega)] &= \frac{-\left(798.3\omega_c^3 + 49.86\omega_c\omega^2\right) p_2 - \left(361\frac{\omega_c^4}{\omega} - 515.83\omega_c^2\omega + 1.63\omega^3\right) p_1}{\left[(-32\omega_c\omega)^2 + \left(361\omega_c^2 - \omega^2\right)^2\right]} \\ &\quad \times \frac{10^3 \omega_c^3 K}{\left(p_1^2 + p_2^2\right) b_0}. \end{aligned} \qquad (17.21)$$

As $\omega \to 0^-$, $p_1 \to 1$, $p_2 \to nT\omega$, and $p_1^2 + p_2^2 = |1/(1 + T\omega i)^n|^2 \to 1$ in the expressions of $\text{Re}\,[G_l(i\omega)]$ and $\text{Im}\,[G_l(i\omega)]$, the terms in the form of multiplying ω diminish to 0, whereas the terms divided by ω tend to ∞. Thus we have

$$\lim_{\omega \to 0^-} \text{Re}\,[G_l(i\omega)] = \frac{(798.3\omega_c^3) - (361nT\omega_c^4)}{361^2\omega_c^4} \times \frac{10^3\omega_c^3 K}{b_0}$$

$$= \left(6.1256\omega_c^2 - 2.77\omega_c^3 nT\right) K/b_0, \tag{17.22}$$

$$\lim_{\omega \to 0^-} \text{Im}\,[G_l(i\omega)] = +\infty.$$

As previously, as $\omega \to 0^+$,

$$\lim_{\omega \to 0^+} \text{Re}\,[G_l(i\omega)] = \left(6.1256\omega_c^2 - 2.77\omega_c^3 nT\right) K/b_0, \tag{17.23}$$
$$\lim_{\omega \to 0^+} \text{Im}\,[G_l(i\omega)] = -\infty$$

Eqs. (17.22) and (17.23) show that there exists $\omega^* = 0$ satisfying the definition in (17.16), so the function of the asymptote is

$$x = \left(6.1256\omega_c^2 - 2.77\omega_c^3 nT\right) K/b_0. \tag{17.24}$$

As it is shown in Fig. 17.3, the right endpoint of the circle of sensitivity locates at $(1/M_{sc} - 1, 0i)$.

Applying the asymptote condition,

$$\left(6.1256\omega_c^2 - 2.77\omega_c^3 nT\right) K/b_0 > 1/M_{sc} - 1, \tag{17.25}$$

and solving (17.23) for b_0, we have

$$b_0 > (2.77\omega_c nT - 6.1256)\,\omega_c^2 K M_{sc}/(M_{sc} - 1). \tag{17.26}$$

Since increasing b_0 can reduce the maximum sensitivity M_s, for a conservative design, let b_0 be m times the lower limit. Then the parameter b_0 can be determined by

$$b_0 = m\,(2.77\omega_c nT - 6.1256)\,\omega_c^2 K M_{sc}/(M_{sc} - 1). \tag{17.27}$$

The coefficients m and M_{sc} can be chosen according to engineering experience. In this study, we choose $m = 1.4$, and the maximum sensitivity constraint M_{sc} is chosen as the maximum of the allowable value 2.5, and thus (17.27) becomes

$$b_0 = (6.4541\omega_c nT - 14.2726)\,\omega_c^2 K. \tag{17.28}$$

In summary, the tuning rules for the second-order linear ADRC controlling high-order processes $K/(Ts + 1)^n$ are as follows:

$$\begin{cases} \omega_c = 10/knT, \\ \omega_o = 10\omega_c, \\ b_0 = (6.4541\omega_c nT - 14.2726)\,\omega_c^2 K. \end{cases} \tag{17.29}$$

Similarly, we can also derive the tuning rules for the first-order linear ADRC:

$$\begin{cases} \omega_c = 10/knT, \\ \omega_o = 10\omega_c, \\ b_0 = (11.1111 \ln T \omega_c - 12.8042)\,\omega_c^2 K. \end{cases} \tag{17.30}$$

Remark: applying the proposed second-order ADRC tuning method to the high-order process, two interesting conclusions can be further inferred.

The asymptotes for the different controlled systems are the same. Substituting expression (17.28) of b_0 into the asymptote function (17.24), we get $x = -0.429$.

It shows the real axis value of the asymptote remains constant, which can be verified in Fig. 17.4(b). Different Nyquist curves of $G_l(i\omega)$ converge to the same asymptote.

The complementary maximum sensitivity $M_t \leq 1$. It is also an important indicator for measuring robustness. It implies the sensitivity of the closed-loop system to the large process dynamic variations and is defined as

$$M_t = \max_{\omega} |G_l(i\omega)/(1 + G_l(i\omega))|, \quad \omega \in (-\infty, +\infty). \tag{17.31}$$

Since $G_l(i\omega) = x(\omega) + y(\omega)i$,

$$M_t = \max_{\omega} \sqrt{\frac{x(\omega)^2 + y(\omega)^2}{1 + 2x(\omega) + x(\omega)^2 + y(\omega)^2}}. \tag{17.32}$$

If $x(\omega) \geq -0.5$, then $M_t \leq 1$. The proposed ADRC tuning method gives the asymptote function $x = -0.429$. Under the proposed tuning method, the Nyquist curve of $G_l(i\omega)$ can be tuned to be located at the right side of the asymptote line, that is, $x(\omega)$ is not smaller than -0.429. Therefore $M_t \leq 1$ is achieved. This conclusion indicates that the proposed ADRC tuning method derived under M_s constraint can lead to satisfactory M_t.

The effect of the tuning parameter k should be clear to users. According to [23], to ensure the closed-loop system stability, b_0 and the process gain K must share the same sign. Then the upper limit of the desired settling time factor k can be deducted from (17.27), $0 < k < 4.5$. To avoid unstable or oscillatory output response caused by nearing the critical values, the range of the tuning parameter k can be further narrowed. In engineering practice, the range of $k = 1.0 \sim 4.0$ is suitable for most high-order processes. However, the adjustment of k is still necessary to achieve a certain robustness level. Consider the cascaded fifth-order process

$$G_p = 1/(8s + 1)^5. \tag{17.33}$$

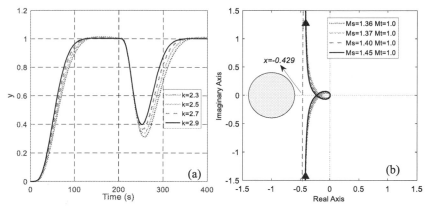

FIGURE 17.4 Performance and robustness under different tuning parameter k: (a) tracking response($t = 0$–200 s) and disturbance rejection ($t = 0$–400 s) of G_p; (b) Nyquist diagram and robustness indices of G_p.

The model parameter $n = 5$, $T = 8$, and $K = 1$ can be directly used to calculate the ADRC parameters by (17.29). The control results and robustness indices under the different tuning parameter k are shown in Fig. 17.4. In general, increasing k results in faster tracking response and better disturbance rejection performance, but higher maximum sensitivity M_s. This influence pattern can be used as a crude guideline to adjust k to a certain maximum sensitivity M_s.

17.3.3 One-parameter tuning rule of first-order ADRC tuning rule

Similarly, the tuning formulas of the first-order ADRC applied to the high-order processes are

$$
\begin{cases}
\omega_c = 10/knT, \\
\omega_o = 10\omega_c, \\
b_0 = (11.1111nT\omega_c - 12.8042)\,\omega_c K,
\end{cases}
\tag{17.34}
$$

where the desired settling time factor k is the tuning parameter for the trade-off between performance and robustness. Since the sign of b_0 should be the same as the sign of the process gain K [23], a reasonable range of k can be determined from Eq. (17.21), $k = 1 \sim 7$.

Because of the conservativeness in the previous design, the tuning formulas in Eq. (17.34) usually result in a lower maximum sensitivity than $M_{sc} = 2.5$. In this subsection the relationship between the tuning parameter k and the real maximum sensitivity M_s is found. The users can then specify the system robustness level by using M_s as a tuning parameter. The expression of maximum sensitivity for first-order ADRC controlling high-order processes is

$$M_s = \max_{\omega} \left| \frac{1}{1 + G_l(i\omega)} \right|$$

$$= \max_{\omega} \left| \frac{1}{1 + \dfrac{1200knT\omega i + 10000}{(111.111/k - 12.8042)\left[(knT\omega)^2 + 210knT\omega i\right]} \dfrac{1}{(T\omega i + 1)^n}} \right|. \tag{17.35}$$

We can see from the expression that the process gain K does not influence M_s anymore, whereas the desired settling time factor k, the process order n, and the process time constant T are still related to M_s.

Simulations have been performed to test how T, n, and k influence M_s in the range of $T/n = 0.01 \sim 100$, $n = 3 \sim 20$, and $k = 1 \sim 7$. The results are plotted in Fig. 17.5.

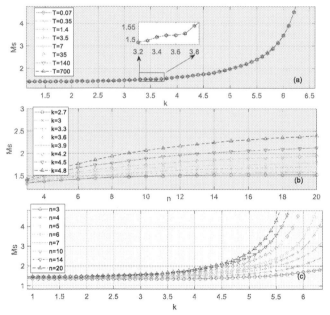

FIGURE 17.5 Influence of n, T, and k on M_s: (a) M_s changes with k under different T/n when $n = 7$ and $K = 1$; (b) M_s changes with n under different k when $T/n = 7$ and $K = 1$; (c) M_s changes with k under different n when $T/n = 1/15$ and $K = 1$.

Fig. 17.5(a) shows that the curves of different T/n ratios coincide, which indicates that the process time constant T has no influence on the maximum sensitivity M_s. This occurs because T and ω appear in pairs in Eq. (17.35). This means that T is a scaling factor of frequency ω, so it does not affect the shape of the Nyquist curve and thus does not affect the value of maximum sensitivity M_s. Figs. 17.5(b) and 17.5(c) also show that M_s changes logarithmically with n and exponentially with k. Therefore we propose that

$$M_s = f(n, k) = a_2 e^k \ln(n - a_3) + a_1. \tag{17.36}$$

Solving for k gives

$$k = \ln\left[\frac{M_s - a_1}{a_2 \ln(n - a_3)}\right].\tag{17.37}$$

Eq. (17.37) calculates a certain k to ensure that the system robustness level at M_s. By using nonlinear fitting techniques the coefficients a_1, a_2, and a_3 will be further determined.

First, the data sets M_S, n and k are required for fitting. For $M_s = 1.4 \sim 2.0$ and $n = 3 \sim 20$, the data set of k can be generated by minimizing the integrated absolute error (IAE) of reference tracking and disturbance rejection response:

$$\min_{k}(\text{IAE}) \text{ such that } M_s = \text{const}, \ n = \text{const}.\tag{17.38}$$

A dataset of k was therefore determined (see dots in Fig. 17.6). Nonlinear fitting gives the estimation of coefficients $a_1 = 1.3966$, $a_2 = 0.0026$, and $a3 = 1.6980$ with an average error $E = 0.165$ and variance $S = 0.0252$. To improve the fitting precision, the regression model for fitting is modified as

$$k = \ln\left[\frac{M_s - a_1 n^{a_4}}{a_2 n^{a_5} \ln(n - a_3 n^{a_6})}\right].\tag{17.39}$$

Thus the corresponding estimated coefficients are

$$a_1 = 1.312, a_2 = 0.002, a_3 = 0.452, a_4 = 0.026, a_5 = 0.48, a_6 = 1.22,\tag{17.40}$$

which gives an average fitting error $E = 0.047$ and a fitting variance $S = 0.0055$. The fitting result is shown in Fig. 17.6.

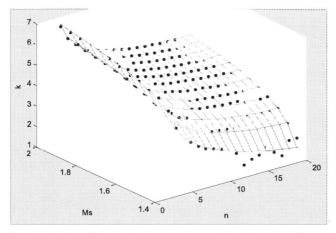

FIGURE 17.6 Multivariable fitting of k. Dots represent the fitting data set, and the mesh represents the fitting results.

To sum up, for a high-order process $K/(Ts + 1)^n$, if a designed maximum sensitivity $M_{sd}(1.4 \sim 2.0)$ is given for the controller design, then the first-order ADRC parameters can be calculated through the following equations:

$$
\begin{cases}
k = \ln\left[\frac{M_{sd}-1.312n^{0.026}}{0.002n^{0.48}\ln(n-0.452n^{1.22})}\right], \\
\omega_c = 10/knT, \\
\omega_o = 10\omega_c, \\
b_0 = (11.1111\ln T\omega_c - 12.8042)\,\omega_c K.
\end{cases}
\tag{17.41}
$$

17.4 Experiment verification

17.4.1 Water tanks

To validate the proposed second-order ADRC tuning method, a laboratory test is performed on a water tank system. Fig. 17.7 shows the experiment setup. The water tank control system, developed by Feedback Instruments Ltd, consists of water tanks, pumps, sensors, a controller, and a monitor. In this experiment, water tanks 1 and 2 connected by a water tube are used. Water level y of Tank 2 is the process variable (PV), and the voltage of the pump u is the manipulated variable (MV).

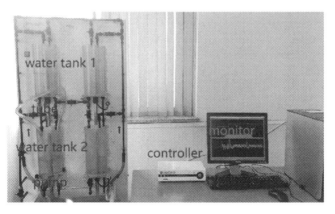

FIGURE 17.7 The water tank experiment setup.

A step input is added in the open-loop control system at the working point $y = 9$ cm, as shown in Fig. 17.8. Note that the water level y is slightly oscillatory before a step input is added, but this does not influence the design of control system because the proposed ADRC tuning does not relay on accurate modeling. A rough high-order model is identified from the open-loop system response data by using the empirical equations in [24]:

$$
G \approx \frac{5.72}{(27.72s + 1)^4}.
\tag{17.42}
$$

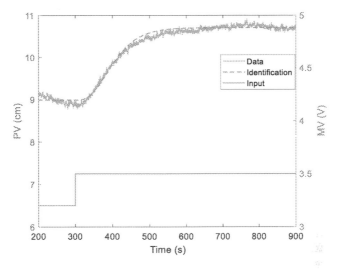

FIGURE 17.8 Open-loop step experimental data vs. identified model response.

For comparison purpose, PID algorithm is also implemented on the water tank. PID controller is tuned by the SIMC method due to its simplicity in use. The choice of $\tau_c = \theta$ leads to the maximum sensitivity $M_s = 1.4$ with the PID controller parameters $K_p = 0.1458$, $K_i = 0.0021$, and $K_d = 2.4249$. The tuning parameter of ADRC, the desired settling factor $k = 4.17$, is manually tuned to achieve the same maximum sensitivity of SIMC-PID, which gives the controller parameters $\omega_c = 0.0278$, $\omega_o = 0.2775$, and $b_0 = 0.0246$.

These two controllers were tested by changing the water level set point from 9 cm to 11 cm first, then an input step disturbance $d = 1$V is artificially added in the system at time $t = 1000$ s. Fig. 17.9 shows the real-time control results. We can see that the proposed ADRC tuning results in slightly slower response during the set-point tracking, $t = 500$–1000 s, but ADRC has much better performance in disturbance rejection. Moreover, the MV chattering under ADRC algorithm is less severe than SIMC-PID. Note that the calculated ADRC parameters are directly used on the plant without retuning. The experimental results demonstrate the reliability and effectiveness of the proposed ADRC tuning method.

Although it is expected that ADRC behaves better than SIMC-PID in terms of disturbance rejection, the ADRC results of the water tank experiment show much better disturbance rejection than the simulated situation. A possible reason is that the working condition has varied, such as water level change of the water reserve, change of pump characteristic, to the direction that is beneficial to ADRC control.

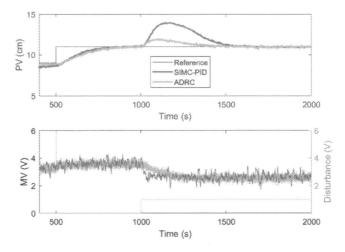

FIGURE 17.9 Experiment results under SIMC-PID and proposed ADRC tuning parameters.

17.4.2 Power plant simulator

To validate the effectiveness of the proposed first-order ADRC tuning method, the method is tested on the total airflow control in a 1000 MWe coal-fired power plant simulator. The simulator and control systems are constructed on a software named Industry Automation Platform (IAP). The calculation step size used in the simulator experiments is 0.1 s, and the data collection step size is 1 s. Excessive or inadequate airflow may decrease the combustion efficiency or even threaten the combustion stability in the furnace. For this reason, the airflow should be maintained at an optimal value. In the DCS the airflow set-point value (SV) is decided by the boiler master demand and fuel feed. Based on the difference between the SV and the measured airflow (or PV), ADRC algorithm gives commands to the forced draft fans to adjust the pitch blade position (scaled in permillage), which is the MV of this control system.

By using data-based modeling the dynamic from the forced draft fans to the total airflow is determined from open-loop step test data. The model parameters K and T are identified by the SIMULINK parameter estimation tool. The optimization method is nonlinear least squares. The model order n is decided by choosing the identified model with the lowest cost function value. The identified result is presented in Eq. (17.43) and Fig. 17.10. Although the identified simple high-order process cannot capture all the complex dynamics of the total air process, the ability of ADRC in estimating and compensating the unmodeled dynamics enables ADRC to provide good control results for complex industrial processes.

$$G_p(s) \approx \frac{3.25}{(2.433s + 1)^5}. \tag{17.43}$$

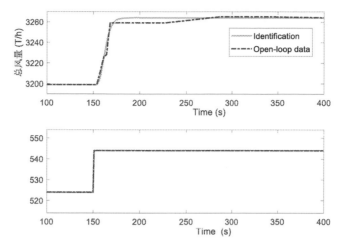

FIGURE 17.10 Open-loop identification of the total air flow control process.

Prior to the test in the power plant simulator DCS, MATLAB® simulations were conducted based on this model to generate suitable parameters. The designed maximum sensitivity is set as $M_{sd} = 1.4$, and the ADRC parameters are calculated using the proposed tuning formulas in Eq. (17.41). The original PI controller used in the total airflow control loop is retuned to achieve the same maximum sensitivity as that in the ADRC tuning.

Fig. 17.11 shows the simulation results of the different control and tuning strategies. The original PI parameters result in a relatively slow response because of the conservative tuning. The retuned PI parameters achieve a maximum sensitivity of 1.4, and it has improved the control performance a lot. The proposed ADRC tuning method results in the real maximum sensitivity $M_s = 1.407$, which is very close to the designed value $M_{sd} = 1.4$.

Fig. 17.12 shows the experimental results in the power plant simulator. Experiments are carried out under a constant load of 1000 MWe. The set-point is changed from $3100T/h$ to $3200T/h$, and a 10‰ pitch blade disturbance lasting about 500 s is added to the system around 1100 s. Experiment results show agreement with the simulation. The retuned PI has improved the control performance of the original PI. The ADRC shows control advantages in set-point tracking and disturbance rejection.

Since the total airflow control is closely related to the power plant load, the system dynamics vary with the operating conditions. It is necessary to test different controllers under varying loads. Fig. 17.13 shows that the proposed ADRC tuning method still maintains good tracking performance when the load changes from 700 MWe to 1000 MWe.

The average settling time $\bar{T}_s$ overshoot σ under constant load condition and the average tracking error $\bar{e}$ under varying load condition have been calculated for different control strategies. The average settling time $\bar{T}_s$ and the average

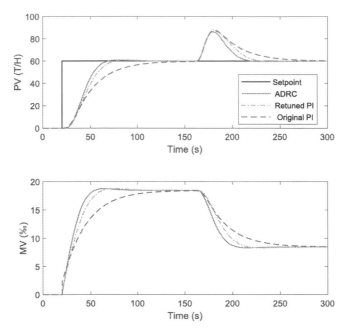

FIGURE 17.11 Comparative simulations based on the identified model.

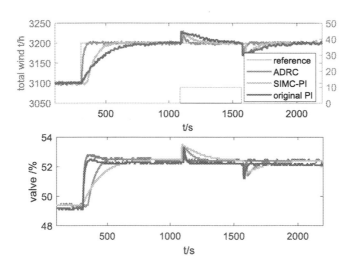

FIGURE 17.12 Set-point tracking and disturbance rejection tests of the total airflow control.

tracking error $\bar{e}$ of ADRC control strategy have been remarkably reduced by about 50%, compared to the retuned PI control. The overshoot σ during disturbance rejection has also been decreased by 5.7%.

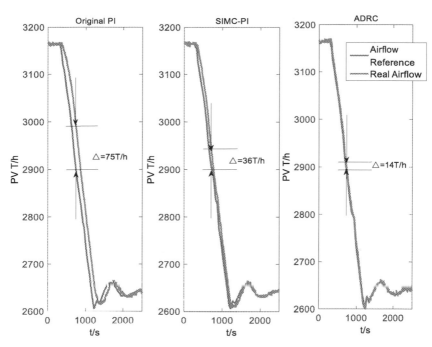

FIGURE 17.13   Reference tracking under varying loads.

The experimental tests in a coal-fired power plant simulator initially demonstrate the feasibility of the proposed ADRC tuning method in coal-fired power plants. It not only provides satisfactory control performance in set-point tracking and disturbance rejection, but also shows a good robustness under varying working conditions.

17.5   Field test

17.5.1   Superheated steam temperature

Encouraged by the positive results from water tank experiment and the power plant simulator, the proposed second-order ADRC tuning method is further applied to the superheater steam temperature (SST) control in a 330 MWe in-service CFB unit in Shanxi, China. The SST control system, which is one of the most important control systems in the power plant, is a typical high-order process. The SST has to be controlled within a certain range, so that the temperature will not exceed the upper limit that is set for safe operation of the steam turbine. At the meantime, the temperature will not drop out of the lower limit that ensures the efficiency of the whole power plant. For this CFB unit, the allowable SST temperature fluctuation range is ±5°C. The steam that comes from the drum is heated by the fuel gas through three sets of superheaters as shown in

Fig. 17.14. Two sets of desuperheaters are deployed to control the steam temperature. Since the control of the second desuperheater directly influences the SST, ADRC control algorithm and the proposed tuning method are implemented on the second desuperheater.

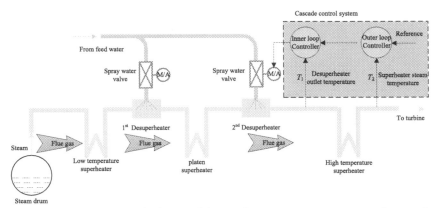

FIGURE 17.14 The schematic diagram of the superheater steam temperature control system in CFB power plant.

For the purpose of controller design, the SST models are identified from the open-loop data. As shown in Fig. 17.15, it is not a standard open-loop step test, because there is a spike in the control input signal and the control input changes before the temperature reaches steady state due to the operation limit. Therefore the superheater models are identified by using an optimization method. The dynamics from the spray water valve to the desuperheater outlet temperature T_1 is denoted as model $G_1(s)$, and the dynamics from the desuperheater outlet temperature T_1 to the SST T_2 is denoted as model $G_2(s)$.

The model identification process is accomplished by Matlab SIMULINK parameter estimation tool. The chosen optimization method is pattern search. The model order is decided by choosing the identified model with the lowest cost function value. The identification results are shown in Fig. 17.15, and the identified transfer functions are

$$G_1(s) = \frac{-1.6165}{(19.363s + 1)^2}, \quad G_2(s) = \frac{1.5528}{(28.234s + 1)^4}. \tag{17.44}$$

For the simplicity of implementing the control algorithm and relevant protective logics in the distributed control system (DCS), the first-order ADRC control algorithm is chosen to enhance the control performance of the SST control system. Compared to the outer-loop process model $G_2(s)$, the inner-loop process model $G_1(s)$ is relatively fast response, and usually the PI controller or even the P controller is enough to eliminate the disturbances in the inner loop. Therefore the inner PI controller remains unchanged, and the first-order ADRC is imple-

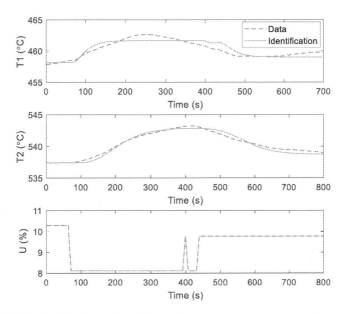

FIGURE 17.15   Identification results of the superheater steam temperature control system.

mented as an outer-loop controller in parallel with the original outer-loop PID controller (see Fig. 17.16).

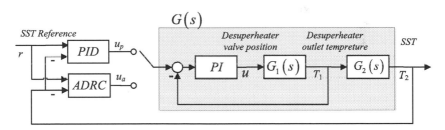

FIGURE 17.16   Cascade control system of the superheater steam temperature.

The tuning of the outer-loop controller is based on the equivalent model $G(s)$ for the combined inner controlled system and the model $G_2(s)$, as shown in Fig. 17.16. The inner PI controller parameters are $K_{p2} = -1$ and $K_{i2} = -1/40$, so the equivalent model for the outer-loop controller is

$$G(s) = \frac{G_{c,\,\text{inner}}(s)G_1(s)}{1 + G_{c,\,\text{inner}}(s)G_1(s)}, \quad G_2(s) \approx \frac{1.5528}{(27.4259s + 1)^5}. \tag{17.45}$$

The outer loop PI is tuned by the experienced field engineer, and the PI parameters are $K_{p1} = 1/3$ and $K_{i1} = 1/240$, which results in the robustness level $M_s = 1.5$. ADRC controller parameters are calculated by using (17.30). The desired settling time factor is manually tuned as $k = 3.9$ to achieve a similar maximum sensitivity of SIMC-PI tuning, which gives the ADRC control parameters $\omega_c = 0.0187$, $\omega_o = 0.187$, and $b_0 = 0.4554$.

Two sets of field test results are shown in Figs. 17.17 and 17.18. Fig. 17.17 shows the SST control result when the outer-loop controller switches between PI and ADRC. Fig. 17.18 compares the control results under reference step change. It should be mentioned that the artificial input disturbance is not allowed for the commercial operation of the power plant, and thus the strict disturbance rejection tests are not performed.

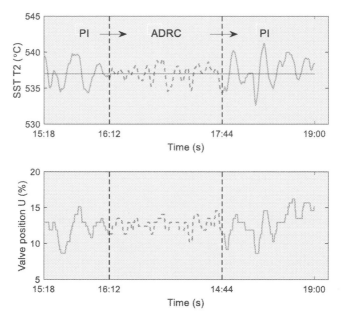

FIGURE 17.17 Field test 1: switch between PI and ADRC with the SST set-point of 537°C (date of test: 15th of March, 2017).

In addition, control performance indices, such as peak positive error e^+, peak negative error e^-, standard deviation σ, integral absolute error (IAE), and the TV of control input, are summarized in Table 17.1.

TABLE 17.1 Performance indices for SST control tests.

Test	Controller	e^+/°C	e^-/°C	σ	IAE	TV
Test1	PID	4.20	−4.34	1.82	8519	44.4
	ADRC	2.08	−2.59	1.05	4645	37.1
Test2	PID	4.20	−5.63	1.93	9836	66.6
	ADRC	2.18	−2.25	1.38	6757	53.5

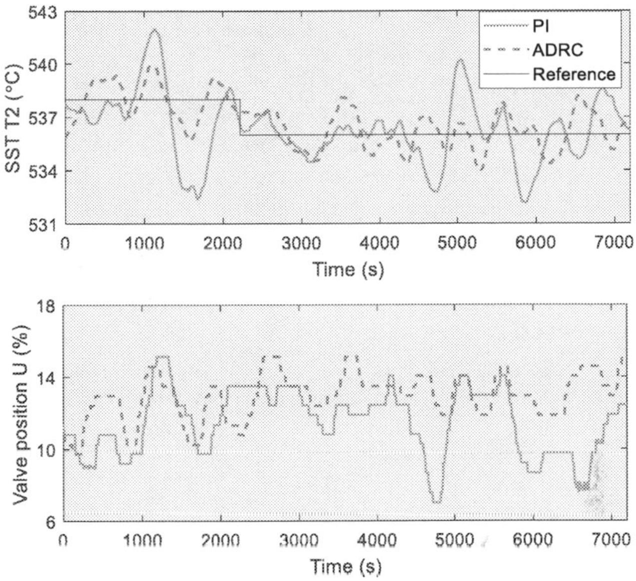

FIGURE 17.18 Field test 2: SST set-point step from 538°C to 536°C (time span of PI test: 14th of March, 2017, 07:00–09:30; time span of ADRC test: 16th of March, 2017, 11:00–13:30).

We can find that the proposed ADRC tuning method reduced the peak error and standard deviation σ by about 50%. The IAE is reduced by more than 30%. The TV of the control input is decreased by about 20%, which means a less overall tear and wear of valves, so the lifetime of valves can be prolonged, and the maintenance cost is therefore reduced.

The field test results show that the proposed first-order ADRC tuning can reduce the SST fluctuation to a large extent. Since the SST is one of the main concerns when the power plant changes its load, the reduced SST fluctuation range indicates the possibility of large-scale load-varying operation and thus also indicates the potential of flexible operation of power plants to integrate more renewables into grid.

17.5.2 Secondary wind

Although the power plant simulator contains the same control system a real power plant, it is not realistic enough to reflect all the characteristics of a real power plant. The control of actual industrial processes is more challenging, especially when unknown multi-source disturbances are often present. The proposed first-order ADRC tuning method is further applied to an actual industrial process. The ADRC control algorithm and the tuning method are applied to the secondary air control system of a 330 MWe circulating fluidized bed (CFB) unit, which is in commercial operation in the Shanxi Province in China.

A combustion system is an essential part of a CFB unit. It is the place where fuel and air are mixed, and the chemical energy of the fuel is converted into thermal energy to heat the working fluid. Compared to a pulverized coal-fired boiler, the solid materials in a CFB boiler must be fluidized and circulated by air, making the combustion air system of a CFB boiler more complicated. Fig. 17.19 shows the schematic diagram of the air and smoke system in the CFB unit. The high-pressure blowers supply high-pressure air to fluidize and overflow the bed material in a loop seal. Primary air fans supply hot primary air upwardly into the furnace to fluidize the solid particles. At the same time, a small proportion of the primary air is sent to the coal feeding system as spreading air. The secondary air fans convey the secondary air to the interface between the lower and upper zones of the furnace, constituting 20–60% of the total air. The secondary air is added to the furnace to create an oxygen-rich environment, which improves the combustion efficiency.

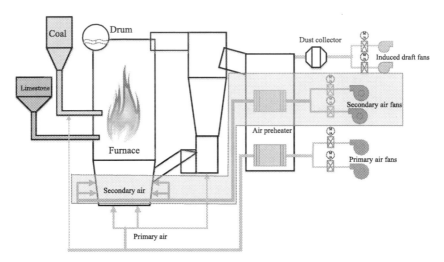

FIGURE 17.19 Configuration diagram of the air and smoke system in a CFB unit.

Because the failure of controlling the primary air and the loop seal air can severely damage the normal fluidization and circulation state of the CFB boiler, whereas the control of the secondary air system mainly influences the combustion efficiency, so it is less risky to test the ADRC tuning on the secondary air system in a commercially operated power plant. Therefore, for safety considerations, the secondary air system is chosen for the field test.

The ADRC algorithm is implemented in the SUPCON DCS in parallel with the original PID algorithm. Implemental issues such as the bumpless transfer between the PID, ADRC, and manual mode, as well as the amplification limit and rate limit, have been carefully solved. The schematic diagram of the control system is shown in Fig. 17.20. The balance module receives control commands from PID or ADRC controller and distributes the control commands to each secondary air fan. Under normal conditions, the control command is equally distributed to the two air fans. If the working efficiency of the two fans changes, then the operators can add an offset to the balance module to even out the power output of the two air fans.

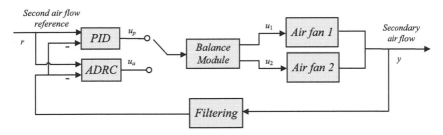

FIGURE 17.20 Schematic diagram of the secondary air control system.

The secondary air flow, scaled in percentage, is controlled by the variable frequency drive (VFD) attached to the air fan motors. The VFD commands are the control input, which is also scaled in percentage. To use the proposed ADRC tuning method to determine appropriate controller parameters, the input-step experiment data are collected for identification purposes. The same data-based modeling method is used as it in the total air identification. The identification results are shown in Fig. 17.21, and the transfer function is

$$G_p(s) \approx \frac{1.9596}{(8.1972s + 1)^4}. \tag{17.46}$$

A set of relatively conservative ADRC parameters are used in the first test. By setting the designed maximum sensitivity $M_{sd} = 1.4$ the initial ADRC parameters are $\omega_c = 0.0969$, $\omega_o = 0.969$, and $b_0 - 4.2721$. No unstable oscillation or frequent control input variation occurs in the first test. After ensuring the stability and robustness, another set of ADRC parameters with a faster response is calculated by increasing the designed maximum sensitivity $M_{sd} = 1.7$ and the retuned ADRC parameters are $\omega_c = 0.0574$, $\omega_o = 0.5744$, and $b_0 = 0.9142$. In addition, tests are also performed with the PID controller for comparison. The derivative part of PID controller is usually not used by engineers in power plant unit, because it may introduce noise to the control system and thus bring oscillation, especially when the reference is frequently changed. The original PID controller parameters are improved by an experienced field engineer with $K_p = 0.3333$, $K_i = 0.0067$, and $K_d = 0$, which have been put in use for a long time to ensure enough robustness.

Reference tracking is the main concern for a secondary air control system, because the secondary air flow must rapidly respond to the varying load reference of the CFB unit. Therefore the air flow reference is step-changed for performance tests. In addition, artificial disturbances are not allowed for this

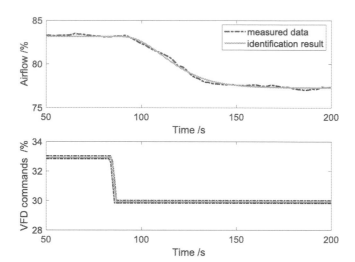

FIGURE 17.21 Identification of the secondary air flow control system.

commercially operated power plant, so strict disturbance rejection tests are therefore excluded. For a fair comparison, reference step tests for the ADRC and PID are performed within the same load range of 230–240 MWe. The test results of 20-min period are shown in Figs. 17.22 and 17.23. The control performance indices including the average settling time $\bar{T}_s$, IAE, ITAE, and the average total variation of two air fans $\bar{TV}$ are summarized in Table 17.2.

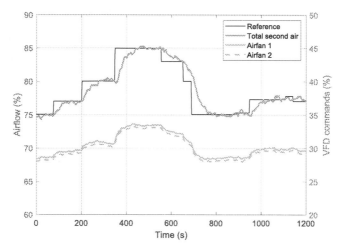

FIGURE 17.22 Reference step test results of the ADRC in the CFB unit (time span of test: 1st of September, 2017, 9:28:50–9:48:50).

We can see from Table 17.2 that the average settling time $\bar{T}_s$ and IAE and ITAE indices have been reduced by more than 40% under the proposed ADRC

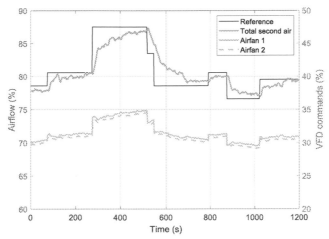

FIGURE 17.23 Reference step test results of the PID in the CFB unit (time span of test: 1st of September, 2017, 9:55:30–10:15:30).

TABLE 17.2 Control performance comparison of ADRC and PID in the secondary air control system of a 330 MWe in-service CFB unit.

algorithm	Average settling time $\bar{T}_s$	IAE	ITAE	Average $\bar{T}\bar{V}$
ADRC	86 s	959	5384	22.95
PID	> 193 s	1644	9041	28.17
Improvement	> 55.4%	41.6%	40.4%	18.5%

tuning method, which means that the proposed ADRC tuning has provided a better reference tracking. In the meanwhile, the average total variation $\bar{T}\bar{V}$ of the control input has also been reduced by 18.5%. A reduction in TV indicates less frequent position changes of the fan blades, which means less tear and wear of the air fans, and thus the maintenance cost could be reduced.

Therefore we can conclude that the proposed maximum sensitivity ADRC tuning improved the control performance. The explicit form of the tuning equations facilitates the calculation of the ADRC parameters. The maximum sensitivity-constrained tuning that provides the desired robustness allows for a safer implementation of the control algorithm. The field test results show that the proposed ADRC tuning method has the potential for further industrial applications.

17.6 Summary

This chapter achieves the goal of deriving maximum sensitivity-constrained ADRC tuning method for high-order industrial processes. To derive the parameters under sensitivity constraint, an asymptote condition is propounded.

Laboratory experiments on water tank and power plant simulator have shown the efficiency of this tuning method. Field tests on an operational large-scale coal fired power plant are further implemented, demonstrating the ability of ADRC in assisting power plant flexible operation. Future research will be continued on the ADRC parameters tuning toward multi-variable process.

References

[1] Y. Wang, C-H. Guo, X-J. Chen, et al., Carbon peak and carbon neutrality in China: goals, implementation path and prospects, China Geology 4 (4) (2021) 720–746.

[2] B. Looney, BP Statistical Review of World Energy: BP Statistical Review, London UK. Retrieved November, 2020, 18 2021.

[3] L. Sun, D. Li, K.Y. Lee, Optimal disturbance rejection for PI controller with constraints on relative delay margin, ISA Transactions 63 (2016) 103–111.

[4] Z. Wu, T. He, Y. Liu, et al., Physics-informed energy-balanced modeling and active disturbance rejection control for circulating fluidized bed units, Control Engineering Practice 116 (2021) 104934.

[5] L. Guo, S. Cao, Anti-disturbance control theory for systems with multiple disturbances: a survey, ISA Transactions 53 (4) (2014) 846–849.

[6] S. Talole, Active disturbance rejection control: applications in aerospace, Control Theory and Technology 16 (4) (2018) 314–323.

[7] Z. Huang, Y. Liu, H. Zheng, et al., A self-searching optimal ADRC for the pitch angle control of an underwater thermal glider in the vertical plane motion, Ocean Engineering 159 (2018) 98–111.

[8] Z. Wu, Z. Gao, D. Li, et al., On transitioning from PID to ADRC in thermal power plants, Control Theory and Technology 19 (1) (2021) 3–18.

[9] Z-L. Zhao, B-Z. Guo, On convergence of nonlinear active disturbance rejection control for SISO nonlinear systems, Journal of Dynamical and Control Systems 22 (2) (2016) 385–412.

[10] S. Shao, Z. Gao, On the conditions of exponential stability in active disturbance rejection control based on singular perturbation analysis, International Journal of Control 90 (10) (2017) 2085–2097.

[11] Y. Huang, W. Xue, Active disturbance rejection control: methodology and theoretical analysis, ISA Transactions 53 (4) (2014) 963–976.

[12] L-J. Wang, Q. Li, C-N. Tong, et al., Overview of active disturbance rejection control for systems with time-delay, 2013.

[13] L. Sun, D. Li, Z. Gao, et al., Combined feedforward and model-assisted active disturbance rejection control for non-minimum phase system, ISA Transactions 64 (2016) 24–33.

[14] S. Pawar, R. Chile, B. Patre, Modified reduced order observer based linear active disturbance rejection control for TITO systems, ISA Transactions 71 (2017) 480–494.

[15] L. Sun, J. Dong, D. Li, et al., A practical multivariable control approach based on inverted decoupling and decentralized active disturbance rejection control, Industrial & Engineering Chemistry Research 55 (7) (2016) 2008–2019.

[16] Y. Xia, M. Fu, C. Li, et al., Active disturbance rejection control for active suspension system of tracked vehicles with gun, IEEE Transactions on Industrial Electronics 65 (5) (2017) 4051–4060.

[17] Y.X. Su, C.H. Zheng, B.Y. Duan, Automatic disturbances rejection controller for precise motion control of permanent-magnet synchronous motors, IEEE Transactions on Industrial Electronics 52 (3) (2005) 814–823.

[18] Q. Zheng, Z. Gao, An energy saving, factory-validated disturbance decoupling control design for extrusion processes, in: Proceedings of the 10th World Congress on Intelligent Control and Automation, 2012, pp. 2891–2896.

[19] Z. Gao, Scaling and bandwidth-parameterization based controller tuning ACC, 2003, pp. 4989–4996.
[20] X. Chen, D. Li, Z. Gao, et al., Tuning method for second-order active disturbance rejection control, in: Proceedings of the 30th Chinese Control Conference, 2011, pp. 6322–6327.
[21] K.J. Åström, H. Panagopoulos, T. Hägglund, Design of PI controllers based on non-convex optimization, Automatica 34 (5) (1998) 585–601.
[22] O. Yaniv, M. Nagurka, Design of PID controllers satisfying gain margin and sensitivity constraints on a set of plants, Automatica 40 (1) (2004) 111–116.
[23] C. Zhao, D. Li, Control design for the SISO system with the unknown order and the unknown relative degree, ISA Transactions 53 (4) (2014) 858–872.
[24] X. Yang, Automatic Control for Thermal Process, Tsinghua University Press, China, 2008.

Chapter 18

Desired dynamic equational proportional-integral-derivative controller design based on probabilistic robustness

Gengjin Shi[a], Donghai Li[a], Yanjun Ding[a], and YangQuan Chen[b]

[a]*State Key Lab of Power Systems, Department of Energy and Power Engineering, Tsinghua University, Beijing, China,* [b]*Mechatronics, Embedded Systems and Automation (MESA) Lab, School of Engineering, University of California, Merced, CA, United States*

18.1 Introduction

Uncertainties usually exist in practical industrial systems such as coal-fired power plants and gas turbines. Their existence, including parameter perturbations, external disturbances, and unmodeled dynamics, brings challenges to modern industrial control systems. For instance, uncertainties caused by working medium can lead to the fluctuation of the shaft speed of a gas turbine and further affect its generation efficiency. It has been pointed out by Brockett that "If there is no uncertainty in the system, the control or the environment, feedback control is largely unnecessary" [3].

To handle the uncertainties in industrial processes, many control strategies were proposed based on disturbance observers in recent decades, such as disturbance observer-based control (DOB) [4,18], perturbation observer-based control (POB) [13,36], equivalent input disturbance-based control (EID) [24], uncertainty and disturbance estimator-based control (UDE) [39], generalized proportional integral observer-based control (GPIO) [10,25], and unknown input observer-based control (UIO) [11]. On the other hand, several robust control strategies are proposed to control uncertain systems such as H_2 control, H_∞ control [20], and the robust controller based on linear matrix inequalities [21]. These control strategies show their strengths on eliminating uncertainties by both simulation and experiments. However, they have strong dependence on accurate mathematical models of processes, which are difficult to obtain in practice, and their mathematical expressions are complex. As a result, these control strategies are hard to build on distributed control systems (DCS) of power plants and gas turbines.

Modeling, Identification, and Control for Cyber-Physical Systems Towards Industry 4.0
https://doi.org/10.1016/B978-0-32-395207-1.00031-7

415

The proportional-integral-derivative (PID) controller is of dominance in modern control systems and widely applied to industrial processes such as thermal engineering, chemical engineering, and mechanics [35]. Because of its simple structure and reliable control performance, it is regarded as the first choice for engineers although advanced control strategies such as model predictive control (MPC) [22], active disturbance rejection control (ADRC) [8], and sliding mode control [16] have been developed rapidly in recent years. The PID controller originates from Minorsky's research on ship steering servomechanism in 1922 [17]. It has a simple feedback control structure of the form "present-past-future", which is independent of the accurate mathematical models of processes [38]. So far, researchers worldwide have devoted attention to the study of PID tuning methods for half a century. Various tuning rules are summarized in the handbook for different processes, including the Cohen–Coon (C-C) method [6], Skogestad internal model control (SIMC) [26], maximum-sensitivity (Ms) constrained integral gain optimization (MIGO) [19], and so forth. Nevertheless, constrained with the structure, the conventional one-degree-of-freedom PID has the problem that tracking and disturbance rejection are coupled. To remove this disadvantage, in 1963, Horowitz [9] designed the two-degree-of-freedom (TDOF) PID. Many tuning rules of TDOF PID were proposed in the past decades, including tuning rules based on closed-loop Ms [2], internal model control (IMC) [7], genetic algorithm (GA) [15], and so forth. However, these tuning rules of TDOF PID are complex, and most of them are derived based on specific models. To simplify the tuning procedure of TDOF PID, the desired dynamic equation (DDE) method was proposed based on the core idea of Tornambè controller (TC) [29]. The basic idea of DDE method is that parameters of TDOF PID are tuned based on the coefficients of the desired dynamic equation. This method is able to tune the parameters of TDOF PID controllers to enable the output approaching the response of the desired dynamic response with less dependency on any specific transfer function.

Nevertheless, most design methods of PID or TDOF PID can only guarantee their control performance of nominal systems (Johnson). For some "worse-case" conditions, this may result in serious oscillation or even non-convergence [32]. Probabilistic robustness (PR), known as one of randomized algorithms, is an effective and powerful method for the tuning and robustness analysis of controllers. Up to now, PR has been applied to the design of PID [28], fractional-order PID (FOPID) [33], and ADRC [31]. These applications show the great potential of PR method.

Aiming at aforementioned problems, in this chapter, we develop a design method of DDE PID based on PR for uncertain systems. The following are the main contributions of this chapter:

- A DDE PID-design based on PR is developed for uncertain systems to enhance the ability of the closed-loop systems to handle with uncertainties. The proposed control strategy has little dependency on the accurate model of the process, and its tuning has been summarized as a flow chart that is easy to follow.

- Stable regions of DDE PID are analyzed in the frequency domain to avoid the divergence of the closed-loop system.
- Superiorities of the proposed DDE PID design based on PR are verified by both simulation examples and an experiment on the level system of a water tank to demonstrate its advantage in robustness.

The rest of this chapter is organized as follows. The problem formulation is introduced in Section 18.2, followed by the principle and the stable region of DDE PID in Section 18.3. In Section 18.4, the DDE PID design based on PR is discussed, and the corresponding design procedure is summarized as well. Then, in Section 18.5 the effectiveness of the proposed DDE PID design based on PR is validated by simulations of several typical processes. Section 18.6 presents the experimental verification of the proposed control strategy on the level system of a water tank. Finally, concluding remarks are presented in the last section.

18.2   Problem formulation

Dynamic characteristics of most industrial systems can be described by differential equations based on physical laws. Consider the time-varying nonlinear system with time delay and external disturbances

$$\begin{cases} \dot{x} = g\,(x, u, t - \tau, d)\,, \\ y = h\,(x, u, t)\,, \end{cases} \tag{18.1}$$

where x, u, τ, d, t, and y are the state variable, input, time delay, external disturbances, time, and output, respectively, and g and h are the state and output functions with respect to related variables in this subsection, respectively. The transfer function model is obtained by linearizing system (18.1) at the nominal working conditions. For practical model establishment, some internal physical properties such as nonlinearities and distributed parameters would be simplified. In addition, the working condition of the system is varying. As a result, the coefficients of the obtained transfer function are not fixed values but rather in a parameter space Q. Then the transfer function family with parameter uncertainties in Q can be depicted as

$$G_p\,(s) = \frac{c_\alpha s^\alpha + c_{\alpha-1} s^{\alpha-1} + \cdots + c_0}{a_m s^m + a_{m-1} s^{m-1} + \cdots + a_0} e^{-\tau s}, \tag{18.2}$$

where $a_i\ (i = 0, 1, 2, \ldots, m)$ and $c_j\ (j = 0, 1, 2, \ldots, \alpha)$ are the coefficients of the denominator and numerator of the uncertain system, respectively. Besides, the time delay τ is a nonnegative real number. Define $q = a_i, c_j, \tau$ as the random parameter vector of Q with probability density function p_r. Hence the parameter space Q of practical industrial processes can be defined as

$$Q = \left\{ \left[a_{i_-}, a_i{}^+ \right] \cup \left[c_{j_-}, c_j{}^+ \right] \cup \left[\tau_-, \tau^+ \right] \right\}, \tag{18.3}$$

where the superscript $+$ and subscript $-$ represent the upper and lower boundaries of Q, respectively. These boundaries are necessary because they determine the perturbation degree of the process. Besides, note that they are flexible. The controller is designed based on uncertainties to meet control requirements for all uncertain systems in Q, which can be formulated as

{the probability to meet control requirements in the parameter space Q}
$$s.t. \text{ control requirements constraints,}$$
$$(18.4)$$

where control requirements can be selected as the settling time T_s, the integral of time and absolute error (ITAE), the overshoot σ, etc. However, the regular design methods of DDE PID in [29,30] are all proposed based on nominal systems with specific robustness constraints which can be formulated as

$$\min \{\text{the control performance indices for the nominal system}\}$$
$$s.t. \text{ robustness constraints,}$$
$$(18.5)$$

where control performance indices can be chosen as ITAE, the integral of absolute error (IAE), etc. Note that with the design method proposed based on expression (18.5), the control performance would be worse when the system varies far from its nominal working condition.

According to expressions (18.4) and (18.5), we can learn that the proposed design method provides a totally new vision where DDE PID is designed for uncertain systems in parameter space Q to meet the control requirements. Therefore, compared with the regular design of DDE PID, the proposed DDE PID based on PR is a practical and effective method, which is possible to check all uncertain systems throughout the parameter space Q.

Finally, the necessity of the proposed DDE PID based on PR can be summarized as follows: the controller designed based on the nominal system, including the conventional DDE PID, is usually designed with the consideration of reference tracking performance and disturbance rejection. As a result, their robustness is generally ignored. However, as for the DDE PID based on PR, its dynamic control performance and robustness are both taken into account for the following reasons:

1. The probability is calculated based on dynamic requirements.
2. The probability reflects the robustness of the controller when the coefficients of the process model are perturbed in a certain parameter space.

18.3 DDE PID principles of DDE PID

Suppose that the process depicted as Eq. (18.1) can be rewritten as

$$G_p(s) = H \frac{b_0 + b_1 s + \cdots + b_{m-n-1} s^{m-n-1} + s^{m-n}}{d_0 + d_1 s + \cdots + d_{m-1} s^{m-1} + s^m} e^{-\tau s}, \qquad (18.6)$$

where m, n, and H are the number of poles, the relative degree, and the high-frequency gain. Besides, d_i ($i = 0, 1, \ldots, m-1$) and b_i ($i = 0, 1, \ldots, m-n-1$) are the coefficients of the numerator and denominator of Eq. (18.6), respectively [27]. a_i, b_i, and H are usually unknown. Eq. (18.6) can be transformed to the state space equations as follows:

$$\begin{cases} \dot{z}_i = z_{i+1}, \ i = 1, \ldots, n-1, \\ \dot{z}_n = -\sum_{i=0}^{n-1} \lambda_i z_{i+1} - \sum_{i=0}^{m-n-1} \varsigma_i w_{i+1} + Hu, \\ \dot{w}_i = w_{i+1}, \ i = 1, \ldots, m-n-1, \\ \dot{w}_{m-n} = -\sum_{i=0}^{m-n-1} b_i w_{i+1} + z_1, \\ y = z_1, \end{cases} \qquad (18.7)$$

where λ_i ($i = 0, 1, \ldots, n-1$) and ζ_i ($i = 0, 1, \ldots, m-n-1$) are unknown parameters. The uncertainties and external disturbances are totally considered as an extended state f:

$$f(z, w, u) = -\sum_{i=0}^{n-1} \lambda_i z_{i+1} - \sum_{i=0}^{n-1} \zeta_i w_{i+1} + (H-l)u, \qquad (18.8)$$

where l is a positive coefficient. Then $\dot{z}_n$ can be rewritten as

$$\dot{z}_n = f(z, w, u) + lu. \qquad (18.9)$$

Consider the desired dynamic equation depicted as

$$y^{(n)} + h_{n-1} y^{(n-1)} + \cdots + h_2 \ddot{y} + h_1 \dot{y} + h_0 y = h_0 . r \qquad (18.10)$$

To reach Eq. (18.10), the corresponding control law should be

$$u = \frac{-h_0(z_1 - r) - h_1 z_2 - h_2 z_3 - \cdots - h_{n-1} z_n - \hat{f}}{l}, \qquad (18.11)$$

where $\hat{f}$ represents the estimate of f estimated by following disturbance observer:

$$\begin{cases} \hat{f} = \xi + k z_n, \\ \dot{\xi} = -k\xi - k^2 z_n - klu, \end{cases} \qquad (18.12)$$

where k and ξ represent the parameter and the intermediate variable of the observer, respectively. Combined with Eq. (18.11), we can rewrite Eq. (18.12) as

$$u = -\frac{\xi + k z_n}{l} - \frac{h_0 (z_1 - r) + h_1 z_2 + \cdots + h_{n-1} z_n}{l}. \tag{18.13}$$

According to Eqs. (18.12)–(18.13), we can learn that

$$\dot{\xi} = k \left[h_0 (z_1 - r) + h_1 z_2 + \cdots + h_{n-1} z_n \right]. \tag{18.14}$$

Integrating both sides of Eq. (18.14) simultaneously, we easily learn that

$$\xi = k \left[h_0 \int (z_1 - r) \, dt + h_1 z_1 + \cdots + h_{n-1} z_{n-1} \right]. \tag{18.15}$$

Combined with Eq. (18.15), we can rewrite Eq. (18.13) as

$$u = -\frac{k \left[h_0 \int (z_1 - r) \, dt + h_1 z_1 + \cdots + h_{n-1} z_{n-1} \right] + k z_n}{l}$$
$$- \frac{h_0 (z_1 - r) + h_1 z_2 + \cdots + h_{n-1} z_n}{l}. \tag{18.16}$$

The set point is the step change in practical processes so that $r^{(i)} = 0$ ($i = 1, 2, \ldots, n - 1$) is bounded and can be set as zero [34]. By defining the error between the set point and the output as $e = r - z_1$ we have $e^{(i)} = r^{(i)} - z_1^{(i)} = -z_{i+1}$ ($i = 1, 2, \ldots, n - 1$). Hence Eq. (18.16) can be written as

$$u = \frac{h_0 + k h_1}{l} e + \frac{k h_0}{l} \int e \, dt + \frac{h_1 + k h_2}{l} \dot{e} + \cdots$$
$$+ \frac{h_{n-2} + k h_{n-1}}{l} e^{(n-2)} + \frac{h_{n-1} + k}{l} e^{(n-1)} - \frac{k h_1}{l} r. \tag{18.17}$$

When n is equal to 2, the parameters of DDE PID are depicted as

$$k_p = \frac{h_0 + k h_1}{l}, k_i = \frac{k h_0}{l}, k_d = \frac{h_1 + k}{l}, b = \frac{k h_1}{l}, \tag{18.18}$$

where k_p, k_i, k_d, and b are the proportional gain, integral gain, derivative gain, and feedforward coefficient, respectively. According to Eq. (18.17), we easily learn that DDE PID has the two-degree-of-freedom structure as in Fig. 18.1, where r, u, d, and y are the set point, control signal, external disturbance, and output, respectively. It has been mentioned in [29] that

$$h_0 = \frac{h_1^2}{4}. \tag{18.19}$$

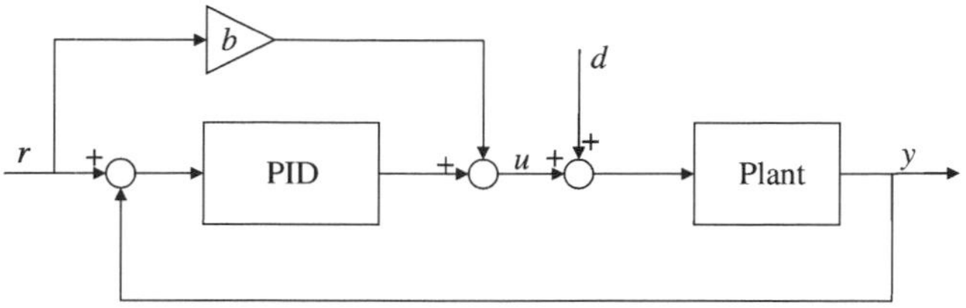

FIGURE 18.1   The structure of DDE PID.

In this chapter, to increase the flexibility of parameter tuning and avoid the oscillation of the output, we modify Eq. (18.19) as

$$h_0 = \zeta \frac{h_1{}^2}{4}, \tag{18.20}$$

where ξ is the coefficient to adjust the value of h_0. Therefore the tunable parameters of DDE PID based on PR (DDE-PR PID) are k, l, h_1, and ξ. Besides, the parameters of DDE PI can be evaluated based on Eq. (18.17) when $n = 1$.

18.3.1   Stable region of DDE PID

D-Partition (DP) method, first proposed by Neimark [14] to regionally partition the zeros of the closed-loop characteristic equation, was developed into solutions of stable regions based on open-loop frequency responses known as Open-Loop D-Partition (OLDP) method [23]. In this subsection, we analyze the stable region of DDE PID based on OLDP.

Proposition 2. *In Fig. 18.1 the feedforward coefficient b does not influence the stability of DDE PID.*

Proof. Denote the transfer function model of PID controller as $G_c(s)$ in Fig. 18.1. According to Fig. 18.1, we can learn that

$$\begin{cases} U(s) = G_c(s)\left[R(s) - Y(s)\right] - bR(s), \\ Y(s) = G_p(s)U(s), \end{cases} \tag{18.21}$$

$$\frac{Y(s)}{R(s)} = \frac{[G_c(s) - b]G_p(s)}{1 + G_c(s)G_p(s)}, \tag{18.22}$$

where $R(s), U(s)$, and $Y(s)$ are defined as the Laplace transformation of r, u and y, respectively. Consequently, the closed-loop characteristic equation of the system is depicted as,

$$L(s) = 1 + G_c(s)G_p(s) = 0 \tag{18.23}$$

From Eq. (18.23), it is obvious that b is irrelevant to the stability of DDE PID. Based on OLDP method, the boundaries of the stable region of DDE PID are depicted as

$$\begin{cases} \partial D_0 : L\,(j0) = 0, \\ \partial D_\infty : L\,(\pm j\infty) = 0, \\ \partial D_\omega : L\,(\pm j\omega) = 0, \\ \partial D_s : \text{The value of } \omega \text{ when } \partial D_\omega \text{ has no solution,} \end{cases} \tag{18.24}$$

where ω represents the frequency in s-domain. Suppose that the process is described by Eq. (18.2) and its frequency response characteristic is simplified as

$$G_p\,(j\omega) = a\,(\omega) + jb\,(\omega). \tag{18.25}$$

Besides, the frequency response characteristic of PID controller is depicted as

$$\begin{aligned} G_c\,(j\omega) &= k_p + \frac{k_i}{j\omega} + k_d\,j\omega \\ &= \frac{h_0 + kh_1}{l} + \frac{kh_0}{lj\omega} + \frac{h_1 + k}{l}\,j\omega. \end{aligned} \tag{18.26}$$

Therefore Eq. (18.23) can be rewritten in the frequency domain as

$$L\,(j\omega) = 1 + G_c\,(j\omega)\,G_p\,(j\omega) = \left[1 + a\,(\omega)\,\frac{h_0 + kh_1}{l} + b\,(\omega)\left(\frac{kh_0}{\omega l} - \frac{h_1 + k}{l}\omega\right) \right]$$
$$+ j\left[b\,(\omega)\,\frac{h_0 + kh_1}{l} - a\,(\omega)\left(\frac{kh_0}{\omega l} - \frac{h_1 + k}{l}\omega\right) \right]. \tag{18.27}$$

Since $h_1 > 0$ and $h_0 = \xi h_1^2/4$, valid boundaries of the stable region of DDE PID are depicted as follows:

$$\begin{cases} h_1 = 0, \\ \partial D_0 : \frac{k\xi h_1^2}{4l} = 0, \\ \partial D_\infty : a_n = 0, \\ \partial D_\omega : \begin{cases} 1 + a\,(\omega)\,\frac{\xi h_1^2 + 4kh_1}{4l} + b\,(\omega)\left(\frac{k\xi h_1^2}{4\omega l} - \frac{h_1 + k}{l}\omega\right) = 0, \\ b\,(\omega)\,\frac{\xi h_1^2 + 4kh_1}{4l} - a\,(\omega)\left(\frac{k\xi h_1^2}{4\omega l} - \frac{h_1 + k}{l}\omega\right) = 0. \end{cases} \end{cases} \tag{18.28}$$

Obviously, there is no parameter of DDE PID in the expression of ∂D_∞. Hence Eq. (18.28) can be rewritten as

$$\begin{cases} \partial D_0 : k = 0, \\ \partial D_\omega : \begin{cases} 1 + a\,(\omega)\,\frac{\xi h_1^2 + 4kh_1}{4l} + b\,(\omega)\left(\frac{k\xi h_1^2}{4\omega l} - \frac{h_1 + k}{l}\omega\right) = 0, \\ b\,(\omega)\,\frac{\xi h_1^2 + 4kh_1}{4l} - a\,(\omega)\left(\frac{k\xi h_1^2}{4\omega l} - \frac{h_1 + k}{l}\omega\right) = 0, \end{cases} \\ h_1 = 0. \end{cases} \tag{18.29}$$

Eq. (18.29) only represents the boundaries of stable regions of DDE PID. However, according to Eq. (18.29), it is unable to decide which side of the boundary is the stable region. The Nyquist criterion is an effective method to determine the area of stable region.   □

Theorem 7. *The necessary condition of the stability of the closed-loop system is that the Nyquist contour does not encircle* $(-1, j0)$ *point when the number of poles of* $L(s)$ *in the right-hand s-plane is zero.*

Based on Theorem 7, the constraint condition of the stable region is depicted as

$$\begin{cases} a\,(\omega)\,\frac{\zeta h_1{}^2+4kh_1}{4l} + b\,(\omega)\left(\frac{k\zeta h_1{}^2}{4\omega l} - \frac{h_1+k}{l}\omega\right) > -1, \\ b\,(\omega)\,\frac{\zeta h_1{}^2+4kh_1}{4l} - a\,(\omega)\left(\frac{k\zeta h_1{}^2}{4\omega l} - \frac{h_1+k}{l}\omega\right) = 0. \end{cases} \tag{18.30}$$

According to Eqs. (18.29)–(18.30), the stable region of DDE PID can be solved. Considering that l dominates the control action of DDE PID, we fix l first when we calculate the stable regions of DDE PID. There is a simulation to analyze the influence on stable regions with different k and ξ. Consider a simple first-order plus delay time (FOPDT) process

$$G_p\,(s) = \frac{1}{10s+1}e^{-s}. \tag{18.31}$$

Based on Eqs. (18.29)–(18.30), the $k - h_1 - l$ stable regions and $k - h_1 - \xi$ stable regions of DDE PID are illustrated in Figs. 18.2 and 18.3, respectively. Note that the range of l is [2, 3], whereas ξ is fixed at 1.4 in Fig. 18.2. Similarly, the range of ξ is [1, 6], whereas l is fixed at 2.

Obviously, a larger l means a wider stable region, and a larger ξ means a narrower stable region as shown in Figs. 18.2 and 18.3. Consequently, for the stability of the closed-loop system, we can first choose a larger range of l. These discussions about stable regions in this subsection are able to offer a rough range of the further parameter optimization for the proposed DDE-PR PID.

18.4 DDE PID design based on PR

Define the parameter vector of DDE PID as $Q = l, k, h_1, \xi$, which should locate in the stable region discussed in Section 18.3.1. When the parameters of DDE PID and the uncertain system in Q are fixed, the dynamic indices of the closed-loop system can be evaluated to check whether they satisfy control requirements. To measure the control performance of the closed-loop system, we define the binary indicator function

$$I_i = \begin{cases} 0 \text{ if the control requirement is not satisfied,} \\ 1 \text{ if the control requirement is satisfied,} \end{cases} \tag{18.32}$$

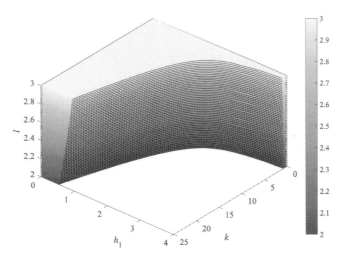

FIGURE 18.2 The $k - h_1 - l$ stable regions of DDE PID ($\xi = 1.4$).

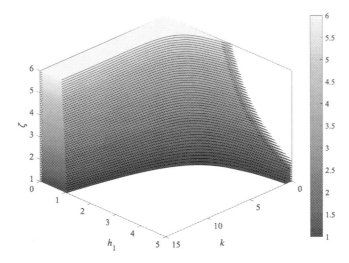

FIGURE 18.3 The $k - h_1 - \xi$ stable regions of DDE PID ($l = 2$).

where the subscript i represents indicator function of the ith control requirement. Then the ith probability P_i of all uncertain systems in Q with fixed DDE PID can be calculated as

$$P_i(\varphi) = \int_Q I_i \left[G_p(q), G_c(\varphi) \right] p_r(q) dq, \tag{18.33}$$

where $G_p(q)$, $G_c(\varphi)$, and $p_r(q)$ are the uncertain systems in Q, DDE PID with fixed parameters, and the probability density function of q in Q, respectively. If there are not only one control requirement, then the PR evaluation index based on the probability meaning can be defined as

$$J(\varphi) = fcn[P_1(\varphi), P_2(\varphi), \ldots], \qquad (18.34)$$

where f_{cn} is the combination function of different design requirements. Note that f_{cn} may be linear, nonlinear, or exponential based on different control targets. The design objective of the proposed DDE-PR PID is to find the optimal controller parameter vector φ^* located in the stable region, which is able to obtain the maximum $J(\varphi)$ for uncertain systems throughout Q. However, it is impossible to calculate the $J(\varphi)$ of all uncertain systems in the parameter space Q. Moreover, in most cases, $J(\varphi)$ is difficult to solve analytically. As a result, the Monte Carlo trial, a practical and effective method to estimate the probability P_i, is applied in design of DDE-PR PID. Based on the Monte Carlo trial, the estimates of P_i and J based on N samples, denoted as $\hat{P}_i(\varphi)$ and $\hat{J}(\varphi)$, can be calculated by Eqs. (18.35) and (18.36), respectively:

$$\hat{P}_i(\varphi) = \frac{1}{N} \sum_{k=1}^{N} I_i \left[G_p(\text{q}), G_c(\varphi) \right], \qquad (18.35)$$

$$\hat{J}(\varphi) = fcn \left[\hat{P}_1(\varphi), \hat{P}_2(\varphi), \ldots \right]. \qquad (18.36)$$

As $N \to \infty$, $\hat{P}_i(\varphi)$ and $\hat{J}(\varphi)$ are able to converge to $P_i(\varphi)$ and $J(\varphi)$, respectively. Actually, the condition of $N \to \infty$ is unrealizable and would lead to estimation errors. Based on the Massart inequality with the given risk parameters ϵ and confidence level $1 - \delta$ [5], the minimal N that can guarantees a certain confidence interval can be depicted by

$$N > \frac{2\left(1 \quad \varepsilon \mid \alpha\varepsilon/3\right)\left(1 \quad \alpha/3\right)\ln\left(2/\delta\right)}{\alpha^2\varepsilon}, \qquad (18.37)$$

where $\alpha \in (0, 1)$. Then N calculated by Eq. (18.37) can ensure that

$$P\left\{\left|P_x - K/N\right| < \alpha\varepsilon\right\} > 1 - \delta, \qquad (18.38)$$

where P_x, N, K, and K/N are the probability of the uncertain system to satisfy control requirements, the sample number, the number of control requirements satisfied in N samples, and the estimate of the probability, respectively. Besides, the confidence interval is calculated as $[K/N - \varepsilon, K/N + \varepsilon]$. In this chapter, we select $\varepsilon = 0.01$, $1 - \delta = 0.99$, and $\alpha = 0.2$. Hence the minimum value of N is 24495. On the basis of Eqs. (18.4) and (18.29)–(18.30), the optimization problem can be redefined as

$$\max\left\{\hat{J}(\varphi) \text{ for uncertain systems}\right\}$$

s.t. σ, T_s, ITAE, et al. constraints, φ locates in Eqs. (18.29)–(18.30)

$$(18.39)$$

As a heuristic optimization algorithm, the genetic algorithm (GA) [1] has a good global convergence. Therefore GA is applied to optimize the index of PR $\hat{J}(\varphi)$ to obtain the optimal parameter vector φ^*. During the optimization, 24495 times of Monte Carlo trials are carried out for each parameter vector φ'. Based on these discussions, the design procedure of the proposed DDE-PR PID is summarized as follows:

1. Define the uncertain system in the parameter space Q and the robustness evaluation index $\hat{J}(\varphi)$ based on Eq. (18.36).
2. Calculate the stable region of DDE PID according to the nominal model of the uncertain system.
3. Produce the initial population of GA in the stable region of DDE PID.
4. Optimize $\hat{P}_i(\varphi)$ and $\hat{J}(\varphi)$ based on GA and obtain the parameter vector φ'.
5. Test the parameter vector φ' by $N = 24495$ of Monte Carlo trials in the parameter space Q. If the test results can satisfy the requirement of the Massart inequality, then $\varphi^* = \varphi'$. Otherwise, return to Step 4.

18.5 Simulation validation

Based on the design procedure in Section 18.4, the proposed DDE-PR PID is designed for the high-order system, non-minimum phase system, integral system, and time delay system. The following factors should be considered when we select the binary indicator function:

1. The indices of control requirements are able to evaluate the control performance of the closed-loop system, e.g., T_s, σ, IAE, ITAE, etc. In addition, the frequency-domain indices can be chosen as well, including the cut-off frequency, gain margin, phase margin, etc.
2. The relationships between different indices should be considered. For instance, a shorter T_s usually means a larger σ. If both of them are selected as the indices of control requirements, then the system can balance their conflicts.
3. The physical constraints of practical systems should be considered such as the actuator saturation.
4. In terms of multiple-input multiple-output (MIMO) systems, the coupling influence from other loops should not be ignored.

On the basis of aforementioned factors, T_s and σ are chosen as the indices of control requirements for uncertain single-input single-output (SISO) systems. Moreover, as for MIMO systems, the recovery time T_{ry} is selected as another index of the control requirement as well to measure the loop decoupling for uncertain MIMO systems. The f_{cn} in Eq. (18.36) may be linear or nonlinear. For simplification of the calculation, the f_{cn} for SISO systems is defined as

$$\hat{J}(\varphi) = 0.8\hat{P}_{T_s}(\varphi) + 0.2\hat{P}_\sigma(\varphi), \tag{18.40}$$

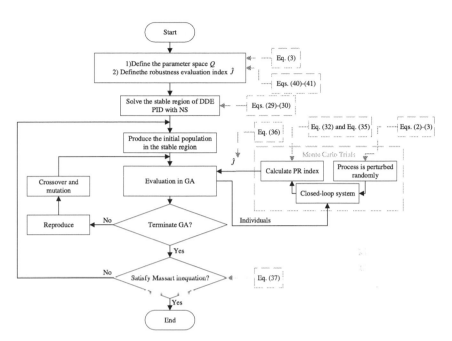

FIGURE 18.4   The design procedure of DDE-PR PID.

where $\hat{P}_{T_s}(\varphi)$ and $\hat{P}_\sigma(\varphi)$ are estimates of the probabilities of T_s and σ, respectively. Similarly, the f_{cn} for a $\gamma * \gamma$ MIMO system can be defined as

$$\hat{J}(\varphi) = \frac{\sum_{i=1}^{\gamma}\left[\hat{P}_{i,T_s}(\varphi) + 0.5\hat{P}_{i,\sigma}(\varphi) + 0.3\hat{P}_{i,T_{ry}}(\varphi)\right]}{1.8\gamma}, \tag{18.41}$$

where i indicates the ith loop. To present the design procedure in Section 18.4 intuitively, a flow chart is summarized as Fig. 18.4. In this chapter, we only focus on the design of DDE-PR PID for SISO systems. As for GA, the number of individuals and the maximum number of evolutionary iterations are set as 200 and 20, respectively.

In this section, to validate the effectiveness of DDE-PR PID, we select six typical processes as the controller objects. These processes are depicted as Eqs. (18.42)–(18.47):

$$G_{p1}(s) = \frac{1}{(s+1)^4}, \tag{18.42}$$

$$G_{p2}(s) = \frac{1}{(s+1)(0.2s+1)(0.04s+1)(0.008s+1)}, \tag{18.43}$$

$$G_{p3}(s) = \frac{1}{s(s+1)^2}, \tag{18.44}$$

$$G_{p4}(s) = \frac{1 - 2s}{(s + 1)^3}, \tag{18.45}$$

$$G_{p5}(s) = \frac{e^{-s}}{(s + 1)^2}, \tag{18.46}$$

$$G_{p6}(s) = \frac{(-0.3s + 1)(0.08s + 1)}{(2s + 1)(s + 1)(0.4s + 1)(0.2s + 1)(0.05s + 1)^3}. \tag{18.47}$$

According to Eqs. (18.42)–(18.47), we can learn that these processes are able to describe almost all types of industrial systems such as the high-order system (G_{p1}, G_{p2}), integral system (G_{p3}), non-minimum phase system (G_{p3}, G_{p6}), and time delay system (G_{p4}). Table 18.1 shows the control requirements and parameters of the Massart inequality of different processes.

TABLE 18.1 Control requirements and parameters of Massart Inequality.

Process model	Control requirements	Parameters of Massart Inequality
G_{p1}	$T_s < 23s, \sigma < 5\%$	$\epsilon = 0.01, \delta = 0.01, \alpha = 0.2, N = 24495$
G_{p2}	$T_s < 3s, \sigma < 5\%$	$\varepsilon = 0.01, \delta = 0.01, \alpha = 0.2, N = 24495$
G_{p3}	$T_s < 30s, \sigma < 5\%$	$\varepsilon = 0.01, \delta = 0.01, \alpha = 0.2, N = 24495$
G_{p4}	$T_s < 28s, \sigma < 5\%$	$\varepsilon = 0.01, \delta = 0.01, \alpha = 0.2, N = 24495$
G_{p5}	$T_s < 30s, \sigma < 5\%$	$\varepsilon = 0.01, \delta = 0.01, \alpha = 0.2, N = 24495$
G_{p6}	$T_s < 15s, \sigma < 5\%$	$\varepsilon = 0.01, \delta = 0.01, \alpha = 0.2, N = 24495$

These control requirements are set based on principles provided in [32]. Table 18.2 lists the parameters of PID controllers based on different tuning methods. Note that PID controllers based on SIMC method [26] approximate MIGO (AMIGO) method [2] and the conventional DDE method [37] and are selected as comparative controllers. Moreover, AMIGO PID has the same structure as that shown in Fig. 18.1.

Measurement noise usually exists in practical processes. As a result, they should be taken into account during simulations. First, we consider the control performance of different controllers when the measurement noise exists. Based on the parameters listed in Table 18.2, the control performance of different PID controllers with measurement noise is illustrated in Figs. 18.5–18.7. According to Eq. (18.40), the overshoot and settling time are calculated to evaluate the probability during the design of DDE-PR PID.

Note that during the design of DDE-PR PID, the random parameter vector q (ai, cj, and τ) is perturbed within the range $[-0.15q, +0.15q]$ randomly, and Monte Carlo trials are carried out $N = 24495$ times for each parameter vector. Besides, the set point of each process has a unit step change at 0 s, and a unit step disturbance is added during simulations. According to Figs. 18.5–18.7, it is obvious that DDE-PR PID has the moderate tracking performance and the

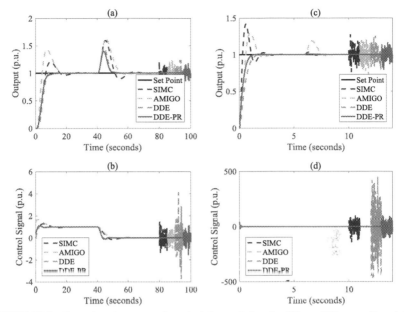

FIGURE 18.5   Control performance of nominal G_{p1} and G_{p2} for different PID controllers: (a,b) G_{p1}; (c,d) G_{p2}.

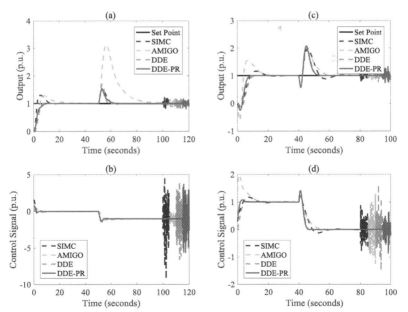

FIGURE 18.6   Control performance of nominal G_{p3} and G_{p4} for different PID controllers: (a,b) G_{p3}; (c,d) G_{p4}.

TABLE 18.2 Parameters of PID controllers based on different tuning methods.

Process Model	K_p, Ti, T_d	b, K_p, T_i, T_d	l, k, h_1, h_0	l, k, h_1, ξ
G_{p1}	0.5, 1.5, 1	0.47, 0.47, 2.0755, 0.8333	12, 9.97, 1.53, 0.59	2.5552, 1.4625, 2.3201, 0.7143
G_{p2}	17.9, 0.224, 0.22	3.5446, 3.5446, 0.5388, 0.0711	1, 10, 8, 16	1.4660, 4.7986, 18.3262, 0.6744
G_{p3}	1.5, 4, 1.5	0.45, 0.45, 12, 0.75	8, 10, 1.13, 0.32	12.5298, 0.6288, 37.9993, 0.0347
G_{p4}	0.3, 1.5, 1	0, 0.3929, 2.4932, 1.0294	31, 12, 1.55, 0.6	9.2132, 0.6632, 3.5021, 1.0761
G_{p5}	0.5, 1, 1	0.65, 0.65, 1.6364, 0.5769	18, 3.36, 2, 1	5.8333, 3.5010, 2.3333, 0.8571
G_{p6}	1.3, 2, 1.2	0.9653, 0.9653, 2.2118, 0.6248	8, 11.28, 1.56, 0.61	3.1769, 1.1549, 5.1989, 0.3338

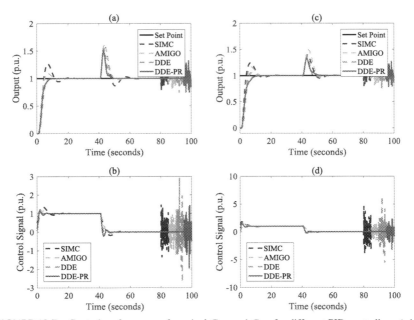

FIGURE 18.7 Control performance of nominal G_{p5} and G_{p6} for different PID controllers: (a,b) G_{p5}; (c,d) G_{p6}.

better disturbance rejection performance than PID controllers based on SIMC, AMIGO, and conventional DDE. Moreover, as for control action signals, DDE-PR PID is less sensitive to measurement noise than SIMC PID and DDE PID and more sensitive to measurement noise than AMIGO PID. Therefore we can learn that although the control requirements are all set based on tracking performance, the disturbance rejection performance and the control signal performance with measurement noise of the proposed DDE-PR PID are satisfactory as well.

The root mean square (RMS) value of the control action signal is an index to evaluate whether a controller is sensitive to the measurement noise. Table 18.3 lists RMS values of control action signals of different PID controllers.

TABLE 18.3 RMS values of control signals of different PID controllers.

Process model	SIMC	AMIGO	DDE	DDE-PR
G_{p1}	0.6616	0.5469	1.3705	0.5285
G_{p2}	51.0286	4.4255	211.6378	47.452
G_{p3}	2.9275	1.1099	2.7107	2.5424
G_{p4}	0.3954	0.5546	0.5413	0.2967
G_{p5}	0.6691	0.5363	0.9727	0.6656
G_{p6}	1.9039	0.0470	2.0404	1.001

A smaller RMS value mean lower sensitivity to the measurement noise. According to Table 18.3, we can learn that DDE-PR PID has a smaller RMS value than SIMC PID and DDE PID. Although the AMIGO PID is less sensitive to the measurement noise than DDE-PR PID, its tracking performance and disturbance performance are sacrificed. Another aspect to be addressed is that the actuator saturation may influence the control performance of controllers in practical processes. As a result, the actuator saturation should be taken into account in this section. Figs. 18.8–18.10 illustrates the control performance of different PID controllers with actuator saturation.

Combined with Figs. 18.8–18.10, it is obvious that DDE-PR PID has more stable tracking performance and better disturbance rejection performance than SIMC PID, AMIGO PID, and DDE PID when the actuator saturation exists. Therefore we can learn that DDE-PR PID can obtain satisfactory control performance when the control action signal is limited even though the actuator saturation is not considered in the design procedure of DDE-PR PID. As for Monte Carlo trials, σ, T_s, and IAE are recorded to evaluate the robustness of different PID controllers. Note that IAE is recorded form 0 s to the terminate time of simulation for each process without the consideration of the measurement noise and the actuator saturation. Fig. 18.11 shows the results of Monte Carlo trials of all perturbed systems with different PID controllers.

The more intensive scatters mean the stronger robustness. From Fig. 18.11 it is obvious that scatters of DDE-PR PID are more intensive than those of SIMC PID, AMIGO PID, and conventional DDE PID, which shows the strong robustness of the proposed DDE-PR PID. To evaluate the robustness of different control strategies quantitatively, the M_s of all processes with different PID controllers are listed in Table 18.4.

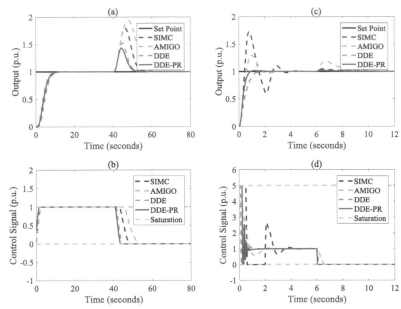

FIGURE 18.8 Control performance of nominal G_{p1} and G_{p2} for different PID controllers with actuator saturation: (a,b) G_{p1}; (c,d) G_{p2}.

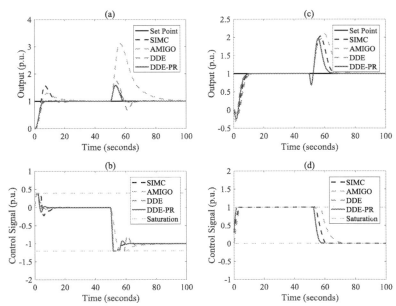

FIGURE 18.9 Control performance of nominal G_{p3} and G_{p4} for different PID controllers with actuator saturation: (a,b) G_{p3}; (c,d) G_{p4}.

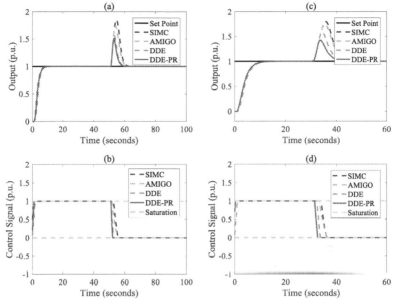

FIGURE 18.10   Control performance of nominal G_{p5} and G_{p6} for different PID controllers with actuator saturation: (a,b) G_{p5}; (c,d) G_{p6}.

TABLE 18.4 The M_s of all systems with different PID controllers.

Process Model	SIMC	AMIGO	DDE	DDE-PR
G_{p1}	1.4856	1.4778	1.3631	1.2662
G_{p2}	1.5199	1.4271	1.3922	1.2884
G_{p3}	1.7396	1.3875	1.303	1.2245
G_{p4}	1.7266	1.5563	1.3441	1.2989
G_{p5}	1.539	1.3647	1.1835	1.2004
G_{p6}	1.4222	1.3947	1.3407	1.2323

If the M_s is closer to 1, then the robustness of a controller is stronger. According to Table 18.4, it is evident that DDE-PR PID has the smallest M_s for most of the processes except $G_{p5}(s)$. In terms of $G_{p5}(s)$, the M_s of DDE-PR PID is close to that of DDE PID. To further illustrate superiorities of the proposed DDE-PR PID, the estimated values of the probability K/N are listed in Table 18.5. Note that the probabilities of SIMC PID, AMIGO PID, and DDE PID are calculated in the same way as DDE-PR PID.

From Table 18.5, the proposed DDE-PR PID always has the largest estimated value of probability K/N, which means that it is able to satisfy con-

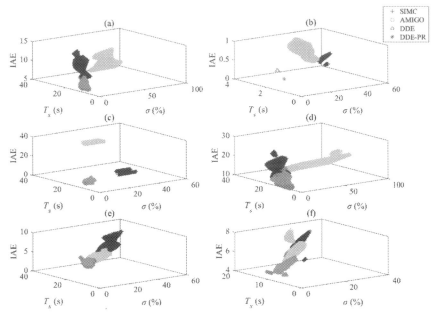

FIGURE 18.11 Results of Monte Carlo trials of all perturbed systems with different PID controllers: (a) G_{p1}; (b) G_{p2}; (c) G_{p3}; (d) G_{p4}; (e) G_{p5}; (f) G_{p6}.

TABLE 18.5 Estimated values of probability for all systems.

Process Model	SIMC	AMIGO	DDE	DDE-PR
G_{p1}	0.4943	0.669	0.9403	1
G_{p2}	0.8	0.7775	0.9996	1
G_{p3}	0.8	0.7097	0.9802	1
G_{p4}	0.7332	0.8	0.9622	1
G_{p5}	0.8	0.8337	0.9995	0.9999
G_{p6}	0.0043	0.7097	0.8707	1

trol requirements with the maximum probability for all uncertain systems in the parameter space Q including "worst-case" conditions. Generally speaking, DDE-PR PID can not only achieve the satisfactory control performance for the nominal systems, but also meet control requirements σ and T_s for all uncertain systems with the maximum probability. This superiority indicates that the proposed DDE-PR PID is suitable for the controller design for thermal systems whose operating condition is changing frequently during daily operations.

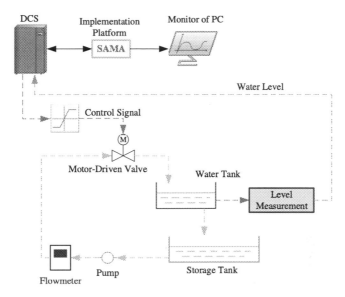

FIGURE 18.12 The schematic structure of the experimental setup of the water tank.

18.6 Experimental verification

In this section, the proposed control strategy is applied to the level control system of a water tank. Fig. 18.12 shows the schematic experimental setup of the water tank. Note that the water level control system is implemented on DCS.

18.6.1 Model identification

In this subsection, the level of the water tank is identified based on particle swarm optimization (PSO), which was proposed in 1995 [12]. The optimal function is defined as the integral absolute error between the real measurement and the model output. During the identification, the parameters of PSO are set as follows: the maximum number of generations $M = 1000$, the number of particles $N = 50$, inertia weight $\omega = 0.6$, learning factors $c_1 = c_2 = 1.2$. Fig. 18.13 illustrates comparisons between the real measurement and model outputs. Note that the process is identified as an FOPDT process. In Fig. 18.13, Δu and ΔH are denoted as changes of the valve opening and the water level, respectively. Based on Fig. 18.13, the transfer functions of the water level can be depicted as

$$G_p(s) = \frac{\Delta H(s)}{\Delta u(s)} = \frac{0.1062}{164.1458s + 1} e^{-8.3872s}. \qquad (18.48)$$

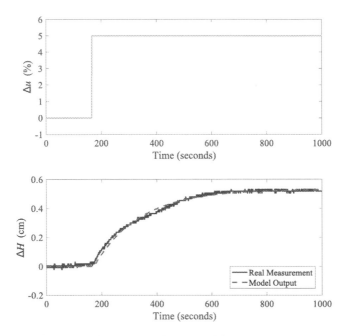

FIGURE 18.13 Comparisons between model outputs and the real measurement.

18.6.2 Preliminary simulations

According to Fig. 18.13, we can learn that the level system has significant measurement noise. In terms of large-scale industrial control systems, PID controllers are rarely used for the reason that the strong derivative action may lead to self-oscillations when there is high-frequency measurement noise. As a result, PI controllers are usually applied to industrial process control. In this section, all controllers, including SIMC PID, AMIGO PID, DDE PID, and DDE-PR PID, are designed based on PI controllers. Considering the range of level variation, the parameter space Q is defined as $[-0.2q, +0.2q]$. In addition, the control requirements are set as $\sigma < 0.05$ and $T_s < 150s$, whereas the parameters of the Massart inequality are set as $\varepsilon = 0.01, \delta = 0.01, \alpha = 0.2$, and $N = 24495$. Table 18.6 lists the parameters of different PI controllers.

TABLE 18.6 Parameters of different PI controllers for the level control of the water tank.

SIMC	AMIGO	DDE	DDE-PR
K_p, T_i	b_f, K_p, T_i	l, k, h_0	l, k, h_0
73.7137,	0, 57.3890,	0.0193, 1.132,	0.0232, 2.0124,
83.8720	69.7687	0.0324	0.0256

Note that the parameters of DDE PI are optimized by GA algorithm, and the optimal objective is IAEsp. Based on the parameters listed in Table 18.6, Fig. 18.14 shows the control performance of the level system with different PI controllers under the nominal water level. During the simulation, the set point has a unit step change at 150 s, and a step disturbance is added at 1000 s. In Fig. 18.14, in contrast to comparative controllers, the proposed DDE-PR PI can track the set point with less overshoot and reject the disturbance more effectively. To quantitatively evaluate the control performance of different PI controllers, the dynamic indices are calculated in Table 18.7.

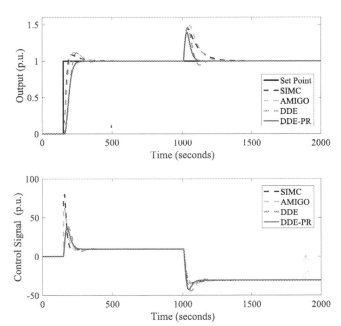

FIGURE 18.14   Simulation results of the level system with different PI controllers under the nominal water level.

TABLE 18.7 Dynamic indices of different PI controllers for simulations of water level control.

Method	σ (%)	T_s (s)	IAE$_{sp}$	IAE$_{ud}$
SIMC	8.83	110	27.8016	45.5124
AMIGO	11.97	148	34.1288	48.6286
DDE	11.31	132	47.2847	26.4296
DDE-PR	0	84	43.3029	18.0133

From Table 18.7, DDE-PR PI has the smallest overshoot, shortest settling time, and smallest IAEud. Besides, its IAEsp is larger than those of SIMC PI

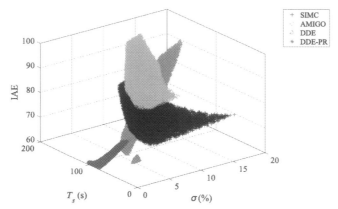

FIGURE 18.15 Results of Monte Carlo trials for the perturbed level system with different PI controllers.

and AMIGO PI, which means that IAEsp has been sacrificed during the design of PR. Similarly, Monte Carlo trails are carried out for $N = 24495$ times to test robustness and calculate the estimated values of probabilities of all PI controllers. Fig. 18.15 shows the results of Monte Carlo trails, and the estimated values of probabilities are listed in Table 18.8.

TABLE 18.8 Estimated values of probability of different PID controllers for simulations of water level control.

SIMC	AMIGO	DDE	DDE-PR
0.7158	0.8	0.7775	0.9999

In Fig. 18.15, it is obvious that scatter points of DDE-PR PI are more intensive than those of other comparative controllers, which means that the DDE PI designed based on PR can obtain strong robustness. Moreover, according to Table 18.8, we can learn that DDE-PR PI can satisfy control requirements with maximum probability for all perturbed level systems.

18.6.3 Experimental results

Based on the parameters listed in Table 18.6, in this subsection, all controllers are applied to the level system of the water tank. The set point of the water level is varying between 5 cm and 6.5 cm during experiments. Figs. 18.16–18.17 illustrate the experimental results of all PI controllers.

From Figs. 18.16–18.17 it is evident that:

1. Compared with SIMC PI and AMIGO PI, the proposed DDE-PR PI can track the set point with little overshoot when the reference signal is changing.

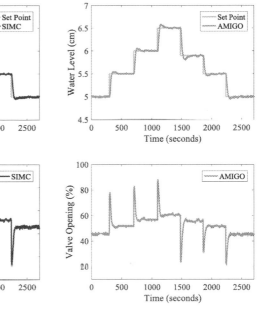

FIGURE 18.16 Experimental results of SIMC and AMIGO. (Date: March 10, 2021; time span of SIMC: 13:12–13:57; time span of AMIGO: 10:38–11:23).

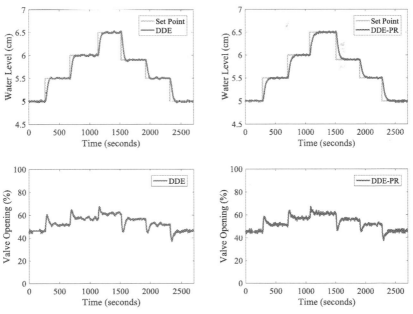

FIGURE 18.17 Experimental results of DDE and DDE-PR. (Date: March 10, 2021; time span of DDE: 9:52–10:37; time span of DDE-PR: 9:05–9:50).

2. Although DDE PI has faster tracking response than DDE-PR PI, its output has obvious oscillations when the set point of water level is varying between 5.5 cm and 6.5 cm.

Similarly to Section 18.5, the overshoot and the settling time of different controllers are calculated for each change of the set point of the water level to test whether DDE-PR PI can meet the control requirements (i.e., $\sigma < 0.05$ and $T_s < 150s$).

18.7 Conclusions

To handle uncertainties existing in industrial processes, DDE PID is designed based on PR. First, the problem formulation is described by defining a transfer function family with parameter uncertainties to introduce the necessity of the proposed method. Second, the principle and the stable region of DDE PID are introduced as fundamentals. Third, the proposed control strategy is designed for several typical processes, and its superiorities are validated by numerical simulations. Fourth, the DDE-PR PID is applied to the level system of a water tank. This successful application indicates the promising prospect in the future field test of DDE-PR PID. Our future work will focus on:

1. The development of the auto-tuning toolbox of the DDE-PR PID;
2. The DDE-PR PID design for MIMO systems;
3. The field application of DDE-PR PID to the practical industrial systems.

References

[1] C.W. Ahn, Practical Genetic Algorithms, John Wiley and Sons, 2006.
[2] K. Åström, T. Hagglund, Advanced PID Control, The Instrumentation, Systems, and Automation Society Press, 2006.
[3] R. Brockett, New issues in the mathematics of control, in: Mathematics Unlimited — 2001 and Beyond, Springer, 2001.
[4] W. Chen, Disturbance observer based control for nonlinear systems, IEEE/ASME Transactions on Mechatronics 9 (4) (2004) 706–710.
[5] X. Chen, K. Zhou, J.L. Aravena, Fast universal algorithms for robustness analysis, in: The 42nd IEEE International Conference on Decision and Control (CDC), 2003.
[6] G. Cohen, G. Coon, Theoretical considerations of retarded control, Transactions of the American Society of Mechanical Engineers 75 (1953) 827–834.
[7] R. Gorez, New design relations for 2-DOF PID-like control systems, Automatica 39 (5) (2003) 901–908.
[8] J. Han, From PID to active disturbance rejection control, IEEE Transactions on Industrial Electronics 56 (3) (2009) 900–906.
[9] I. Horowitz, Synthesis of Feedback Systems, Academic Press, 1963.
[10] C. Johnson, Optimal control of the linear regulator with constant disturbances, IEEE Transactions on Automatic Control 13 (4) (1968) 416–421.
[11] C. Johnson, Accommodation of external disturbances in linear regulator and servomechanism problems, IEEE Transactions on Automatic Control 16 (6) (1971) 635–644.
[12] J. Kennedy, R. Eberhart, Particle swarm optimization, in: 1995 IEEE International Conference on Neural Networks, 1995, pp. 1942–1948.

[13] S.J. Kwon, W.K. Chung, A discrete-time design and analysis of perturbation observer for motion control applications, IEEE Transactions on Control Systems Technology 11 (3) (2003) 399–407.

[14] R. Lanzkron, T. Higgins, D-Decomposition analysis of automatic control systems, IRE Transactions on Automatic Control AC-4 (3) (1959) 150–171.

[15] D. Li, F. Gao, Y. Xue, C. Lu, Optimization of decentralized PI/PID controllers based on genetic algorithm, Asian Journal of Control 9 (3) (2007) 306–316.

[16] H. Li, P. Shi, D. Yao, L. Wu, Observer-based adaptive sliding mode control for nonlinear Markovian jump systems, Automatica 64 (2016) 133–142.

[17] N. Minorsky, Directional stability of automatically steered bodies, Journal of the American Society of Naval Engineers 34 (2) (1922) 280–309.

[18] K. Ohishi, M. Nakao, K. Ohnishi, K. Miyachi, Microprocessor-controlled DC motor for systems for load-insensitive position servo system, IEEE Transactions on Industrial Electronics IE-34 (1) (1987) 44–49.

[19] H. Panagopoulos, K. Astrom, T. Hagglund, Design of PID controllers based on constrained optimization, in: American Control Conference (ACC), 1999, pp. 3858–3862.

[20] I.R. Petersen, R. Tempo, Robust control of uncertain systems: classical results and recent developments, Automatica 50 (5) (2014) 1315–1335.

[21] B.T. Polyak, R. Tempo, Probabilistic robust design with linear quadratic regulators, Systems & Control Letters 43 (5) (2001) 343–353.

[22] S. Qin, T. Badgwell, A survey of industrial model predictive control technology, Control Engineering Practice 11 (7) (2003) 733–764.

[23] Z. Shafiei, A. Shenton, Tuning of PID type controllers for stable and unstable systems with time delay, Automatica 30 (10) (1994) 1609–1615.

[24] J. Shi, M. Fang, Y. Ohyama, H. Hashimoto, M. Wu, Improving disturbance-rejection performance based on an equivalent-input-disturbance approach, IEEE Transactions on Industrial Electronics 55 (1) (2008) 380–389.

[25] H. Sira-Ramirez, From flatness, GPI observers, GPI control and flat filters to observer-based ADRC, Control Theory and Technology 16 (4) (2018) 249–260.

[26] S. Skogestad, Simple analytic rules for model reduction and PID controller tuning, Journal of Process Control 13 (4) (2003) 291–309.

[27] A. Tornambè, P. Valigi, A decentralized controller for the robust stabilization of a class of MIMO dynamical systems, Journal of Dynamic Systems, Measurement, and Control 116 (2) (1994) 293–304.

[28] C. Wang, D. Li, Z. Li, X. Jiang, Optimization of controllers for gas turbine based on probabilistic robustness, Journal of Engineering for Gas Turbines and Power 131 (5) (2009) 054502.

[29] W. Wang, D. Li, Q. Gao, C. Wang, A two-degree-of-freedom PID controller tuning method, Journal of Tsinghua University 48 (11) (2008) 1962–1966.

[30] X. Wang, X. Yan, D. Li, L. Sun, An approach for setting parameters for two-degree-of-freedom PID controllers, Algorithms 11 (48) (2018) a11040048.

[31] Z. Wu, T. He, L. Sun, D. Li, Y. Xue, The facilitation of a sustainable power system: a practice from data-driven enhanced boiler control, Sustainability 10 (4) (2018) 1112.

[32] Z. Wu, D. Li, Y. Chen, Active disturbance rejection control design based on probabilistic robustness for uncertain systems, Industrial & Engineering Chemistry Research 59 (40) (2020) 18070–18087.

[33] Z. Wu, D. Li, Y. Xue, T. He, S. Zheng, Tuning for fractional order PID controller based on probabilistic robustness, IFAC-PapersOnLine 51 (4) (2018) 675–680.

[34] Z. Wu, G. Shi, D. Li, Y. Liu, Y. Chen, Active disturbance rejection control design for high-order integral systems, ISA Transactions (2021).

[35] Z. Wu, J. Yuan, D. Li, Y. Xue, Y. Chen, The PI controller design based on a Smith-like predictor for a class of high order systems, Transactions of the Institute of Measurement and Control 43 (2020) 875–890.

[36] K.S. You, M.C. Lee, W.S. Yoo, Sliding mode controller with sliding perturbation observer based on gain optimization using genetic algorithm, Journal of Mechanical Science and Technology 18 (4) (2004) 630.

[37] M. Zhang, J. Wang, D. Li, Simulation analysis of PID control system based on desired dynamic equation, in: The 8th World Congress on Intelligent Control and Automation, 2010, pp. 3638–3644.

[38] C. Zhao, L. Guo, PID controller design for second order nonlinear uncertain systems, Science China. Information Sciences 60 (2) (2017) 022201.

[39] Q. Zhong, A. Kuperman, R.K. Stobart, Design of UDE-based controllers from their two-degree-of-freedom nature, International Journal of Robust and Nonlinear Control 21 (17) (2011) 1994–2008.

Index

Printed in the United States
by Baker & Taylor Publisher Services